Love, Power and Knowledge

RACE, GENDER, AND SCIENCE

Anne Fausto-Sterling, General Editor

Love, Power and Knowledge

Towards a Feminist Transformation of the Sciences

Hilary Rose

Indiana University Press
Bloomington and Indianapolis

Published in Great Britain by Polity Press, 65 Bridge Street, Cambridge
CB2 1UR, UK and in North America by Indiana University Press, 601 North
Morton Street, Bloomington, Indiana 47404

Manufactured in Great Britain

Library of Congress Cataloging-in-Publication Data

Rose, Hilary, date–
    Love, power and knowledge : towards a feminist transformation of
the sciences / Hilary Rose.
        p.   cm.
    Includes bibliographical references and index.
    ISBN 0–253–35046–8 (cloth). –– ISBN 0–253–20907–2 (pbk.)
    1. Women in science.   2. Women scientists.   I. Title.
D130.R673   1994
305.435––dc20   94–1883

1   2   3   4   5   00   99   98   97   96   95   94

Contents

*To Steven for (mostly)
sharing dreams and being
practical*

Plates

Between pp. 178 and 179

Prologue

The genesis of this book lies a long way back in personal biography and its continuing intersection with history. As with any pre-occupation which has developed over a period of years there is no one single moment, no clear unfolding, just a lumpy thread in the fabric of everyday life, a lumpiness which insisted that I look at it more closely. Genesis becomes less a tidy chronological account than a series of troubling memories; generation as well as gender, class and 'race' frame my thinking. The starting point is memories of war, the horrifying enigma of the Nazi death camps, for these were simultaneously real and unbelievable. How could anyone systematic-ally exterminate an entire people? Other cultures, not least that to which I belonged, had massacred and killed; the new dimension was the meticulous book-keeping of murder. Thus the specific obscenity of the death camps was this 'rationality'. At the time I think I understood this as perverse, for my sense of rationality stood on the side of freedom and justice. But history was to render this sense problematic.

It was perhaps not until the fifties, as a very young woman, that I became intensely aware that the nuclear bomb might well mean no future for my own or for any other child. How was it that science, which seemed to promise so much, was also so deadly that it threatened the human experiment itself? With many others of that generation I walked at Easter to Aldermaston, the centre of bomb research in Britain. Science's collusive relationship to militarism, and scientists' liking for the corridors of power, were untidily entangled with the social optimism of that postwar generation which believed that full employment and an improving welfare infrastructure were its birthright. I remember reading a book in the late fifties which spoke of the authoritarianism of science; it was like being told about a key which might unlock the puzzle.

The 1960s brought the Vietnam War and the explosive appearance of an international student movement which wanted both an end to an imperialist war and a beginning to a new and more democratic society. An enraged opposition to a genocidal technoscience was integral to the refusal of a genocidal and racist culture; the visible and international network of those who were both in and against this technoscience formed the radical science movement. Being part of an immense social movement gives courage, not least courage to look closely at science and at its self-representation – even to begin to smell and see the possibility that not only was science these things but also both it and its critics were profoundly androcentric.

As I grow older I feel that trying to capture criticism in words, by writing and publishing, is like trying to put salt on the tail of a devil. The book that Steven (Rose) and I wrote in the sixties called *Science and Society* – which at the time seemed just the right title – later made both of us feel rueful. For such a title reinforced the very idea we were trying to overcome, namely that science and society were distinct. Collected essays grouped under the banners of *The Radicalisation of Science* and *The Political Economy of Science* seemed fine in the mid-seventies until my growing sense of the conceptual and political obliteration of gender brought discomfort. Adding women to the marxist political economy of science and stirring was no longer enough.

Trying to comfort research students who tell me that their theoretical framework has changed and that they are finding it difficult to finish their theses, I sometimes suggest looking critically but kindly at oneself over time, as, rather concretely, the 'younger and misguided Hilary Rose'. They and I know that it is not quite as easy as this; I still feel myself accountable for what I have written before (taking responsibility for that allegedly generic 'he' I know I have used) and that I have to make a reckoning with that younger self. So writing, as I did, a paper called 'Hand, Brain and Heart' in the early eighties (*Signs: Journal of Women in Culture and Society*, 9 (1) 1983, pp. 73–90) was for me a way of seeking the reconciliation of a number of different selves and above all of this new self, which had been able to come more clearly into existence within second-wave feminism. In my essay the metonym of the Heart stood in for the caring labour of women, left out by marxist political economy, and for the responsible thinking that arose from this labour which was left out of a marxist theory of knowledge. Such a new feminist knowledge might, I argued, re-vision rationality itself, fostering representations of nature which were more pacific to women and nature alike.

Teaching social work and women's studies students at the University of Bradford, with their immense respect for women's everyday lives, has made me very conscious of the relationship between knowledge and love. Yet what were the connections between that everyday responsible rationality, that thinking from caring, and the power/knowledge couple which has dominated thinking about science from Francis Bacon to Michel Foucault? There were theoretical difficulties too. Although I felt myself to be a marxist feminist, within Britain the feminism which took gender and the body most seriously was that of radical feminism. How could I admit the body without biological reductionism, and still make connections to class and race? My precarious solution was to think of myself as a materialist feminist and to engage with the feminisms of Scandinavia and the US, as these placed gender as central and took the body as real. For that matter so did the strong tradition of British feminist research on human reproduction to which I also felt indebted.

However, for those of us living in Britain, an old industrial society with a problematic economy and a growing culture of social indifference, the changed context of the 1980s and 1990s has seen those fierce divisions of radical and socialist feminisms diminish; the body and gender are now central issues for feminism. The significant difference is that now feminist materialism is itself having to compete for intellectual space against a strong poststructuralist current. It has been in and against this changing context that the present book has been all too slowly written. I wanted to explore and listen to the many different voices within the feminist science debate. I have felt more than uneasy at some of the new developments but have had no intention of returning to that tradition of fierce polemic in which I was constructed and which it has been a source of satisfaction to resist. (Though I am not so good a feminist that I do not chuckle over robust and witty denunciations, usually from within that self-same marxist viriculture of androcentric poststructuralists.)

The nine chapters of *Love, Power and Knowledge* are organized around three broad concerns: first, the content, context and history of the feminist critique of science as it has developed since the 1970s (chapters 1–4); second, the situation of women within the institutions of science (chapters 5–7); and third, the culture of science – both actually existing science, and science as feminists might reconstruct it (chapters 8–9, and the epilogue).

The book thus begins with a focus on theoretical issues. Chapter 1 surveys feminist science criticism and theory as they have developed (primarily but not only) in the West, tracing their origins as in part the disobedient daughters of the radical science movement, and in part the daughters of the women's liberation movement and of academic feminism. These were to become powerful voices within and of feminism in the eighties and nineties. The second chapter explores feminist constructions of a responsible rationality as shaped by the everyday lives of women and by feminist values; the belief which is central to my book is that such a revisioning of rationality is crucial to the reconstruction of science. Such a feminist project is both politically realist and utopian; realist because the contemporary culture of technoscience is so deadly that it must be reconstructed; and utopian because the gap between this reality and any gentler one is still immense. Chapter 3 explores the institution of academic feminism as the means through which feminism as a social movement is seeking to change the knowledge system. This chapter is preoccupied with the tension between academic feminism, its cultural and political projects, and its location in diverse national contexts within a global production system of knowledge. I wrote it as a first stab at a feminist sociology of feminist knowledge, as it seemed to me that this might help academic feminists in our many and manifold struggles against that old and appropriately gendered adage of 'he who pays the piper calls the tune.' My fourth chapter turns to the debates within feminist science theory. I read these debates between realists (or standpoint theorists) and postmodernists as very different from those within the mainstream culture, because of the overt commitment of all their participants to feminism as a political project, but also as subtly different from the debates in other areas of feminist knowledge and cultural production. The common preoccupation with nature and with representations of nature frames the debate in ways that are different from discussions of literature, film or the psyche.

My second theme occupies the next three chapters, which are concerned with the structure of the scientific knowledge system and where women scientists are within it. Chapter 5 is thus a structural counterpart to chapter 4; where the latter looks at ideas, the former looks at how far different patriarchal academies have admitted, or been forced to admit, women. Just how far has feminism achieved its goals of equality of representation within the academic labour force? How near is the objective of 'nothing less than half the labs'? Chapter 6 examines the story of the admission of women scientists into the Royal Society, that bastion of British scientific eminence which for three centuries managed to exclude women. The interest of this particular account is the dramatic contrast between the self-representation by this elite body of how Fellows are customarily elected, and the quite extraordinary treatment accorded the first woman candidate to be proposed in the light of the anti-discriminatory legislation passed some two decades previously. The archives of the Royal Society provide a marvellous insight into the ways men have managed to exclude women and how actively they 'man'aged their admission into elite institutions. It is to the credit of the Royal Society that unlike the British government it does not seem to weed its archives, so that the mechanics of the patriarchal scientific power elite are exposed to view. The third chapter in this group focuses on women scientists at the apex of the prestige sytem of science, the nine women Nobel Laureates there have been over the nine decades since the institution of the prize. As well as honouring the extraordinary achievements of these women I wanted to show how their biographies could also be understood as like those of other women scientists of their time. (Working on this chapter in Sweden, where governmental papers are, especially to a British social scientist accustomed to a culture and law of official secrecy, remarkable for their openness, was peculiarly frustrating, as the Nobel archives are closed for fifty years.) Chapters 5–7 thus seek to reinforce the need to battle for space for women within the organizational structure of the production of knowledge even while feminism struggles to reconceptualize the knowledge system itself.

Chapters 8 and 9 are in very different ways about threats and hopes. Chapter 8 focuses on the powerful emphasis given to the new genetics within the life sciences. Increasingly consuming a significant section of the life science budget, its highly reductionist explanations of human bodies and behaviour alike seek to dominate the biomedical culture and bring particular challenges to women. The chapter brings together a recognition that science is socially shaped with a critical analysis of the cultural content and implications of that knowledge. Chapter 9, for me one of the most enjoyable to write, explores some of the texts of feminist science fiction in an extensively revised version of a paper, 'Dreaming the Future', originally published in *Hypatia*, 3 (1), 1988, pp. 119–37. Here, in a laboratory of our own, feminists can explore and experiment with other ways of knowing, other sciences and other futures than those offered by the seeming inevitability of an androcentric technoscience. Finally, in the epilogue, I address the unfinished business of moving beyond the one-sided rationality of masculinist science, to ask how, within our everyday lives, we can begin to create sciences which bring together love, power and knowledge.

Thinking and writing this book in a changing socioscape has for me been a protracted process, a mixture of isolation and feeling part of a continuing and immensely creative conversation. At the birth of modern science in seventeenth-century England the men and tiny numbers of women who corresponded with one another nationally and internationally, sharing and arguing over ideas about nature, felt themselves to be part of an invisible college. Over the past two decades a new invisible college, this time of feminist critics and theorists of science, has come into existence. Initially no more than a handful, the numbers have grown quite rapidly. Making my acknowledgements is thus, for very welcome reasons, hard. But the particular invisible college to which I am indebted, both individually and collectively, and which I think has never corporeally and completely met, includes: Lynda Birke, Tarja Cronberg, Anne Fausto-Stirling, Donna Haraway, Sandra Harding, Nancy Hartsock, Ruth Hubbard, Evelyn Fox Keller, Maureen McNeil, Nellie Oudshom, Vandana Shiva, Kate Soper and Ethel Tobach. Death has taken some of the most wonderful participants; I think with sadness of Ruth Bleier's and Wendy Farrant's premature deaths; yet different others enter. A central pleasure in this feminist symposium is heterodoxy, pleasure in contention rivalling delight in discovered agreement.

The University of Bradford has provided an extraordinarily rich feminist milieu, especially since we established the women's studies degree in 1981, an act of creative resistance in a year of brutal and stupid cuts in British higher education. Colleagues and friends there have been a precious resource; two with very different approaches to feminism, who live a caring responsibility of knowledge and have been particularly important to me, are Sheila Allen and Jalna Hanmer. Errollyn Bruce, Pauline Brier and my students within the West Yorkshire Centre for Research on Women have been a valued source of stimulation and friendship. Conversations with feminists over the years working in and on human reproduction have been important, notably Ann Oakley, Frances Price, Wendy Savage, Meg Stacey, Michelle Stanworth and Gail Vines.

In addition to drawing general support and encouragement from being part of a rich feminist culture I have many directly book-related debts owed to an amazingly multidisciplinary network of friends and colleagues who read draft chapters and sets of chapters: Lynette Hunter, my literary friend; historian Diana Long; and biologists Ann McLaren, Clare Woodward and Val Woodward. I owe very special intellectual political and personal debts to that heroic band who read and commented on the entire book: Sandra Harding, Donna Haraway, Ruth Hubbard and my Polity editor Michelle Stanworth. Last I must thank Steven Rose, who read and discussed many drafts at different stages and whose sustained emotional and intellectual support was crucial to my finishing.

Financial support which made possible time to think and write was provided by a fellowship at the Swedish Collegium for the Advanced Study of the Social Sciences during 1990–1. Intellectual stimulation from within SCASSS came especially from Bjorn Wittrock, Tinne Vannen and Allan Pred, and from outside through the wonderful Scandinavian network of feminists whose thinking and conversations I have been privileged to share. Among them are: Sylvia Benkarts, Jolke Esseveld, Elizabeth Gulbrandsen, Harriet Holter, Eva Lundgren, Ingun Moser and Hildur Ve.

I am also indebted to TMV for a stimulating month in Oslo in May 1992 which fostered the first draft of chapter 8, and to the University of Minnesota for a Hill Professorship, attached to the Center for Advanced Feminist Studies and to the College of Biology for the fall quarter of 1992, which enabled me to finish the first draft of the book. I should also acknowledge an earlier grant from the UK Economic and Social Research Council which supported my study, carried out with Helen Lambert, of a particular public within the Public Understanding of Science programme.

1

Introduction:
Is a Feminist Science
Possible?

Science it would seem is not sexless; she is a man, a father and
infected too.

Virginia Woolf, *Three Guineas*

For the master's tools will never dismantle the master's house.
They may allow us to beat him at his own game but they will
never enable us to bring about genuine change.

Audre Lorde, *Sister Outsider*

To ask, 'Is feminist science possible?' is to return to our own history of
struggle and the contradictory relationship of feminism to science and
its changing definition.[1] For second-wave feminism, science and
technology have not – with the almost single and certainly exceptional voice of Shulamith Firestone – been seen as progressive for
women's interests. There has been little chance of invoking the
metaphor, unhappy or otherwise, of courtship and marriage that was
widely used to foster the hoped-for relationship between marxism
and feminism. Where the radical science movement of the 1960s had
to free itself from the progressivist claims of science – to show that
science was not even neutral but often oppressive and antithetical to
human liberation – many women, already outside such progressivist
claims as a result of their very exclusion from science, had a hunch
that modern science and technology served all too often as means of
domination and not liberation.

Overtly relegated to nature by the recrudescence during the seventies of the patriarchal determinism of sociobiology, feminists learnt to uncover and contest the practices of an androcentric science. In claiming a place in culture, feminism has had to think much more deeply about both social relationships and the relationship of women to nature. Indeed feminist biologists, in contesting the boundaries of nature and culture laid down by sociobiology, understood in a direct and practical way that as women we, our bodies and ourselves, are part both of nature and of culture. Political and cultural struggles waged by feminists within and without science have contested a patriarchal science's right to determine those boundaries. For the most part feminist struggles have resisted biological determinism, which reduced women to nothing but their wombs, hormones, genes, or whatever was the bodily part in biological fashion,[2] but there is also a record of feminists using nature – even essentialism – as a resource in the defence of women.[3]

The recurrent mood, as and when the feminist movement preoccupied itself with science, has been one of anger. This anger extended from a sense of injustice at being shut out of an activity that some women, despite the engendered rules of the game, always wanted to take part in to an overwhelming sense of fury that masculinist science and technology are part of a culture of death. The ideology of science, proclaiming objectivity, freedom from values, and dispassionate pursuit of truth, has excluded women and been integral to our cultural domination, has harmed women's bodies (in our best interests, of course), and has threatened the environment itself. That science claimed its ideological purity, leaving by implication its partner technology to carry the responsibility for the dirty side of the relationship, was part of science's skill at conveying a culture of no culture.

Second-wave feminism began relatively slowly to analyse and contest science, to see the connections between this entity called 'science' and those issues that the movement defined as its own.[4] There were good reasons why the movement was slow; its central preoccupation was with women's shared experience, to reclaim what had been denied or trivialized out of existence and return it to social and political existence. The feminist movement has developed and changed in many ways since those early, path-breaking years of the late 1960s and early 1970s. Then, to consider housework, abortion, sexuality, love, birth control, motherhood and male violence as central social issues was to work against the grain of an arrogant and naturalizing masculinism. Feminism necessarily embraced body politics; the struggle for the repossession of our bodies, including knowledge about them, was to become central to the movement. The very process of examining these everyday aspects of women's lives, learning to speak about them, forged new concepts, new names.

Naming – conceptualizing – has been rightly seen within feminism as empowerment.[5] Naming brings into consciousness phenomena and experiences hitherto denied space in both nature and culture. In the fierce opposition to new concepts, it becomes clear that often these are not merely unacknowledged aspects of reality waiting to be discovered, but are actively erased by the values of the dominant culture. Even today feminism's concept of gender meets strong resistance from androcentric social theorists, or it is used as a euphemism for women, thus denying relationality and so diminishing the political and cultural claims. Naming, above all when the words become part of the language of new historic subjects seeking to take their place in society, simultaneously contests existing hegemony and affirms a changed consciousness of reality.

Feminists both constructed new knowledge, new accounts of the world from the perspective of women's everyday lives, and also tore down existing hegemonic ideas. Central concepts which had organized thought and culture, not least sacred reason itelf, were interrogated and found to be far from some timeless universal thought form, but instead a gendered, historically and geographically specific construct. The intense abstractionism of masculine thought came into visibility.[6] To catch the distinctive character of women's and feminist thought, feminists evoked alternative metaphors of spinning and quilt-making, reconstructions of a responsible rationality, of an ethic of care.[7] As Adrienne Rich wrote: I am convinced that 'there are ways of thinking that we don't yet know about. I take these words to mean that many women are even now thinking in ways which traditional intellection denies, decries or is unable to grasp.'

Although feminism has touched women's lives the world over and draws increasing numbers of women into its vortex, it is none the less true that the movement has been strongest within the old capitalist societies – and it is here that the discussion of science has been most intense. This is not to say that feminists in what were the societies of 'actually existing socialism' and third world or sometimes black feminists within advanced industrial societies have experienced science and technology in a particularly favourable way; rather that, for necessary reasons, their attention has been primarily focused elsewhere. It has been the unremitting struggle to produce enough food without further green revolutions harming people and land alike, the struggle against disease, not least the AIDS which sweeps Africa, and other crises of the environment which have placed science and technology on the agenda of third world women's struggles to survive.[8]

From the earliest days of the radical science movement of the 1960s, the critique of science and technology has focused attention on the ways in which existing science and technology are locked into the contemporary forms of capitalism and imperialism as systems of domination. This denunciation has served two functions. Negatively, it has facilitated the growth of an antipathy to science that rejects all scientific investigation carried out under any conditions and at any historical time.[9] Within feminism this took the form of denouncing all of science and technology as monolithically and irretrievably male. More positively, the denunciation has fostered the difficult task of constructing, in a prefigurative way, both the forms and the content of a different, alternative science – one that anticipates the science and technology possible in a new society and, at the same time, contributes through innovatory practice to the realization of that society.[10] But from its inception, with its false starts as well as real achievements, its perilous balancing between atheoretical activism and abstract theoreticism, the project was not without its contradictions and difficulties. Feminism is just beginning to recapture the full force of Virginia Woolf's compelling aphorism; science, it would seem – to rephrase – is neither raceless, sexless nor classless; she is a white man, bourgeois, and infected too.

The trouble with science and technology from a feminist perspective is that they are integral not only to the systems of domination of late capitalism and its new forms of imperialism, but also to one of patriarchal domination; yet to try to discuss science under these structures of domination or to argue that they constitute one social formation has proved peculiarly difficult. The present chapter serves to open that discussion by looking, first, at the radical critique of science of the 1960s and 1970s, and then at the growing body of feminist scholarship which developed partly in co-operation with, and partly against, the androcentric voice of the radical science movement.[11]

The radical critique of science

The critique of science was to explode into practice and to struggle into theory during the radical movements of the late 1960s and early 1970s. The rich and complex issues contained in the class and social struggles of those movements were frequently narrowed and constrained as the theoreticians filtered the wealth of lived experience through the abstract categories of theory. From an early rhetoric which attacked with a certain even-handedness the class society, imperialism, racism and sexism (those who were black, colonized or women might well have had doubts about their equal prioritization in practice as well as in rhetoric), two main lines of analysis were devloped. The first considered the political economy of science, and the second took up the relationship between science and ideology. While the two are linked at many points, work in political economy was more coherently developed; work on the debate over science and ideology was and remains more problematic.[12]

The need to reply immediately to the renewed biological determinism of the 1970s and 1980s was urgent as scientific racism sustained a growing political racism. In Britain the movement did not manage well the double task of opposition and maintaining internal coalitions. As I discuss later, despite the potential alliance between the critics of IQ theory, in which social constructionists and those who argued that it was 'bad science' ideologically organized around race and class interests shared a common project of overturning a would-be canonical IQ theory,[13] the movement split. In a larger country with a larger movement this might have been less significant. As it was the split did tremendous harm, making it very difficult for radical working scientists and radical social constructionists to co-operate. At the time my own feeling was that such radical relativism, such hyper-reflexivity, aided the monolithic rejection of science which was simultaneously being proposed by the counterculture.[14]

The new left came into existence opposing the old left analysis which claimed that there was an inevitable contradiction between the productive forces unleashed by science and the capitalist order. Within the old left account science was seen as uninfluenced by class, race, gender, nationality or politics; it was the abstract accumulation of knowledge – of facts, theories and techniques – which could be 'used' or 'abused' by society.

In the chill early years of the cold war the only space open to left scientists was to criticize science's use for militaristic purposes, a space epitomized in Britain by the organization of Science for Peace. While this movement failed to criticize the content of science, one of its abiding offshoots has been the continued struggle against the scale and proportion of the British science budget spent on military purposes – a struggle no less urgent in the 1990s. Despite the collapse of the former Soviet Union there has been little in the way of resetting of research objectives, so that half the UK science budget is still directed towards military ends.[15]

But the experiences of the sixties and seventies overthrew notions of reharnessing actually existing science. What the sixties' radicals discovered in their campaigns against a militarized and polluting science was that those in charge of 'neutral' science were overwhelmingly white and male occupants of positions of power within advanced industrialized society – whether the project of that society was capitalism or state socialism. The anti-human (and as feminists were increasingly to demonstrate, the specifically anti-women) technologies that science generated were being used for the profit of some and the distress of many. Thus the politics of experience brought the radical movement's attitudes toward science into a confrontation with the old left analysis of science, in particular in an effort to recover those hopes of a second science, a science for the people, which had been a striking feature of the early days of the Soviet revolution but had subsequently been brutally destroyed. Hope for that lay buried in the cupboard of the Lysenko affair, and disinterring and coming to terms with this denied past was critical for the radical science movement.

The Lysenko affair epitomizes the period from the 1930s to the 1950s in the Soviet Union, during which there was an attempt to develop a specifically proletarian interpretation of all culture, including the natural sciences.[16] This interpretation of the history of science, with its thesis of the 'Two Sciences' (bourgeois and proletarian), had been raised by the theorists from the young Soviet Union and introduced to the West at the 1931 International Conference for the History of Science in London. The thesis was strongly attractive to a group of young British marxist scientists, who wanted to revolutionize their science along with their society. Such hopes died in the Lysenko controversy. Against the genetic consensus, but apparently in accord with dialectical principles, the plant breeder Trofim Lysenko advanced the thesis that acquired characteristics are inherited. Initially it was merely a scientific dispute, but Lysenko also set his social origins as a peasant (and thus his experiential knowledge as a proletarian) against the aristocratic origins (and therefore abstract knowledge as a bourgeois) of his leading opponent, Nikolai Vavilov. The debate was resolved by Lysenko's presentation of falsified statistics on the amounts of grain produced and by the direct intervention of Stalin on the side of this fraudulent, but proletarian-claiming, science. In 1940 Vavilov was arrested and Lysenko became director of the key Agricultural Research Institute.[17] Marxist scholarship at the time, as for example expressed in the debates within the British Communist Party, particularly in the natural science group (the Engels Society), tore itself apart on the issue, which was ultimately presented starkly as a matter of loyalty to the Soviet system at the height of the cold war. Many biologists and geneticists distanced themselves from the Party, leaving non-biologists, above all the distinguished crystallographer J. D. Bernal, a leading figure within the Communist Party, to support Lysenko's claims in loyalty to the Soviet Union. The Engels Society soon ceased to exist.

Thus, when the radical movement of the 1960s and 1970s turned to marxist analyses of the natural sciences, it found either the terrifying language of 'mistakes' and a desire to repress all mention of the past or an insistence, by for example the biologist and historian of Chinese science Joseph Needham, a figure from the old left but who was felt to be more sympathetic to the aspirations of the radical science movement, that there is only one universal modern science.[18] Nor was the movement helped by the special status of science within the history of marxism – from Marx's and Engels's claims for a scientific socialism, Engel's tendency to claim scientists as natural allies of socialism, and Lenin's enthusiasm for the Taylorist scientific management of industrial production to Althusser's structuralist project to remove the human agent from marxism so as to make it truly scientific. Indeed the enthusiasm for structuralism of marxist social sciences – not least cultural studies[19] – was not shared by the radical science movement, which was struggling both to restore agency and responsibility into the impersonal deterministic voice of science and more generally to locate science in social context.[20]

The myth of the neutrality of science

While the movement was forging its own politically engaged critique, within the academy there was also a parallel and dramatic shift in the history, philosophy and sociology of science. A sophisticated form of 'externalism',[21] holding the thesis that scientific knowledge is structured through its social genesis, had by the early 1980s become common to all three, so that one major strand of research has become aimed at demonstrating how interests construct knowledge while another has focused on the deconstruction of the language of science.[22]

The academy's recognition of the changed universe that modern science inhabited was signalled by Thomas Kuhn's 1962 publication *The Structure of Scientific Revolutions*, which presided over the steady thaw of an epistemology that had seemed forever frozen in the timeless certainties of positivism and the Vienna circle. More or less concurrently the historian Derek De Solla Price pointed to the distinction between a past era of 'Little Science' and the modern trend towards 'Big Science'.[23] Later Jerry Ravetz[24] developed this distinction through an examination of the circumstances in which scientists actually produced scientific knowledge. Abandoning the internalist and very abstract Popperian theory of 'bold conjectures and refutations'[25] which had come to dominate mainstream philosophy and history of science, Ravetz showed that whereas in its early period science was considered a craft,[26] by the beginning of the twentieth century scientists increasingly adopted industrialized methods of production.

More oriented towards contesting existing science in practice, others within the radical science movement were none the less pursuing the same theoretical concerns. Revolted by the genocidal technoscience that the United States was employing in its war in Southeast Asia and by the expanding new technologies of urban repression at home, they asked how science can claim to be ideologically pure, value-free, and above all neutral when it is torn from the context from which it is constructed and within which it will be used. Slowly, from a simple 'use-and-abuse' model in which science, though open to abuse by political others, was seen as itself fundamental, basic and pure (created by scientists who by implication shared in the purity and disinterest of their creation), the new critics of science – to the equal concern of both the scientific establishment and the old left – laid siege to the myth of the neutrality of science itself.[27]

Advocates of the new political economy of science[28] argued that in bringing science into the capitalist mode of production, knowledge itself, as the product of scientific labour, had been made a commodity. The history of patenting within science and technology was one of steady encroachment, beginning within physics and chemistry[29] but now enveloping the burgeoning area of biotechnology – and indeed life itself. In the 1990s the marxist analysis has been matched and replaced by the market language of 'intellectual property' designed to police ownership patterns in the interests of capital. Even for the basic sciences, seemingly remote from technological exploitation, the rewards and prestige go to those who publish the knowledge first. The very process of diffusion reduces the value of the knowledge (typically produced in the elite institutions of the metropolitan countries) as it is transferred to the weak and isolated institutions in the periphery. The value of the knowledge as it passes from the centre of production to the periphery declines as surely as that of a car as it moves from second to third hand. Susantha Goonatilake gives support to this thesis of 'dependent knowledge', drawing on the third-world experience of Sri Lanka and India.[30]

The change in the mode and place of scientific production, and its subjugation to the laws of commodity production, are features of the sciences most closely integrated with the reproduction of social and economic power. The physical sciences, above all physics itself, are at once the most arcane and the most deeply implicated in the capitalist system of domination. The means of producing new knowledge based on experimentation are symbolized by the giant machines (above all the particle accelerators, whose costs are so immense that they are restricted to the US, the former Soviet Union, Japan and, as a shared facility at CERN – Centre Européen pour la Recherche Nucléaire – the reinvigorated capitalism of Europe), and by the international collaboration/rivalry of the Human Genome Project. Experiments now take immense teams of researchers, so that a single paper may have some thirty authors, who acknowledge the support from unnamed cadres of technical staff without whom the experiments could not take place. At the same time, the physical sciences, particularly in the old capitalist countries, more or less successfully exclude all but extremely small numbers of women.[31] Industrialized sciences – Big Sciences – have been highly resistant to feminist reconceptualization; the successes of feminist re-visioning have lain in sciences such as sociology, history and ethology – all characterized by little capital equipment per worker and by craft methods of production.[32]

The social origins of science as alienated knowledge

While many within the radical science movement were influenced by the writings of the Frankfurt school, which alone within the Western marxist tradition saw science as a social problem,[33] it was Alfred Sohn Rethel, as part of that tradition, who was to seek to explain the social origins of the highly abstract and alienated character of scientific knowledge.[34] He suggested that while abstraction arose with the circulation of money, the alienated and abstract character of scientific knowledge has its roots in the profound division of intellectual and manual labour integral to the capitalist social formation. Scientific knowledge and its production system are of a piece with the abstract and alienated labour of the capitalist mode of production itself. The Chinese Cultural Revolution, with its project of transcending the division of mental and manual labour and their associated knowledges, was seen by Sohn Rethel – and indeed by many or most of the New Left – as offering a progressive model of immense historical significance. They saw not only the possibility of transcending hierarchical and antagonistic social relations, but also the means for creating a new science and technology not directed toward the domination of nature or of humanity as part of nature. It was politically significant that the hope for this new science came from the East, decisively breaking that Eurocentric and class story of the birth of science.

Today, it is questioned how far the radical impulse of Maoism was constrained and deformed by the continued practices of Stalinism, yet at the time the attempt to create new knowledge drawing on both the everyday experience of peasants and workers and the academic knowledge of the intellectuals was embraced.[35] (It goes without saying that the Cultural Revolution was an ungendered project, so that both peasants and intellectuals in an entirely naturalized way were understood as the necessarily masculine harbingers of change.)

The Cultural Revolution was not an isolated phenomenon, and was reflected by and influenced struggles in a number of third-world as well as first-world countries. In an early article the US black mathematician Sam Anderson[36] reported the struggles within Guinea Bissau and Mozambique to build a new science and technology *with* the masses. He also drew attention to the related task of recovering the erased history of African and Asian scientific achievement, pointing to the 40,000-year history of iron smelting in Zimbabwe and the history of the systematic destruction of cotton production in India and Africa so that production could be relocated in Manchester, where child labour could profit the British imperial trade. (By the 1980s, cotton along with other textiles was relocated once more, this time to the newly industrializing countries, as a new global – and sexual – division of labour was established by the alliance of footloose patriarchal capital and new technologies.)

Anderson's is an almost solitary voice, for he writes compassionately about the pain and danger of being inside and outside, of being, as a black scientist, one of 'America's peculiar beings'. He proposes that each black scientist could ensure that 'at least two sisters or brothers' get into college and pursue the sciences for black people'. Because today history seems to be erasing the pioneer voices trying to make such links between movements, it is important to set the record straight. Well before the powerful wave of black feminism of the eighties there were exceptional voices insisting on the connections between feminism and black militancy. The poet and political activist June Jordan, who taught and advanced black writing, was among the numbers of early black feminists criticizing the 'ostensible leadership' of the black movement for only advocating the liberation of black men.[37]

In a world where the costs of the new technosciences confront us in the pollution of the seas, the cities, the countryside, and in the fear of nuclear holocaust, such longings for alternative knowledges encompassing both the sciences and the arts and whose purpose is to serve the people cannot be dismissed as merely romantic. The realization of such longings has become a contributor to survival itself. Such hopes lay behind the mobilization of white male aerospace workers in Britain in the 1970s[38] – people not easily equated with romantic intellectuals – who were driven to conclusions very similar to those of Sohn Rethel. Beginning with their opposition to the threat of redundancy and with a moral distaste for being so deeply involved in the manufacture of war technology, the workers went on to design, and in some cases to create, alternative, socially useful technologies such as a vehicle which could run on both road and rail.[39] Although such projects could only be seen as prefigurative, and did not outlast the arrival of Thatcherism at the end of the decade, in their contestation of the division of mental and manual labour in the production of technology through the unity of hand and brain, they were part of the long struggle to transform technology itself.[40]

Although it is retrospectively easy to criticize the radical science movement for its shortcomings, not least its androcentricity, its preoccupation with the global political economy of science did help weaken the Eurocentricity of the history of science.[41] Above all the radical science movement had restored, through its political demands for a science to 'serve the people' without the need for corrupt statistics or Stalinist terror, the epistemological possibility of a 'two sciences' thesis. The movement had laid powerful siege to claims that science and technology transcend history, and made plain the class character of science within a capitalist and imperialist (or for that matter state socialist) social formation. The ideology of science was 'demystified', the myths that had served to gloss over the class structure of scientific production were exposed. From outside science it has become quite difficult to remember the hegemonic grip of science, the taken-for-granted internalism of the academic history, philosophy and sociology of science in the 1960s and 1970s, and to appreciate the transgressive practices and analyses which de-stabilized the old categories and created space for new alternative accounts – including those of feminism.

There is, however, a paradox. While today's social studies of science take for granted the social context of science, their practi-tioners none the less typically tell their own origins story so as to emphasize the internal development of their history, and to neglect any version of externalism, whether the historical materialist question of 'What conflict outside us was within us the reflex of thought?' or any social constructionist account. There has been a tendency to focus on Kuhn as founding father, single-handedly opening the doors to the possibility of a fully social account of science.[42] To question this account is not to diminish Kuhn's contribution, nor to neglect the importance of intellectual development, but rather to insist that attention is paid both to theories and to their historical location – not least our feminism's own theorizing and our own contexts of production.[43]

Thus while feminists discuss the relationship of feminism as a social movement to other such movements – in the past the New Left, and today the peace and environmental movements – the connec-tions between the feminist critique of science and the radical science movement with its primarily class but also anti-racist concerns have been often left in some obscurity.[44] Indeed, as the radical critique of science developed, the disjuncture between the politics of practical struggle and the politics of theorizing seemed to increase. Looking back over the writing of the sixties and early seventies, it is difficult not to feel that, as the critical work became more theoretical, more fully elaborated within a marxist viriculture, so the theorists' willingness to engage with the complexity of social relations – not least of those between women and men activists, which had been thrown into visibility through political struggle – was reduced.

The birth of the feminist critique of science

The willingness to engage with feminist questions within science, or rather technoscience (to give full weight to that iron-bound marriage of science and technology in the West), had to wait until the new wave of feminism was ready; initially there were other more pressing issues of women's daily lives to respond to. In the rest of this chapter I introduce the strands within the feminist critique of science with

which the remainder of this book is concerned, but I am conscious
that the price of selecting these is that I leave out other connected
areas such as the work of feminist science educators,[45] feminist
studies of technology in both employment and the domestic contexts,[46] feminist critics of militarism and feminist environmentalists.[47]

I have named my five strands as follows: first, 'Why so few?',
recalling Alice Rossi's provocative question concerning the paucity of
women in the university and research system in 1965;[48] second,
'Recovering Hypatia's sisters', to evoke that patient historical and
biographical work to recover the history of women in science, from
Hypatia herself to present-day women scientists; third, 'Contesting
patriarchal science', to bring out the committed resistance by feminist
biologists against the 1970s' wave of biological determinism, and also
the work of feminist historians of science in exposing science's
construction of women's nature. The fourth strand introduces 'The
feminist critique of epistemology', which speaks both to the strong
hopes of the early 1980s for an alternative feminist epistemology as a
successor science but also to the challenge posed by the postmodernist turn to any meta-narrative, and how this debate has been
managed among the feminist critics of science.[49] Finally, my fifth
strand I call 'Dreaming the future'. A number of friends remain
surprised that I here discuss feminist science fiction, as this seems to
me to have a special relationship both to feminist culture and to
technoscience. This relationship is both playful and serious; 'we' read
it – certainly more of us than follow the feminist science criticism. My
feeling is that, as a critical, wide-reaching engagement in feminist
debates around technoscience, feminist SF is quite simply far too
important to be left out.[50]

The ordering of these strands is fairly arbitrary, not least because
they start from different places in different countries, for example in
Britain and the US. Thus concerns about science in the British
feminist movement reflect the dominant radical feminist and socialist
feminist currents, and therefore began by contesting patriarchal
science. In the US, by contrast, the dominant radical feminist and
liberal feminist currents meant that their critique also began by
contesting patriarchal science but spoke strongly about the under-
representation of women in science. Within the framework of liberal
feminism, science and technology were simply occupations where
women were particularly thin on the ground.

Why so few?

'Why so few?' can be read as one of the central questions around
which the feminisms of the second wave have organized their critical
analyses of science. As the sixties come into historical perspective, it
becomes even clearer what a well-posed question this was, as the

problem in higher education and research was rarely that of the total and legal exclusion which first-wave feminism was to challenge so effectively, but of material and ideological practices which served to exclude all but a handful of women. The distinctive contribution of feminists who have survived in what Anne Sayre's[51] pioneering biography accurately called 'an especially male profession' has been pointed accounts of the men-operated exclusion mechanisms of science, from physics to psychology.[52] Naomi Weisstein's 1977 paper is a classic in this genre.[53] She wrote:

> I am an experimental psychologist, doing research in vision. The profession has for a long time considered this activity, on the part of one of my sex, to be an outrageous violation of the social order and against all the laws of nature. Yet at the time I entered graduate school in the early sixties I was unaware of this. I was remarkably naive.

Later, Weisstein was to add a footnote observing that she had since realized how exceptional it was for a woman to have survived as far as graduate school. Evelyn Fox Keller, writing of her experiences as a student physicist and later as a research worker, echoes this theme – the continuous, subtle and not-so-subtle exclusion mechanisms deployed against women scientists. She writes that as a student, she had to be careful to enter a lecture room with or after other students; if she entered first and sat down, men students found it threatening to sit near this low-status person – a woman student – and she was often surrounded by a 'sea of seats'. On one occasion when she solved a mathematical problem the male university teacher was so incredulous that Keller, like Naomi Weisstein in a similar situation, was quite gently asked who (i.e., which man) did it for her, or where she got (i.e., stole) the solution. Keller's experiences were not, however, unique; what was new was that they, and the sexual harassment that often accompanies them, are now discussed.[54]

As Diane Narek, teaching physical sciences, wrote, 'The only reason that there aren't any more women scientists and technicians is because the men don't allow it.'[55] Since the sharp insights of these pioneering voices, the careful historical work of Margaret Rossiter has detailed both what US men scientists would not allow and also how the feminist scientists of the time understood the nature of the problem, and what they were able to achieve in the task of change.[56]

Current liberal feminist attempts to attract attention both to the under-representation of women and to the lack of promotion for them have made but few inroads in Britain.[57] Indeed, despite some gains, such as the increase of women in medicine, losses are also evident and the proportion of women is now lower in some areas of science than it was in the interwar period. Further, it is not simply difficult to get into science, it is difficult to stay in. The labour process of an experimental scientist is even more in conflict with the demands of child care than that of, say, a historian or a sociologist. What Airlie Hochschild in a brilliant pioneering article characterized as 'the clockwork of male careers' ticks in an even more pronounced way in the laboratories.[58] While both the woman laboratory scientist and the woman historian may have in common the problem of the double day, the former has much less flexibility in choosing when or where to work.

Women who manage to get jobs in science have to handle a peculiar contradiction between being women and not women (i.e. scientists) at the same moment. Many have resolved this by withdrawing or letting themselves be excluded from science; others become essentially honorary men, denying that being a woman creates any problems at all. Long before the postmodernist language of multiple fractured identities enriched feminist analysis, it was understood that a woman scientist is 'cut in two'. Ruth Wallsgrove wrote, 'A woman, especially if she has any ambition or education, receives two kinds of messages: the kind that tells her what it is to be a successful person; and the kind that tells her what it is to be a "real" woman.'[59] In a later article Keller observes that 'any scientist who is not a man walks a path bounded on one side by inauthenticity and on the other by subversion.'[60] Arguing for the continuing need to build two-way streets between feminism and science, biologist Anne Fausto Sterling agrees with this depiction of catch twenty-two for women scientists, and adds with some feeling that 'being a feminist scientist makes matters worse'.[61] Men scientists, as we see from Sharon Traweek's study of physicists, like to get marriage sorted out quickly so that once their needs for love, sex and a pleasant domestic life are resolved they can put all their energies into science.[62] Small wonder that women, let alone feminists, working in physics are still rare. It is difficult enough to conceal part of oneself to pursue knowledge of the natural world; it is even more difficult to develop a feminist practice in the competitive world of science.[63]

Rita Arditti, noting how common it was for women scientists to marry men scientists – often in the same field, saw that, 'All had secondary positions to their husbands regardless of ability; their loyalty as wives had led them to accept precarious work situations in which their research was dependent on their marriages.'[64] Nor is the problem only about love and personal relationships; it is also about motherhood. 'Can I be a geneticist (or whatever) and have children?' is a not a question from the past, but a painful and very practical question for young intellectual women interested in the sciences today. For women who wish to contribute to the frontiers of knowledge in the US, the answer seems pretty much to be 'Probably no', whereas in the UK it is still, 'Well, maybe.'[65]

What were the conditions through which the few survived and in some cases made important contributions to knowledge? Censoring out problems is particularly evident in the autobiographical accounts of highly successful women in the sciences, and the attempt to ask distinguished women scientists (who rarely see themselves as feminists) to reflect on their lives has not been particularly successful. Two major collections were made, one by the New York Academy in the 1960s, and one published by UNESCO.[66] (It is not without irony that the male editor of this latter collection, an erstwhile biologist, had never appointed a woman scientist to the laboratory he directed – but the tone of the meeting which led to the volume was too polite for such blunt political comment.) These autobiographical accounts are in themselves by and large unrevealing; the difficulties of personal life are ironed out; the emphasis is on the excitement of science. Yet, reading across these autobiographical accounts typically shows a highly privileged class origin and the unusual support and encouragement of a scientist father or husband. While I return in a later chapter to this matter of the linkage between privileged class origin and access to science, it is important to note that it was in the craft areas of science that the daughters and wives of leading scientists enjoyed a certain privileged access to laboratories.

I remember asking one of the very few women Fellows of the Royal Society, Dorothy Needham, why the biochemistry laboratory of Gowland Hopkins had served, in the interwar period, as a refuge not only for brilliant Jewish scientists fleeing Nazi Germany (e.g. Hans Krebs and Fritz Lipmann), but also for so many brilliant women. These included Dorothy Wrinch, Dorothy Needham, Marjory Stephenson (who was to succeed Hopkins in the chair) and Barbara Holmes. Needham explained that it happened through Hopkins's daughter, Barbara (Holmes), whose love of science he had actively encouraged. The class element was important, for these women were not given any income; most of the work for which Needham herself received scientific recognition, including Fellowship of the Royal Society, was done without a 'proper job'. One woman scientist of this generation described how she was paid exactly the cost of replacing her own labour in child care – not a way in which male scientists' salaries have been arranged.

Recovering Hypatia's sisters

While it was an early and widespread political objective of feminist scholarship to find the women written out of history,[67] the erasure of women by the masculinist account of science was dramatically put onto the political and research agenda, not by a professional historian or by the general impulse to recover women, but because of the deep friendship between the crystallographer Rosalind Franklin and the writer Anne Sayre. That Sayre was married to the crystallographer David Sayre gave her a personal link into Franklin's research community. The shabby treatment of Franklin was widely known and deplored among crystallographers, but the issue was brought to a head by the publication of James Watson's highly personal and macho account of the discovery of the double helix structure of DNA on the basis at least in part of photographs stolen from Franklin.[68] But it was the context of the rising feminist movement which turned a story of professionally shabby treatment into a public issue of the gender politics of science. Sayre's passionate defence of Franklin probably did more than any other single book in the seventies to demonstrate the erasure of women scientists by men. That she challenged the accreditation system of science in the particular area of DNA, that macho molecule of the biological revolution, was an act of profanity. No fewer than two (James Watson and Maurice Wilkins) of the three male Nobel prizewinners felt it necessary to modify their accounts of the history of the discovery of the structure of DNA to accommodate Sayre's critique. Even more importantly, it enabled similar accreditation scandals to surface, such as the doubts about Hewish receiving a Nobel prize for work leading to the discovery of quasars based on observations made by his graduate student Jocelyn Bell,[69] the questionable passing over of Candace Pert for the Lasker Award for work related to neurotransmitters, and the erasure of Lise Meitner's contribution in the award of the Nobel Prize to her collaborator Otto Hahn. While there is little evidence that the accreditation system has become less loaded, there is greater realism about the extent of the bias built into the system. Equally, historians of science generally seem now to be more sensitive to gender issues in their accounts of scientific developments: Hypatia's sisters have come into increasing visibility, if not into justice.

From early new-wave biographical work, such as Laura Osen's[70] study of women in mathematics from the third-century Alexandrine Hypatia onwards,[71] came a common theme of the significance of sympathetic men family members in providing encouragement and practical support to enable these women's talents and interests to develop. And in using that word 'talent' we should remember both Pierre Bourdieu's characterization of it as learnt skills esteemed within bourgeois culture, and also Anne Phillips's[72] characterization of it as learnt skills esteemed within a patriarchal culture, rather than considering it as some lingering expression of essentialism[73]. Some support came from mathematical families, where women were able to draw on the cultural capital available to them as family members. Mathematics, like literature, could be and was seen as an appropriate activity for ladies, for like the novelists they only needed a 'room of their own'. The craft mode of production in specific cultural activities, then as now, makes them more accessible to socially privileged women.

The task of recovering Hypatia's sisters, of making the distinctive history of women scientists visible, was paralleled by the theory-driven historical investigation of the gendering of science itself. No single scholar made this double-faceted enquiry more apparent than did Evelyn Fox Keller, in two widely read books: the biographical study of the plant geneticist Barbara McClintock, *A Feeling for the Organism*, and her theoretical essays, published as *Reflections on Gender and Science*. The title of the biography was a clue to both McClintock's method and Keller's own, for this was no conventional biography. In part Keller's theoretical insights came through a re-reading of Plato, Bacon and the debate surrounding the birth of modern science as marked by the founding of the Royal Society. Listening to these gendered origin stories of science she focuses on their strongly sexual imagery, contrasting Plato's homoerotic sexuality with Bacon's mainly heterosexual and frequently violent prose. In her account of the debate to define the new science she argues that mechanical philosophy competed with hermetic, and that the triumph of the former both reflected and was constitutive of the polarity between femininity and masculinity – a polarity which was crucial to the formation of early capitalism.

Whereas the classical (i.e. ungendered) reading of Bacon claimed him as the father of the scientific revolution and emphasized the persuasive claims for the scientific method of enquiry as yielding the most reliable knowledge, the feminist readings during the early 1980s were radically revisionist.[74] Carolyn Merchant[75] pointed to the crucial shift from the central Renaissance metaphor of the earth as a nurturing mother to the seventeenth-century conception of nature as a disordered female demanding mastery. Following swiftly on the heels of this dramatic reinterpretation was that of Brian Easlea, who saw the struggle over the dominant metaphor of science as central to the knowledge system. Whereas Merchant insisted on the relationship of masculine science to nature, Easlea also emphasized political economy and the imperialist racializing connection.[76] He sees the physicist Robert Boyle, a founding father of the Royal Society and governor of the New England Company, as acutely aware that the American Indian conception of nature as a mother had to be overthrown. It was, as Virginia Woolf wrote, a 'discouraging impediment to the empire of man over the inferior creatures of God' and should be replaced by the sole claims of the Christian God the Father. This new metaphor of nature pointed not just to difference but to racial and gendered domination.

Such racializing and gendering of science was not confined to the eighteenth and nineteenth centuries but is present today, despite the ideological claims of the neutrality of science. But while in the seventeenth century only a few isolated voices (a prominent example was the polymath science fiction writer Margaret Duchess of Cavendish)[77] had sufficient social power even to begin to contest this conception of either themselves or nature, both first- and second-wave feminism were to expose the linkages betweeen science, masculinity and violence.[78] In the late twentieth century the languages of nature and natural science are still sexualized, but today's 'woman in the metaphor' is also modernized, as when the physicist Richard Feynman in his 1965 Nobel address spoke of scientific theory as a beautiful young woman to be wooed and won; as the theory aged, it remained to be honoured merely as an old mother who has produced children. Less constrained by the metaphor of marriage and motherhood, the philosopher Paul Feyerabend[79] sees nature as the compliant mistress whose sole function is pleasuring the (man) scientist. It has been the deconstructionist historians who have, through their detailed attention to the language of science, explored the changing constructions of femininity and masculinity and who have thus been in no small part responsible for opening the epistemological debate.[80]

Contesting patriarchal science

Overtly patriarchal theorists, responding to the question 'Why so few?', claimed that 'anatomy is destiny', and that this expression of the division of labour was and is biological in origin. Thus the debate about biological determinism was forced by feminism's enemies, who were not content to permit arguments concerning modifiable social discrimination to explain why there were so few women in science, or indeed anywhere very visible in society and culture. A virulent masculinism – with Steven Goldberg, E. O. Wilson and David Barash among the more conspicuous protagonists – threatened by the rising challenge of the seventies' women's movement, set itself the project of insisting, once again, that women's biology is destiny. Feminists generally and feminist biologists in particular found themselves centrally engaged in resisting such arguments.

But of course misogynist science was no new phenomenon within the history of science. Both the historians of science and the biologists began to explore this long past, increasingly developing a picture of science as usually androcentric and under particular conditions moving into active misogyny. The text and title of *Women Look at Biology Looking at Women*[81] summed up this increasingly complex analysis of a deepening research programme, gradually shifting the focus away from women in science to a view of women as produced by science. Increasingly the new scholarship drew on the concept of gender to illuminate a double process of a gendered science produced by a gendered knowledge production system. Was the seemingly taken-for-granted androcentricity, even misogyny, of science a matter of 'bias' which good, unbiased science carried out by feminists and their allies would correct, or was the problem more profound, one that only an explicitly feminist science could displace, so as to become, in the language of the Enlightenment, a 'successor science'?

The recrudescence of biological determinism during the seventies was committed to the renaturalization of women; to an insistence that, if not anatomy then evolution, X-chromosomes or hormones were destiny; and to the *Inevitability of Patriarchy*.[82] Such views fed upon the work of IQ advocates, whose views had become an important location for social and political struggle around issues of 'race' and class.[83] Within the US these interventions were greedily taken up by a government looking for ways to justify the withdrawal of resources from the Poverty Programme, as a *laissez-faire* approach to welfare was more in accord with nature. Despite resistance by the Welfare Rights movement, scientific racism helped justify cutting welfare benefits of poor – primarily black – women and their children, thus enabling more resources to be committed to the Vietnam war. In Britain IQ theory was extensively cited by the racist campaign for immigration restriction and fed racist sentiment that genetic inferiority explained high levels of unemployment and thence excessive demands on the welfare system by black people. The critical counterattack mounted by anti-racists, discussed earlier in this chapter, helped prevent the new scientific racism spreading unchallenged.

In the prevailing political climate, the relationship between biological determinists – especially in the guise of the new sociobiology – and the New Right was a love match. In Britain a New Right government happily seized on biological determinism as a scientific prop to their plan to restore women to their natural place, which at that point was not in the labour market. (By the mid-eighties the view changed and part-time women's work became the ideal solution, achieving unpaid labour at home and cheap labour in employment. From then on we heard little about women's natural place.) No one put the government's view in the early 1980s more succinctly than the Secretary for State for Social Service, Patrick Jenkin, in a 1980 television interview on working mothers: 'Quite frankly I don't think mothers have the same right to work as fathers. If the Lord had intended us to have equal rights, he wouldn't have created men and women. These are biological facts, young children do depend on their mothers.' While it was perhaps overkill to draw on both creationism and biology to make his point, in the political rhetoric of government ministers and other New Right ideologues, the old enthusiam for biological determinism was given fresh vigour by the fashionable new sociobiology. Thus at the height of the struggle of the feminist movement to bring women out of nature into culture, a host of greater or lesser sociobiologists, their media supporters and New Right politicians joined eagerly in the cultural and political effort to return them whence they came.

Although the early feminist movement – not least because of the Eurocentricity of much of its theorizing and politics[84] – had tended to dismiss science as peripheral, black activists, including feminists, were aware of the threat of racist science and the need to resist it. It was the distinctive theoretical and political contribution of the

growing numbers of black feminist writers and theorists in the
eighties to insist that 'race', gender and class form a crucial
'trialectic'.[85] By the end of the seventies the ideological avalanche of
the new sociobiology meant that the critical opposition by feminist
biologists became more relevant to the wider movement. Among the
feminist scientists themselves, many of whom had cut their political
teeth within the radical science movement (and the metaphor
indicates the painfulness and growth in that process), the wave of
biological determinism directed against women resulted in a conscious coming together to develop the scientific arguments against
sociobiology's claims. The result was a host of pamphlets, confer-
ences, symposia and books.[86]

Feminist ethologists, psychologists and biologists argued that the
claims to ground the sex-gender system in hormones, in evolutionary
sociobiology, or in terms of just-so stories derived from ethological
observations of other species were based on bad science in the
classical sense of the term: weak theory, inadequate and mis-
interpreted data, poor experiments, and inadmissible extrapolations
from observations made on rats, ants and ducks to humans. Such a
cavalier approach to the limits of scientific method would not be
acceptable in any less ideologically charged task than the legitimation
of male (and white) domination and female (and black) subordination
as rooted in biology and, therefore, natural.

The debate was and is waged in both the popular and the scientific
domains. While sociobiologists have repeatedly claimed that the
media have vulgarized the scientificity of sociobiology, it is difficult,
reading their own claims, to believe that further vulgarization is
practicable. Thus

> It pays males to be aggressive, hasty, fickle and undiscriminating. In
> theory it is more profitable for the females to be coy, hold back until
> they can identify males with the best genes . . . Human beings obey
> these principles faithfully.[87]

Or

> Rape in humans is by no means as simple, influenced as it is by a
> complex overlay of cultural attitudes. Nevertheless mallard rape and
> blue bird adultery may have a degree of relevance to human behaviour.
> Perhaps human rapists, in their own criminally misguided way, are
> doing the best they can to maximise their fitness. If so, they are not very
> different from the sexually excluded bachelor mallards.[88]

When sociobiologists themselves are writing like this, it seems no
great leap for *Playboy* to inform its readers that male promiscuity is a
biological part of every man's birthright, or for *Science Digest* to claim
that rape is genetically programmed into male behaviour. Other

props to biological determinism were provided by scientists who were more cautious in their texts but less so in their press interviews, giving the message that they were not unhappy with the strongly sexist and racist constructions being placed on their work.[89] Certainly the would-be theoreticians of neo-Nazism within both France and Britain welcomed both IQ theory and sociobiology.[90]

Contesting such patriarchal and racist science required two distinct moves on the part of the feminist biologists and their allies.[91] The first was to challenge the truth claims of the account within the scientific canon itself. The second was to create alternative accounts. This strategy of using 'good science' to drive out 'bad', with an alternative account claiming a stronger truth claim, was a more powerful move than merely pointing to the biased nature of androcentric science or even to the social determinants of science. While the latter could situate the discourses of 'biology as destiny' within their time–space contexts, it could not replace them; thus it could criticize but not offer a new life science.

The feminist critique of epistemology

It was partly by reflecting on these debates about 'good science' and 'bad science', which I have described here as contesting patriarchal science and which Sandra Harding speaks of as feminist empiricism,[92] that the fourth theme of the new scholarship developed. This, during the course of the eighties, came to be spoken of as the epistemology debate. Before I say where I stand in the current theoretical debate, I must emphasize that I do not want to be read as suggesting that these other strands are in some sense closed or entirely bypassed in the current hot debate. In large measure all these strands will continue to offer important new knowledge. Intensity of theoretical debate in one area does not indicate the desirability of closing down other approaches.

The epistemology debate was framed less by the feminist biologists located within science who had taken a major part in contesting contemporary patriarchal science than by the fast-developing feminist social studies of science. These, like their masculinist counterparts in the main/male/stream social studies of science, were primarily drawn from the disciplines of sociology, history and philosophy – but unlike the former were openly committed to the political project of feminism. Feminist theorists of science thus had at that stage two sets of tools which they could draw on: those of the radical science movement from which a number of the feminists had come, and those being developed by the new mainstream social studies of science. However, plundering tool-kits is not unproblematic. It raises Audre Lorde's troubling question of whether it is possible to use the master's tools to take down the master's house. One fundamental response has been negative, and has developed strategies such as *écriture feminine*, or sought to create a new language and culture of women. However, because I think feminists like any other community of resistance are both inside and outside culture, I want only to restate the warnings of the difficulties and the dangers.

The problem for feminist materialists is to admit nature, particularly the body – that is, a constrained essentialism – while giving priority to the social, without concluding at the same time that human beings are infinitely malleable. The dialectical relationship between two systems of production – the production of things and the production of people – seemed to hold the explanation not only of why there are so few women in science, but also, and equally or even more importantly, of why the knowledge produced by science is so abstract and disembodied. The very fact that women are, by and large, shut out of the production system of scientific knowledge, with its ideological power to define what is and what is not objective knowledge, paradoxically has offered feminists a fresh page on which to write. Largely ignored by the oppressors and their systems of knowledge, feminists at this point necessarily theorized from practice and returned theory to practice.

While it would be false to suggest that all work claiming to be feminist achieved such a dialectical synthesis, there is a sense in which theoretical writing looks and must look to the women's movement rather than to the male academy. Thinking from the everyday lives of women necessarily fuses the personal, the social and the biological. It is not surprising that, within the natural sciences, it has been in biology and medicine that feminists have sought to defend women's interests and advance feminist interpretations. To take an example: menstruation, which so many women experience as distressing or at best uncomfortable, has generated a tremendous amount of collective discussion, study and writing. A pre-eminent characteristic of these investigations lies in their fusing of subjective and objective experience in such a way as to make new knowledge. Cartesian dualism, biological determinism, and social constructionism fade when faced with the necessity of integrating and interpreting the everyday experience of bleeding, pain and tension. Taking pain seriously has not meant collapsing into the arms of biomedicine's imperializing strategies.[93] Any reading of the abundant literature of the grass-roots level of the movement reveals a feminism with a taken-for-granted embodiment combined with an understanding of the immense power of the social construction of knowledge.

The early feminist critique of science used language as a means of going behind the appearance of things, to reveal what the sciences were saying about the changing construction of gender. However, until the middle eighties the critics felt no necessity to refer to the rapidly expanding poststructuralist currents emanating particularly from France. Feminism spoke as an engaged discourse looking primarily to the movement rather than the academy. In the work of neither Merchant nor Keller, nor the early Donna Haraway, nor the influential standpoint collection *Discovering Reality* edited by Sandra Harding and Merrill Hintikka in 1983, nor Ruth Hubbard and her various collaborators, to take some of the early well-known names, are there references to Foucault as the theorist of discourse analysis, Derrida for deconstructionism or Lyotard for postmodernism. For that matter Merchant, Easlea and Keller made no reference to the influential work by Brian Vickers using rhetoric theory. Neither did the feminist critiques of science within the Harding and Hintikka collection (notably Nancy Hartsock and Jane Flax), nor my own work, which drew on US feminist object relations theory to support a realist project, show any interest in the linguistically oriented work of Lacan. Yet by the end of the eighties these, together with Lyotard as the theorist of postmodernism and the anti-feminist Baudrillard, to say nothing of Nietzsche himself, came increasingly to frame the debates around knowledge. It is not of course that postmodernism was not influential within other areas of feminist scholarship, but simply that within the feminist critique of science these androcentric voices and their feminist revisionings were until the middle eighties pretty much absent. Today, in the concern within feminism for a theory of knowledge, it is not so much the postmodern turn as the postmodern deluge.

Integral to feminism's struggle to gain power over our bodies and our lives has been a claim for a distinctly feminist science. But while there is general agreement that the first move is to challenge and overthrow existing canonical knowledges, the question of what we might replace them with produces, broadly speaking, two responses. The first is feminist standpoint theory, which looks to the possibility of a feminist knowledge to produce better and truer pictures of reality; the second is feminist postmodernism, which refuses the possibility of any universalizing discourse, but which argues instead for localized reliable feminist knowledges. The debate between these positions occupies chapter 4. Initially standpoint theorists shared much of that successor science tradition which sought to make over science, to locate it in proletarian, radical and now, feminist lives and knowledges. The crucial difference between those alternative epistemologies, which are direct descendants of the Enlightenment tradition, and the re-visioned projects of feminism is that where the former claimed both reason and objectivity as the historic allies of the oppressed masses, the latter echo the claim but simultaneously give new meanings to the categories of reason and objectivity themselves.

I want to sustain standpoint project theory, not least because I share Harding's political understanding that it 'empowers all women in a world where socially legitimate knowledge and the political power associated with it are firmly located in white, Western, bourgeois, compulsorily heterosexual men's hands'; but I also want to underline the connections between those who argue for a localized plurality of discourses and for whom the main enemy is hegemony itself[94] and the feminist successor science project, for both share a hostility to the dichotomous choice of any theoretical 'either/or'. Part of our energy comes from the empowering visions offered us by feminist science fiction writing – that fifth strand – that enables us to feel, like Marge Piercy's Dawn, that not only do we 'want to do something very important' and move beyond masculinist conceptions of reality, but we can actually achieve this. Through her we feel that, 'Some day the gross repair will be done. The oceans will be balanced, the rivers flow clean, the wetlands and the forests flourish. There'll be no more enemies. No Them and Us. We can quarrel joyously with each other about important matters of idea.'[95]

The question which gives its title to this chapter, 'Is a Feminist Science Possible?', makes clear that I ground my answers within one side of the current theoretical debate and a commitment to a fully historicized critical realism, even while I acknowledge that the postmodernist turn has provided much more sophisticated means to analyse the texts within which both science and its critique are necessarily located. The traffic between feminist realists and feminist postmodernists has been richly productive, and even though this book is pitched on one side of that debate, I want to make clear my debts to the other.

Most negatively I read feminist postmodernism, at least in its 'strong' or 'hyper' forms, as denying the possibility of empowering knowledge about either the social or the natural worlds, and in the last analysis as parasitical on those who continue, like most feminist natural and social scientists, to make 'truth claims'. None the less postmodernism has, along with realism, a shared conception of knowledge as being historically and geographically located and produced. Adherents of both see science as increasingly pervasive, so that experts and expertise consistently invade daily life.[96] But there the similarity stops, for whereas historicism and critical realism continue to distinguish between 'true' and 'false' knowledge – reflected within the natural sciences in the realists' insistence that it is possible to distinguish between 'good science' and 'bad science' – postmodernism makes a further move. It insists that the criteria through which goodness and badness, truth or falsity are determined are themselves integral to modernism, and cannot be legitimized outside it. The possibility of epistemology is thus dissolved. The distinction between 'good science' and 'bad science' becomes simply a matter of preferred belief, or as Richard Rorty, author of the influential *Philosophy and the Mirror of Nature* puts it, of 'science as solidarity'. Rorty argues that 'We pragmatists, who wish to reduce objectivity to solidarity, do not require a metaphysics or an epistemology. We do not need an account of a relation between beliefs and knowledge called "correspondence" nor an account of human cognitive abilities which ensure that our species is capable of entering into that relation.'[97] In this way postmodernism points to the situatedness of thought, relegating truth claims as having the same status as fictions: 'they are stories we choose to believe.'[98]

However, except in the stories told by their postmodernist critics, few contemporary scientists actually think they are making absolute truth claims, although they do believe that the claims they make, the stories they tell, have a different kind of reliability to those of fiction. Perhaps truth in the strong sense used by Rorty et al. never exists outside the certainty of 'true for me', which I get when I read a poem or a novel. Contemporary scientific knowledge is simultaneously hedged around with conditions, probability and provisionality – except when making claims to government or industry for more resources, when the ideology of science is wheeled out to provide the rhetoric of certainty. (Nowhere is this more apparent than in the history of the last fifty years of nuclear research, and today in the Human Genome Project, which I discuss in chapter 8.)

Thus while it is unquestionably the case that there is a deep crisis in the theory of representations within the arts and the humanities[99] and that Agnes Heller's question *Can Modernity Survive?* admits of no easy dismissal, it is difficult not to see postmodernism as a nihilistic assertion of the primacy of the humanities and arts in the face of the almost overwhelming powers associated with modern science and technology. Postmodernism dominates the cultural journals and the debates of the humanities intelligentsia; science and technology as usual continue with the imperializing and frequently lethal agenda of the real. Postmodernism's relegation of the contest between more and less truthful accounts to one about different stories that we may choose to believe or not has immense and politically weakening implications for social criticism. To give a recent and bloody example, the remorseless media campaign conducted during the war in Iraq to insist that the new military technology was 'smart' and in conse-quence only eliminated military objectives cannot be falsified by postmodernism, only deconstructed as an account which we may choose to believe or disbelieve. While both postmodernists and realists may point to the long association of white European masculinity with the slaughter of third-world people and the violent 'rationality' of modern science and technology, only realists (and solid liberal empiricists) can and do construct an alternative and truth-claiming account that the bombing was indiscriminate.

Yet even while I want to hold on to the possibility of making and claiming truth for critical accounts, unquestionably societies charac-terized as modern are undergoing profound changes, and there is much less certainty about whether, how, if and when scientific knowledge can be used to increase personal or social rationality. For example, women are simultaneously given the epidemiological news that to have two alcoholic drinks a day decreases the chance of a heart attack, and also that this increases the chances of stroke and that their babies will suffer should they be pregnant. In these circumstances scepticism, even cynicism, about scientific knowledge claims becomes widespread, making its own connection with the fashionable and conservative nihilism of mainstream postmodernism. The issue for feminism is whether political commitment can be integrated with postmodernism in such a way as to empower women, or whether its anti-realism weakens the feminist project.

There are in a number of feminist postmodernist writers troubling signs of avant-gardism for its own sake. Their not infrequent references to Nietzsche's 'gay science' do more than merely forget the danger of using the master's tools to dismantle the master's house but go on to ignore that theorist's profound misogyny. Unquestionably my distaste for social theorists explicitly associated with misogyny and anti-semitism and implicitly with the Nazism once more stalking Europe gives me particular difficulties. My commitment to defending what can and should be defended of modernity, to feminist realism and to the feminist re-visioning of the concept of rationality stems in part from being part of a Western European feminism. By contrast postmodernism – or anti-realism – seems since the mid-eighties to have formed a powerful current within North American feminism.

The opening lines of two recent and well-received feminist theory texts exemplify these cultural and political differences. From the US, feminist philosopher Linda Nicholson writes:

> From the late 1960s to the mid 1980s, feminist theory exhibited a recurrent pattern: its analyses tended to reflect the viewpoints of white, middle class women of North America and Western Europe. The irony was that one of the the the powerful arguments feminist scholars were making was the limitation of scholarship which falsely universalised on the basis of limited perspectives.[100]

Against this US account of Western feminism, which in itself makes an appeal to a reality which I have very great difficulty in recognizing (or, to put it in postmodernist terms, Nicholson tells a story I do not entirely believe), the British sociologist Sylvia Walby opens her book *Theorizing Patriarchy*:

> Why are women disadvantaged compared to men? Has this inequality reduced in recent years? What difference, if any, does the increase in women's employment make to other areas of women's lives? Is the sexual double standard a thing of the past? Are contemporary forms of femininity as restricting as those of the past?[101]

While Walby, in her claims to critical realism, quickly appeals to the evident empirical 'facts', she fails to grapple with the crises of representation which Agnes Heller[102] or Joan Scott[103] confront so bravely. None the less her stance as a feminist critical realist is still dominant among Western European feminists within the social and natural sciences, though not within literary studies, philosophy, psychoanalysis and cultural studies, for there the postmodernist currents are very strong. (At least, for these matters of theoretical current change rather rapidly, that is how it seems to me as I write.)[104] By contrast my impression is that in third world feminism, where the issues of survival are too stark to admit such debates, realism holds sway, except in literary and philosophical circles. Despite Rorty's assertions, the feminist realist does not hold up a mirror to either the social or the natural world. Her task is to go behind the mirror, to go behind the appearance of things. Alice's project of going through the looking glass is closer to feminist critical realism than the 'glassy mirror' of either Francis Bacon or Richard Rorty.

At the same time, because feminism as a *political* project has a fundamentally critical relationship to theory, as a distinctively postmodernist feminism as against a disengaged (and often both covertly and overtly androcentric) postmodernism has come more clearly into view, the divisions between realists and postmodernists become less sharply drawn. Nowhere is this more distinct than within the feminist critique of science.

2

Thinking from Caring: Feminism's Construction of a Responsible Rationality

Not everyone learns from books.
Sheila Rowbotham, *Resistance and Revolution*

The absence of 'women' as a social category was one of the distinguishing features of 1960s political and academic culture. By contrast with the previous edition of the *International Encyclopedia of the Social Sciences* in the late thirties, which had twenty-five pages addressing women's political cultural and economic concerns, the 1968 edition had no references to women. Even as late as 1974, when feminists made sexual divisions the theme of the annual conference of the British Sociological Association, they precipitated the response from a leading social theorist that 'women' were not a sociological category. 'You might', he said, showing a double lack of sociological imagination 'as well debate the colour green'.[1]

It was against this swamping naturalism that second-wave feminism's initial concern was to make women's lives visible, so that that which was hidden and naturalized could be seen by the movement itself as socially and historically constructed. Women's groups explored their common experience[2] of daily life, in the process building a different knowledge which sought to transcend the division between the personal and the political. Providing social explanations for women's oppression was crucial in enabling women to cease being mere objects of history, and instead (though as usual not necessarily in circumstances of our choosing) becoming historical subjects who made our own history. Smashing the oppressor's glassy mirror which reflected and sustained the status quo, and making an alternative knowledge which explained who women were and might be in the world, was integral to becoming a social movement.

By the 1990s a self-consciously postcolonial feminism was actively reinterpreting the social category of women to emphasize the immense diversity while seeking to hang on to commonality. But attending to diversity or difference could only follow from an initial project of rescuing the category of women from the realm of nature and relocating it in the realm of culture. Gayatri Spivak writes that the academic disciplines are unable to do much to right the wrongs of colonialism but that they can contest the colonialism of the mind.[3] She speaks of undoing 'the effects of colonial history on the production of knowledge' by 'the retrieval of information to restore the balance of historical knowledge'. Much of this was achieved by focusing politically and theoretically on women's daily lives – to find out what women in all their diversity actually did in both paid and unpaid work.[4]

The early concern with the latter – the unpaid, unacknowledged labour of housework, child care and sexually servicing men – was crucial in exposing the social nature of the sexual division of labour.[5] Such collective self-scrutiny also brought the social nature of the sexual division of labour into visibility. The fact that within the dominant culture of the West this division of labour was recognized, but in a profoundly naturalized way, meant that the task of breaking through the construction of womanliness could only be carried out by a social movement so unincorporated in the existing social and cultural order that it was free to develop its own accounts of the world. Within the dominant ideology the suitability of a woman's paid work was determined by its compatibility with her other nature-given tasks of being a wife and mother, so that schoolteaching was a classic example of a job which accommodated what was spoken of, at best, as 'woman's two roles'.[6] Most occupations are so deeply gendered that unless 'woman' preceded the occupational category, it went without saying (and still often does) that its holder was a man. Scientists are not the least among this group; women who are scientists are the marked other. Breaking through the intense naturalism which surrounded and defined women's lives demanded little short of a cultural revolution.[7] At a time when the 'false universalism' of early second-wave feminism is widely decried, it is important to insist on its achievement in beginning to mobilize women in all our diversity.

This present chapter returns to the discussion of women's work and women's knowledge or, as Bettina Aptheker[8] speaks of it, 'women's work, women's consciousness and the meaning of daily experience'. To get at this afresh while drawing on feminist materialism I want to return to the painful contortions of particularly the British marxist feminist debate over domestic labour. This is partly because (and I entirely agree with Aptheker here) when faced with a choice between staying faithful to women's lives or faithful to marxist theory, influential British marxist feminists seemed to chose the latter, weakening if not losing a feminist agenda in the process. As I argue below, US marxist feminism by contrast kept its feminist agenda sharper, partly by its greater willingness to use and grapple with the concept of patriarchy and partly by its creative use of object relations theory. However, during the seventies and early eighties marxism was unquestionably a powerful resource to feminism. What feminism took from Marx was his critical method, which enabled him to go behind the appearance of things to what he spoke of as their 'essence'. Above all he was able to go behind the appearance of freedom in the labour market, in which buyers and sellers freely bought and sold, to reveal the systematic relations of domination and subordination which are located within the capitalist mode of production. For a materialist feminism the task was to go behind – the appearance of love, the naturalness of a woman's place and a woman's work – to reveal the systemic relations of domination and subordination within patriarchy. Marx himself, despite a flicker of interest in his early more philosophic writings in just social relations between men and women (and which I guess is why some of us were always more attracted to the 1844 manuscripts than to the mature work of *Capital* or the *Gundrisse*), was the theorist of revolutionary change achieved through class struggle around the means of production.

The main weight of Marx's theorizing locates the first social division of labour as between mental and manual, taking the division of labour between men and women, not least within the family, as entirely natural. In his insistence on the social division of labour – between that of the 'hand' and that of the 'brain' – Marx entirely misses that of the 'heart'. Yet women's work is of a particular kind. Whether menial or requiring the sophisticated skills involved in child care, it almost always involves personal service. To make the nature of this caring, intimate, emotionally demanding labour clear I use the ideologically loaded term 'love'. For without what we call love, without close interpersonal relationships, human beings, especially young human beings, cannot survive.[9] Emotionally demanding labour requires that the carer gives something of themselves to the person being cared for,[10] so that even while child care is capable of immense variation within societies, across societies and across time, it remains the case that nurturance – a matter of feeding and touching, comforting and cleaning bodies – is cross-culturally primarily the preserve of women.[11] To say this is not to suggest that every woman necessarily takes part in such activities; simply that culturally the production and reproduction of people are allocated primarily to women and that this activity is qualitatively different from the production of things or ideas.[12]

Imperialism, and above all slavery, brought the terrible dimension of racism to this reproductive labour of women. Under slavery many black women were compelled to work like beasts of burden in the fields, both beaten as harshly as male slaves and also sexually abused as women. Others were set to work in the house under the intimate surveillance of white women and as sexual objects for white men. bell hooks, writing about the process of enslavement of black women, observes: 'The slaver regarded the black woman as a marketable cook, wet nurse, housekeeper, it was crucial that she be thoroughly terrorized so that she would submit passively to the wills of the master, the mistress and their children.'[13] The dictionary reminds us that 'servant' was used in seventeenth-century North American colonies as the usual designation for a slave. The long imprint of the colonial past means that domestic service in private homes remains a substantial labour market activity for US black women, and that cleaning and menial work in public institutions is still disproportionately allocated to black women in the UK.[14] Paid psychological care work is 'raced' as well as gendered. This placing in the labour process has its effects on the production of knowledge.

Thus, although that strand within feminist theorizing which claimed critical realism for its accounts was grounded in the materialism of Marx, to sustain a feminist agenda it necessarily continues in a spirit of such radical revisionism that it no longer makes sense to ask whether the theory is or is not 'marxist'.[15]

For materialist theory, human knowledge and human consciousness are not abstract or divorced from experience or 'given' by some process separate from the material reality of the world. As Marx observed in his critique of methodological individualism, 'Man does not squat outside the world', nor do women. Knowledge comes from practice, from being in, working on and changing the social and natural world. As people work on nature and transform it, they gain knowledge of how nature – including their own nature – is organized and may be explained.[16] The birth of modern science in the seventeenth century was above all the equation of the new knowledge with experiment, with intervening in nature, in order to go behind the appearance of things. Yet even while modern science rested on experiment, thus uniting hand and brain, and was anti-elite in the sense that all (men) could engage equally and fruitfully in making the new knowledge, by the nineteenth century the ideology of science came to celebrate theory, to cast something called 'pure' science as of greater value and higher status than 'applied' science.

Not only is the history of science told as a story of theoretical development but even today the theoretical physicist is understood as being hierarchically superior to the mere experimentalist.[17] Within Britain the hierarchy is intensified by a class division, with science understood as more genteel than engineering, given the latter's associations of manufacture and social contact with working-class people.[18]

Even within feminism there is a growing celebration of the 'feminist theorist' who in feminist discourse seems to be located as somehow on a higher, more prestigious level than the empirical researcher or the community activist. Such hierarchical divisions are not separable from the class and race differences between women; only a minority of women have that leisure which is crucial for space to reflect and for the creation of theory.[19] The kinds of debate which have from time to time erupted within feminism, concerning the problems of expertise and a commitment to flat organizational forms, have counterposed three kinds of knowledge: (1) the production of knowledge about women, (2) knowledge for women and (3) knowledge by women. As an academic feminist I feel myself caught in that tension between my political commitments and my paid location within the academy.

Such divisions of esteem within the production system of knowledge, between the mental labour of the theorist and the fusion of mental and manual skills of the experimentalist, are not unconnected with the division of labour within capitalist patriarchy. Science, as organized knowledge, is constituted through material practices, through labour and access to the means of knowledge production – whether the latter is the computer and the books of the writer or the vast accelerator at CERN for the nuclear physicist. The deep divisions of gender, race and class thus structure both the labour process of science and its knowledges. The ideas which come from the perspective of bourgeois white men, whether about the social or the natural worlds, serve to preserve the status quo or – as Dorothy Smith puts it – are part of the relations of ruling.[20] Transformative or empowering knowledge comes within a materialist theory from the political struggles of the subjugated. In consequence its standpoint is more to be trusted, is more truthful, than the standpoint of the dominant group whose view of the world is shaped by their need to retain mastery. Reliable knowledge is knowledge from below.

To understand the specificities of masculinist knowledge and the exclusion of women from it, and the transformation of knowledge made possible by a feminist epistemology, it is therefore necessary to return to the particular nature of women's labour in the world and the division of labour between the genders. Masculinist knowledge in the West has taken the form of an intense emphasis on the domains of cognitive and objective rationality, on reductive explanation, and on dichotomous partitioning of the social and natural worlds.[21] Increasingly, as the global economy structures production including the production of knowledge, the new international division of labour allocates the processing of both material things and nature to the periphery. It is here that women labour at less than even subsistence wages. By contrast, a feminist epistemology which derives from women's lived experience is centred on the domains of inter-connectedness and caring rationality, and emphasizes holism and harmonious relationships with nature, providing links to that other major social movement of our time, the ecological. At times the ecological movement also takes on gender and challenges the new forms of imperialism.[22] To borrow from Maurice Bazin, against 'their science' we range 'our sciences'. To understand feminist knowledge, a 'feminist science', one key place to begin is with women's everyday lives as both paid and unpaid workers.

The social origins of the segregated labour market

Examining this problem within contemporary capitalist patriarchy, Western feminists point to the 'family wage' in which the male breadwinner receives from capital an income sufficient to reproduce not only his own labour power, but also that of his wife and dependents. Indisputably, the family wage as it emerged during the nineteenth century in the most unionized and better-paid sectors of the economy served to improve the conditions of a particular class fraction – but at the price of enforcing women's and children's dependence on men. Feminist historians have shown that men's wages elsewhere within the economy never enabled them to achieve family breadwinner status, but the ideological goal of the family wage was held before them.[23] Protective legislation removing women and children from certain kinds and conditions of work (ostensibly for their sake) led to their systematic exclusion from the leading sectors of the economy, where the organized (male) breadwinners and (male) capital could battle out the higher wage levels together.[24] During the nineteenth century, although many women were, in fact, bread-winners (not least because of the high rates of widowhood, to say nothing of the large numbers of single women), ideologically they were marginalized in their claims for equal participation and equal pay within the labour market.[25]

Yet while in Britain today the ideology of the family wage has weakened, women's wages have not taken a sudden upward bound. Indeed as the numbers of single parents continue to rise, many women and their children are finding themselves both in poverty and with new emotional pressures, as biological fathers simultaneously retreat from their commitments as financial providers,[26] and make new emotional claims, backed up by law, as fathers. Most women's lives are precariously balanced at the intersection of family, labour market and social policy. Thus while it is true that more women with children are in paid work than ever before, most are also in the most casualized and worst paid occupations,[27] with little union protection.[28] The labour market of science and technology is part of this changing but still profoundly segregated form.[29] Nor, despite some gains here and there, have the educational reforms at the turn of the century which admitted women to study, or the expansion of higher education in the sixties, or legislation for equal opportunity and equal pay since the 1970s overcome this structuring of the scientific labour market. Despite national variations between countries, men and the interests of men have long commanded science.[30]

Although in conditions of boom, such as those that characterized Western capitalism up to the mid-seventies, capital looks to women (and other marginal workers) as a source of labour power, and talks of opening up the entire labour market to women, in practice it remains intensely segregated.[31] Even in years of expansion women have remained in an exceedingly narrow range of clerical and service occupations. In Britain, for example, segregation within the labour market was more marked in 1971 than it was in 1901.[32] Sweden, whose equality legislation constitutes an immense historical achievement, none the less has one of the most segregated labour markets in Europe. Nor is a segregated labour market necessarily to capital's advantage; in the post-Sputnik years, the United States, anxious to boost its numbers of scientists and engineers, looked to women as a possible supply source. Focusing on resocialization strategies and publicizing successful 'role models', the state was none the less largely unsuccessful in opening the scientific labour market to women. It seems to take little less than a nation state at war to modify the segregated market significantly.[33] A buoyant economy plus pressure from below, which existed in the sixties and early seventies, produces some, but not radical, concessions. In the present conditions of recession – even slump – the massive casualization of the British labour market finds its reflection in the laboratory. The trade union movement has been profoundly weakened by right-wing social and economic policies, beginning the eighties with a much stronger commitment to defending men's than women's jobs, and closing the decade by retreating from an agenda of struggle towards a re-discovery of its origins in mutual benefit employee associations. More positively, the increasing presence of feminists and black sections within the unions is moving them towards policies of gender and racial justice.

Science as a global production system is simply part of the segregated labour market, excluding women – other than those in exceptionally favourable circumstances – from occupying elite positions within the production of knowledge.[34] Many, even most, women in natural science and engineering are relegated to those tasks that most markedly parallel their 'primary' task as wife and mother. If we examine the full labour force – not just the scientists but also the technicians, secretarial staff and cleaning personnel – we see that the majority of women are still carrying out manual and personal service work. Neither chance nor biology explains men's occupancy of the leadership positions within science. The analysis of the segregated labour market, nationally and internationally, shows that women are concentrated not only in low-paid work but frequently in work of a particular kind – human service work and menial work, a remarkable echo of women's work within the home. It is precisely in those societies where the largest proportion of women are employed in the paid labour market that we see the sharpest expression of segregated occupations and the greatest extent of part-time work. The latter, which is ideologically proposed as 'choice' and as a gain for all, is for women in reality structured through the greedy time demands of the double day. Despite the ideology of science being above gender, this holds within the scientific labour market as much as any other.

In science therefore it is true, but only in a particular sense, that there are few women. In the United States, as in Britain, the academic staffs of science and engineering departments are predominantly men, though the pattern varies from engineering (where there are almost no women academic staff and under 10 per cent women students) through physics and chemistry (which are slightly more mixed) to biology (where there is the greatest proportion of women as academic staff, although still a minority and at the lower levels). But, while relatively few women are in evidence in advanced science and technology education, that is not to say that none are to be seen at all. Women clean the floors, under the supervision of men supervisors; women act as technicians, under men senior technicians; they work as waitresses under men catering officers; and they work as secretaries, typing letters dictated by men, and generally smoothing interpersonal relations. The point is – and it has to be made again and again – that women's paid work, even in the science or technology laboratory, echoes what they do at home. The laboratory is simply part of the segregated labour market.

Time for theory?

Although women are just over half of humanity, women do much more than half of the total labour of the world. There is a depressing consistency in the international time-budget studies which have over the last decade or so mapped out the gross inequality between the hours worked by women and by men.[35] While the double shift of paid and unpaid labour seems, at an everyday level, sufficient explanation as to why women have been denied access to the time-demanding arena of public life, not least science, it does not explain the silence over the gendered politics of time.

Even now, a quarter of a century after the advent of second-wave feminism, we are faced with the contradiction that while the history of the organized labour movement acknowledges the reduction of hours of work as one of its key objectives, and now increasingly sees it as the creative response to the restructuring of employment, the cruel hours of labour extracted from women are neglected by the state and the male-dominated labour movement alike. André Gorz, in an imaginative political response to a postindustrial capitalism, writes:

> Nowhere is the line separating left and right clearer than on the question of time; the politics of time. According to whether it is a politics (and policy) of the right or the left, it may lead either to a society based on unemployment or to one based on free time. Of all the levers available to change the social order and the quality of life, this is one of the most powerful.[36]

Gorz echoes Marx's point about the politics of time and class domination: 'In a capitalist society, spare time is acquired for one class by converting the whole life time of the working class into labour time.' But Gorz fails to add the dimension of gender which could well argue that 'In a patriarchal society, spare time is acquired for one gender by converting the whole life time of the women into labour'.[37]

Indeed, with the recrudescence of market economics and the retreat from the welfare state which has characterized the past decade, the state is intent on making sure that for the majority of women their time will not be their own.[38] The collapse of 'actually existing socialism' has reduced the social supports to women in the former Soviet Union and Eastern Europe. In the countries of the third world, afflicted by water shortages in no small measure created by the first world, women may spend up to ten hours each day simply collecting water for themselves and their households.[39] Reflection on experience and the acquisition of organized knowledge require time. This time has been appropriated by one gender, and with time, the means for theorizing. For great numbers of women, theory has to be done in the cracks.[40]

The labour of love

Getting hold of the labour process of women's work – remembering that it exists as 'not men's' – gives us a way of looking at words like 'intuition', 'sensitivity', 'interrelatedness' and all those other feminine words about which many of us have held complex and contradictory feelings.[41] It has been feminism, aided by the crisis in the welfare state, which has not only named and thus brought into visibility the distinctive labour of women, but also insisted that we understand its double-sidedness both as labour and as love.[42] This combination of menial labour, often involving long hours, boring repetitive house-work and very complex emotional work with children, husbands and dependent elderly people, has been patiently unravelled by femin-ists. It has been salutary to read Charlotte Perkins Gilman's turn-of-the-century classic *Women and Economics*[43] and reflect how long it took my generation to recapture the clarity of her distinction between the morally legitimate claims of mothering children and those of servicing men.[44]

Feminism's rethinking of women's work and learning to recognize women's skill had to tackle a keystone within the patriarchal ideology of work, namely that where 'skill' is, women are not.[45] The under-valuation of unpaid labour is of a piece with the undervaluation of women's employment. To use Diane Elson and Ruth Pearson's term,[46] women are inferior bearers of labour and their presence in any significant way within an occupation signals that it is of low status, requiring only modest financial reward. Black and third-world women are racialized within the labour market, and their skills even further devalued.

Skills which are acquired by women through practice within the home are both undervalued and systematically denied their social origins.[47] This is so whether they are utilized in paid or unpaid labour. In the new electronics world factories, women's skill at microcircuitry and patience with repetitive tasks are seen as biological attributes; that these are third-world women underlines the racist as well as patriarchal assumptions.[48] Within public caring labour – what Leila Simonen[49] speaks of as 'social or occupational mothering' – even where the relational skills of women are acknowledged, it is not in terms of status or financial reward. In a nursing context the patient's wellbeing, even life, may turn on close personal support (TLC – tender loving care) which women are widely seen as being able to supply in abundance, yet women's pay cheques do not acknowledge this as a socially acquired skill. In the domestic context women's nurturing qualities are simultaneously praised and seen as pre-scientific practices awaiting the emancipatory certainty of scientific knowledge.[50]

Experiential knowledge is thus dismissed and trivialized, while an arrogant, objectivizing science seeks to instruct women in its own practices. Not for nothing does the woman in the feminist cartoon say, 'Well, if I get my instincts biologically I'm not having you tell me what to do!' The increasing tendency to make caring 'scientific' has eroded women's confidence, delegitimizing the knowledge they have gained individually and intergenerationally from the practice of caring.[51] This professionalized conception of scientific knowledge, as arcane knowledge handed down by gendered experts, appoints itself to lead women into emancipation, for when it has succeeded in making a definitive map of mothering, it will be entirely possible for men to carry it out.[52] As with other professionalizing theories, this one turns on the denigration and disempowering of those it purports to aid. For example, it refuses to understand the emotional complexity and particularity of child rearing; as Carol Gilligan[53] puts it, this 'intuitive ability' is not an innate faculty but one 'that comes only with a certain sort of training'.

While the first phase of feminist research on women's labour was concerned with claiming it as work and as largely evaded by men, the second phase began to explore the labour process of caring and what caring work meant to the givers of care. Lacking the language to explore the tacit knowledge derived from caring, the practice and the knowledge have been treated with even less social esteem than is accorded to the tacit knowledge of manual labour.[54] Women themselves discussing childrearing emphasize the accumulation of skill – the second baby is easier than the first. At the same time, each infant is unique and requires a special and highly flexible response. Sara Ruddick's discussion of 'thinking as a mother' deepens this focus and argues that the labour of motherhood requires a distinctive sensitivity, which demonstrates a 'preservative love . . . fosters growth and values change . . . values open rather than closed structures, and refuses sharp divisions between inner and outer-self and the other'.[55] As a materialist Ruddick sees thought as arising out of social practice; she both admits the body and opens up the possibility of a distinctive kind of thinking arising from the labour of mothering.

The problem for women has been how such collective knowledge may be shared and developed when experiential knowledge has been dismissed as purely subjective. Feminist studies of birthing consistently report that women's sense of self-confidence has been eroded by the medicalization of reproduction.[56] Yet those areas of caring where the direction is almost entirely in male hands and claims to be guided by the achievements of science are precisely those where fad and fancy seem to have been most free. The sorry history of medicine, above all psychiatry and gynaecology, is full of representations of women, their bodies and their minds, from perspectives which range from the paternalist to the misogynist.

The intervention of scientific experts into areas of domestic and people work has not infrequently harmed practice which has been built up through careful observation. Waerness[57] examined cookery books and showed how those inspired by the latest scientific nutritional thinking led to unsound advice, while the practical guides offered by women cookery writers stood the test of time. She makes a parallel argument for nutrition to that made by Barbara Ehrenreich and Deirdre English, Margaret Versluysen and Jean Donnison,[58] on the replacement of the female midwife by the – in reality more ignorant – male and rapidly professionalizing doctor. Knowledge born of practice was frequently more securely founded than the proposals from a fragile – and often arbitrary – science.

Alienated and non-alienated caring

It has been both a theoretical and an empirical problem that even where feminists tried to separate housework from peoplework, the two continually merged. Caring, despite the best efforts of social policy research and psychology, requires much more than the abstraction of words. It is possible to feel or recollect the satisfaction of caring for someone, of finding all the little pieces of comfort that were important to that small child, that very elderly person – a mixture of words and silences, of favourite food and drink, of hard work in cleaning up a wet or dirty bed, of special ways of doing things. All the senses were involved; the cared-for looked good, felt good, sounded good, smelled sweet. Yet the pleasure did not just belong to the carer; it belonged also to the cared-for; at best it was mutual.

Is it all a con? Is this part of the emotionalization of housework? For in an entirely negative way emotion as integral to caring labour has become historically linked through the processes of mass consumption to a degrading emotionalization of housework.[59] At its nadir in television advertisements women are invited to feel that love is superwhite shirts for their husbands and children. A woman's feminine identity as madonna/whore is beamed out as the sexually attractive, perennially young woman celebrating her immaculate laundry. It is important to see that this emotionalized housework within industrialized countries is a relatively new phenomenon associated with the emergence of middle-class houses without servants, and of mass working-class housing. Providing the breadwinner/housewife division of labour of welfare capitalism is realizable, then unprecedented standards of domestic comfort can be achieved through the emotionalization and mechanization of housework.

But this does not mean that the pleasure of caring for someone is unreal, nor that it involves no work, nor that taking part in the relations of caring labour does not yield understanding. Indeed both feminist psychologists and philosophers have proposed that the scrutiny of women's caring could yield an ethic of care,[60] even to a theory of citizenship in which caring, like other duties, becomes a citizen's public obligation.[61] In consequence the task has been to analyse caring as labour, meaning and relation. Under what conditions do women freely care and under what conditions is caring extracted from them? How far is women's caring part of what Hilary Land and I have called compulsory altruism?[62] For caring, whether paid or unpaid, like other forms of labour, exists predominantly in its alienated form but also contains within itself glimpsed moments of an unalienated form. It is important with all forms of labour to insist that the experience of the unalienated form is located – however fleetingly – within the alienated, as otherwise we have no means of conceptual-izing – however prefiguratively – the social relations and labour processes of a society which has overcome alienation. Reflecting on caring labour offers clues as to why such work can on one occasion offer great satisfaction and on another be the site of tremendously hostile and painful feelings, in which the cared-for person and the carer confront one another as hostile beings. The same reflection also speaks of the pleasure and satisfaction to be found in reciprocal care.[63]

Less visibly, at least because it is masked by the ideology of romantic and maternal love, husbands and grown sons extract caring labour from wives and mothers. In an early article Heidi Hartman[64] estimated that for every grown man in her home a woman has to provide an additional seven hours' labour a week. She also made the point that the reproduction of people is very much more complex than the production of things. While the issue of housework has been forced back from the frontiers of feminist debate and struggle by the recession and the ascendance of the right, and the image of the new man strongly proselytized, the politics of time have not changed significantly.[65]

In similar vein Kari Waerness,[66] beginning with a common under-standing of caring as 'taking responsibility' and providing nurtur-ance, distinguishes between three kinds of caring: the mutual caring reciprocally exchanged between equals; enforced caring extracted, above all, from the woman; and caring for dependants – by these she means those who by age or disability need help to care for them-selves. She sees mutual caring as offering no problems, only pleasure. Her strongest strictures are directed towards enforced caring in which women are coerced into doing caring work for, typically, male others. Her third category of caring for ('natural') dependants she sees as necessary and hence acceptable labour. It is above all, as Hilary Graham[67] puts it, 'when labour outlasts love' that the recognition of caring as labour is inescapable. The accounts of women caring for their mothers, or for husbands who have become totally dependent through accident, stroke and the like (a phenom-enon aided by the age imbalance between husbands and wives), describe a world of unrelenting labour in which women, enchained by the social expectation of what the neighbours and the health and professional workers think, but also by what they themselves feel as duty, cannot escape. Elizabeth Cady Stanton may have announced at Seneca Falls in 1848 that 'Women's self-development is a higher duty than self-sacrifice', but the ideology of self-sacrifice appears to be remarkably robust. Even if Carol Gilligan[68] is right and women are increasingly trying to balance the claims of personal self-development and the just claims of others, unless this balancing takes place in a context of adequacy of resources – not least social support – then for many women it will fail. Hard choices are made in hard contexts.[69]

Criticizably, both British and Nordic literature has researched unpaid caring as if it was only an issue for white women, but where there is some solidarity between the carers and the cared for in much of the Nordic literature, the British has been marked by a strong voice of middle-aged and middle-class women who are confronted by the prospect of caring for their elderly mothers. There is little or no reflection that they themselves will at some point be old women. Jenny Morris has with justifiable anger drawn attention to the same silencing of the perspectives of disabled women.[70] In Britain it has been the feminist activists working in community groups, in trade unions and in town halls, not the academic feminists, who have brought issues of race and racism into the feminist politics of care. It has been primarily from the US that the literature from black and postcolonial feminism has been written which grapples with the complexity of domestic labour.[71] Far from being 'simply' menial we learn from Aptheker 'how a domestic worker has to be able to see into her employer's mind regardless of the external posture that the employer may adopt'.[72] There are differences but there are also commonalities with the managed heart of Airlie Hochschild's airline hostesses. As the old manufacturing industries of the nineteenth and earlier twentieth century collapse/are relocated in the South, people-work proliferates in the advanced industrial countries, and work which involves the production and reproduction of people is qualitatively different from the production of things.

Alienated or unalienated, freely exchanged in reciprocal caring, given as a labour of love or enforced by an individual man or by the state, internalized by duty or the fear of gossip, women's caring labour is much more than the formation of a (white) feminine identity. As a profoundly sensuous activity, women's labour constitutes a material reality which structures a distinctive understanding of the social and natural worlds. As Aptheker puts it,

'The point is to suggest a way of knowing from the meanings that women give to their labours. The search for dailiness is a method of work that allows us to take the patterns women create and the meanings women invent and learn from them. If we map what we learn connecting one meaning or invention to another we begin to lay out a different way of seeing reality. This way of seeing is what I refer to as the women's standpoint.[73]

A feminism which is interested in the possibility of thought arising from the distinctive social practices of women has to be willing to think about the distinctive labour of women, not least the labour of women birthing, for despite the massive medicalization of human reproduction in the West, to say nothing of the advent of new reproductive technologies, it remains the case that birthing is something that women do and men do not. To be willing to consider the labour of birthing within a discussion of women's knowledge is to admit the body, to accept what Janet Sayers usefully spoke of as a limited essentialism and a constrained social constructionism.[74] Despite the attempts on the part of a number of male social theorists[75] to suggest that the relationship between social reproduction and biological reproduction was merely accidental, many feminists not only remained unconvinced but also disentangled the polysemic character of 'reproduction'.[76] Mary O'Brien[77] insists that childbirth makes nonsense of Marx's division between the labour of the architect and the bee, where he argues that architects imagine even the worst building in their minds before they build, whereas the bee produces its always beautiful structures instinctually. Instead she sees a woman birthing as a moment when bodiliness and culture mingle boundaries.[78]

But there is also no compelling reason to continue to accept Marx's conception of animals as natural and thus without culture and human beings as transcending nature and thus having culture. The nineteenth-century passion for classifying and erecting boundaries, and the belief in the great chain of being, has receded. Even the high tide of behaviourism of the 1960s has passed, and a rather different conception has become possible of both animals and people making and even sharing culture.[79] The moves within biology that take down the barriers between animals and people are echoed within a political culture which is increasingly uneasy with casting animals as simply the instruments of larger human purposes. I do not wish to endorse the mutant liberal democratic theory of 'animal rights', but the language marks a dramatic shift in popular conceptions of nature and culture.

Breastfeeding a child similarly refuses to be located in that masculinist division between the natural and the cultural. While the health educationalists claim it as natural and every baby's birthright, in the West breastfeeding is more likely to be chosen by educated middle-class women. If a woman and her baby are fortunate, they may discover that breastfeeding can be not only caring labour but a deeply sexual pleasure. The boundaries of nature and culture, of self and non-self, of caring and pleasure, soften and merge. Such trafficking pushes feminist theory making beyond being a mere speciality within social sciences or literary studies and towards a thoroughly anti-disciplinary feminism, which seeks to overcome the old and oppressive dichotomies between the natural and the social, between caring and thinking.

Discipline and unfree labour

While doing women's work brings little recognition, 'failure' to do it can cause criticism, anger, even violence on the part of an individual man. If that failure takes place around children, it may result in intervention by the state, ranging from psychotherapeutic support for the woman to ECT and/or the forcible removal of her children. Unpaid the labour of housewifery and child care may be, but as a form of unfree labour it is carried out against a backdrop of extraordinarily powerful sanctions. Such sanctions are the penalties for breaches of the sexual contract, that invisible text between men and women written in the same moment of birth of liberal democratic society when men wrote their social contract.[80] Its silence about the level of violence acceptable within the family is an expression of the division in the control of women between the public patriarchy of the state and the private patriarchy of the home. Men's violence can be understood simply as the savage labour discipline imposed by men on women. Yet these often brutal punishments (a beating for a burnt meal, murder for infidelity) police the boundaries of women's familial labour as surely as factory fines and overseer beatings have policed the labour of free labourers and slaves, or for that matter as surely as sexual harassment polices women's paid employment. The massive resistance to men's violence, denying its naturalness and denying its inevitability, is simultaneously an immense achievement of the women's movement and the other side of a political struggle which values care.

Yet sociobiology today naturalizes the sexual contract, and sees violence[81] against women as rooted in biology, both his and hers, rather than arising from the unfree labour of women within patriarchy. The stubborn resistance that sociobiology offers to moving violence out of nature and into culture speaks of the enormous task which feminists and their allies face in defeating male violence, either practically or even in terms of a culture which continues to celebrate violence as desirable masculinity.

While it is true that housework and personal caring work are now much more widely seen as 'work' rather than as the natural expression of femininity itself, the savagery of the punishments for inadequate performance or labour refusal are so great that there is a tendency to think of men's violence as a problem of a different kind, to be considered in different terms to the sex–gender division of labour. These 'different terms' invoke a vague biologism, an unarticulated sense that it is natural for men to be violent to those who care for them. The difference in the sentences given to men and to women who kill speaks of the depth to which this naturalism is institutionalized within the state.[82] Yet the naturalism of the connection between violence and caring only holds when they are between men and women, and, being natural, may be deplored but impossible to oppose. It is only in the context of caring and violence in relationships between men that explanations are sought in culture rather than in nature.

The violence issue cannot simply be naturalized, even though the relationship of men and women to interpersonal violence is very different. Men may be violent to one another, to women and to children. Women are rarely violent to other adult people, whether women or men, but their relations with children and elderly dependent people can involve violence. Radical feminism with its preoccupation with men's violence has sometimes been too silent about women's violence to dependents, even though this can be explained, though not explained away, as stemming from the enforced nature of much of their caring work. Compulsory unpaid and very badly paid caring not only diminishes the freedom of the care-giver, but also threatens the safety of the cared-for. Timothy Diamond's study of US residential homes and the lack of psychological and material security for the first generation of Americans who paid into Social Security speaks of this ever-present danger. Yet the very specificities of the limited violence of women serves to highlight the contrast with the all too generalized violence of men.

The revolutionary and psychiatrist Franz Fanon[83] described the torture used by the French against the Algerians at the height of their struggle for national liberation in which the colonized colluded – and were meant to collude – in their own oppression. Fanon speaks of how a French police officer who worked as a torturer sought psychiatric help from him, so that the torturer could continue in his work without experiencing personal discomfort. The possibility that something very similar goes on every day between men and women is a matter few wish to discuss. Men rely on the emotional support of women to sustain them, even in their violence. Even where individual men play no part in violence, at least at the interpersonal level, active opposition is for the most part left to women. Thus even non-violent men benefit from the violence of others.[84] The lack of resistance simultaneously denies the existence of violence and colludes with the naturalistic justification of violence. Men's violence is rendered natural and normal, so integral to masculinity that it becomes difficult to connect it to other aspects of men's lives, let alone to place violence within a theory of knowledge.

Given that the potentiality of the modern state for collective violence is at an unparalleled historic level, feminism has begun to trace the connections between the everyday violence of men's culture and the inbuilt militarism of so much of modern science and technology. At the very birth of modern science in the seventeenth century, Francis Bacon used rape as his central metaphor, to invoke the process whereby the scientists forced nature and 'wrested her secrets from her'.[85] In the nineteenth century at the beginnings of physiology, Claude Bernard spoke of 'Nature as a woman, who must be forced to unveil herself when attacked by the experimenter, and who must be put to the question and is subdued.'[86] Except for a minority among scientists,[87] nature was seen as something separate from humanity, to be 'dominated'.[88] That a pacific relationship between humanity and nature could offer an alternative metaphor was a possibility that, until the rise of ecology (particularly deep ecology) as both a subject and a social movement, could not be seriously considered. The West's Judaeo-Christian cultural inheritance shored up this conception of Man as made in the image of God and as therefore having 'dominion' over nature.[89] Biotechnologists who are also active Christians explicitly draw on this tradition to justify genetic manipulation. When scientists describe their laboratory practice they frequently use a language soaked in militaristic and aggressive sexual metaphor.[90] The masculinist values of violence and domination are embedded within science; here as elsewhere Virginia Woolf illuminates: 'the values of men are different from the values of women . . . it is however the values of men that prevail.' Ideologically men's violence and women's caring are locked together, each integral to the ordering of patriarchal society not least in terms of civic and domestic duty.

Explaining women's labour

How was feminist theory to interpret women's daily lives? The power of early feminist insights, like Pat Mainardi's wonderful aphorism 'His resistance is the measure of your oppression',[91] as she analysed all the weird and wonderful ways men buckpassed housework, were lost, at least in Britain, in the problem of defining what domestic labour did for capital.

Women, argued the marxist feminists Selma James and Maria Rosa Dalla Costa,[92] do housework because it benefits capitalism, and therefore it should be paid. Positively, the ensuing debate around wages for housework served to accelerate the struggle against the naturalization of women in both bourgeois and socialist politics. Negatively, it understood domestic labour as a relationship between women and the capitalist system and failed to grasp that women came into society as 'not-men'. It thus conceptually lost both men, and patriarchy, from its analysis. Instead of answering Marx's crucial question 'cui bono?', British marxist feminism in the seventies let men off the hook.

But perhaps the most negative by-product was that the power of marxist tools to define the problem erased women's experience. The anger and wit which had sharpened the earlier feminist critique gave way to increasingly Talmudic exchanges.[93] Compared with the exhilaration of the rediscovery of housework as work, these

refinements to theory were often experienced and criticized as a new
and unwanted separation between the theoreticians and the activists.
(Some famous cartoons bore witness to the conflict, not least the
image of two women cleaners sweeping up polysyllabic words,
complaining that if they – the feminist theoreticians – had to clean up
afterwards, they would not use all these long words.)[94]

In dramatic response to this loss of a feminist agenda Christine
Delphy's pamphlet *The Main Enemy*[95] offered a radical re-visioning of
feminist materialism. She argued that women's domestic labour can
only be understood in the context of the domestic mode of
production, in which men benefited from and controlled the labour of
women. This re-visioning stemmed from France. It was from the US
that Heidi Hartman, examining women's material relationship to
patriarchy and to capital in two influential papers,[96] analysed the
allocation by sex of occupations within the labour market, and argued
that it was within the capitalist mode of production that men were
able to exclude and marginalize women, thus forcing them into
relations of subordination within the factory and the home. She went
on to examine the division of labour within the household, arguing
that here also was a locus of struggle, through which men were able
to force women into weaker places within the labour market.

Dual systems theory, such as Hartman and others proposed, was
criticized from a number of perspectives. For example, it was argued
that it is not appropriate to speak of more than one mode of
production. Yet third-world studies clearly document the coexistence
of different modes of production. A more serious difficulty was
contained in Iris Young's[97] criticism that if the problem is set up in
this way, the analysis of the sex–gender system becomes auxilliary –
and subordinate – to the analysis of class relations. An even more
devastating critique came from Gloria Joseph with her insistence on
adding race into the 'ménage à trois'.[98] While dualism was theoretically and politically flawed – for it divides the world only by sex and
by class and is silent on race, sexuality, age, differently abled bodies –
it did make space for an autonomous social struggle which was not,
in both the first and last analysis, entirely reducible to class.

Troubles with patriarchy

At the centre of the desire to move beyond dualism, which saw
capitalism and patriarchy as relatively autonomous systems of
domination, lay a deep unease on the part of many socialist feminists
with the concept of patriarchy.[99] The most serious objection
concerned its ahistoric character; it seemed to suffuse all relations
between human males and females. There was some force in such
criticisms, made cogently by a number of writers including Sheila
Rowbotham and Michèle Barrett, although the latter was by the end

of the eighties to revise her position. [100] Thus, Zillah Eisenstein is criticized for her use of the concept of patriarchy, which she sees as 'universal in Western society', so that patriarchy 'changes historically, but universal qualities of it are maintained even if they are specifically redefined'.[101] Throughout history men have retained their power by dominating the public realm and relegating women to the private. Hartman, too, speaks of patriarchy as predating capitalism and persisting within it, but also refers to it as 'universal'; not surprisingly, she is less than optimistic about the prospect of changes in the economic organization of society doing anything other than changing the forms through which sexual hierarchies are organized.

In so far as the concept of patriarchy does embrace a sense of universality and timelessness, then Joan Smith justly points to the political danger that feminists unwittingly provide support for their clear enemy – sociobiology.[102] The naturalistic thesis of the 'inevitability of patriarchy' is strengthened if feminists themselves claim that it is everywhere, all the time, simply changing in form. None the less, subsequent developments, in which US, Scandinavian and British feminists theorized the move from private to public patriarchy in the analysis of the welfare state,[103] served to historicize the concept.

There were ironies in this debate, as historicizing the new feminist concepts was crucial to both sides. Despite the massive research effort into family history, which has produced complex, contradictory and diverse understandings in which 'the' family disappears, at a political level the research interest in family history leads to a belief that there is such an entity. Thus despite their elegant demolition of left patriarchs and their passionate opposition to the bourgeois family, Michèle Barrett and Mary McIntosh[104] showed little sympathy with the diversity of families – not least as lived within by white working-class and black women – where family is often a source of both oppression and strength.[105] Right-wing ideologues such as Ferdinand Mount[106] understand this very well, arguing that despite all intervention by external powers, 'the subversive family' waits to spring back into existence with all its 'natural' resilience. Over the course of the eighties, feminism was to develop a more compassionate understanding of the complexities of everyday life, so that the monolithic anger of an equal opportunities feminism (particularly characteristic of British approaches[107]) which saw caring labour solely from the perspective of the carer was joined by the perspectives and the anger of the movement of differently abled people. Disabled feminists, together with black feminists and older feminists began to push against a conceptualization of caring which had produced a picture of the world solely from the perspectives of white, middle-aged, able-bodied and heterosexual women.[108] Because most of this caring literature was developed by socialist feminists – ranging from fabian to marxist in orientation – it had paid sensitive attention to issues of class. But this sensitivity to the intermeshing of class and gender on the part of academic feminism with a taken-for-granted commitment to dual systems theory was increasingly coming under question from community feminism, which was confronted by many more social divisions demanding a much greater willingness to live with complexity. New understandings of caring were being forged in practice by feminist activists whose everyday worlds were surrounded by far greater complexity than the prevailing theoretical models permitted. Gradually the awareness of the gap erupted into feminist debate, which took into account the perspectives of the cared-for as well as the carers, and recognized these relationships as being located in very diverse social contexts. Within these different contexts it became increasingly possible to explore caring as a labour process, as a fusion of often hard physical work, bodily intimacy and close psychological attention, and to appreciate the difference between caring for a child, who is growing into independence, and caring for an elderly or differently abled person, who may well be moving towards increased physical and perhaps psychological dependence.[109]

The different political and cultural contexts of caring labour became increasingly evident as the comparative literature developed. Thus Nordic feminism, while not holding a neutral concept of the state none the less saw it as relatively friendly and supportive to women's everyday lives (acknowledging the origins of the state in opposition to aristocratic power), whereas white US and British feminist, radical and anti-racist analyses interpreted social welfare personnel as largely coercive over women's, and black and working-class people's lives,[110] and therefore worked to develop distinctively feminist practices in social work, medicine and nursing. By contrast, Scandinavian analysis was less preoccupied with control, and both found more solidarity between professional carers and unpaid carers and was more sensitive to the cared-for.[111] The US black feminist Patricia Hill Collins argued even more generally that connectedness and caring for others are reflected in Afrocentric knowledge and practice: 'The parallels between Afrocentric expressions of the ethic of caring and those advanced by feminist scholars are noteworthy.'[112]

Thus, while the early criticisms of patriarchy as an ahistorical concept have yielded as feminist research has documented the complexity of patriarchal relations between and within societies, a situated concept of patriarchy has increasingly entered theorizing. But there are no easy answers even for those reluctant to enter the pluralistic project of postmodernism. Thus the initial optimism of influential theorists such as Sylvia Walby, that feminist materialism could be characterized as dual systems theory then subsequently revised to include 'other patriarchal structures', seems to have opened the way to a piling on of other structures, which echoes the pluralism of the postmodern turn even while resisting it.[113] Patricia Hill Collins's attempt to find a path between materialism and postmodernism also has some difficulties, for the concept of a matrix carries with it a notion of equality between the differences – yet the matrix is located on a map of social divisions where no such equality prevails. These difficulties in theory find their echo in practice in moving beyond the rhetoric – however longed for – of a rainbow alliance to specifying the nature of possible alliances between old and new forms of political struggle, between the old labour movement and the new social movements of anti-racism, peace, ecology and feminism, above all in a period of right-wing ascendance. Like 'wave' and 'particle' theories feminism may simply need different explanations for different purposes and maybe should be less concerned about the totalizing capacity of feminist theory.

What I want to argue is that women's caring practices, even in the alienated forms generated by the social division of labour, foster a more relational understanding both socially and bodily. Women's sense of the body is grounded in the real and material practice of taking care of both our own and the bodies of others: small babies, children, and sick, differently abled, and very elderly people. From the perspective of caring the body is no grand linguistic abstraction, but is very concrete, constantly fluctuating, sometimes dramatically and sometimes very subtly. Caring labourers have to learn to read the body, to understand from the muscle set of a face the strain within, to learn the labour discipline of caring, for bodies make time-specific demands which cannot be scheduled to some external conception of time. The book which falls to the ground and cannot be reached, the incontinence which demands the practical assistance of others, make their own demands. Caring demands empathy and affection which honour the autonomy of the cared-for; effective support demands complex practical and emotional labour, skills developed actively through the carers' lives.

Building a responsible rationality

This radical re-visioning of the concept of labour, so that emotion is restored within work and within knowledge, has accompanied a feminist reconstruction of rationality. A rationality of responsibility for others becomes central in this feminist reconceptualization.[114] Nor is this rationality limited to the understanding of the social world; indeed it is central to my argument that a feminist epistemology redraws lines between the social and the natural in a better, more accurate way, for emotions are also needed in non-violent understanding of the natural world. As Alison Jagger,[115] in her discussion of the epistemic potential of emotion, observes of Jane Goodall and Barbara McClintock, the former's work with chimpanzees demanded an extraordinary level of empathy (an empathy which turned to protective love when their survival was threatened by the demands of

AIDS research), and the latter's work as a maize geneticist required an empathic feeling for the organism. Evelyn Fox Keller describes McClintock's relation to her research as one of affection, empathy and 'the highest form of love: love that allows for intimacy without the annihilation of difference'.[116] Jagger suggests that the claim that emotion is vital to knowledge both challenges positivism's construction of knowledge, with its split between feeling and knowledge, and is part of the move to overcome the historical separation of the faculties: of reason and emotion, thought and action, evaluation and perception. The faculties, which have been constructed as separate and arranged in hierarchical dualities, need bringing together in a way which is both non-hierarchical and anti-foundationalist. All the faculties need developing, for knowing requires them all.[117]

Where Bacon's origin story for science spoke of the intimate connection of knowledge and power, the feminist critique of science, from Mary Shelley onward, has spoken of the danger of knowledge without love. It is the admission of love, a recognition that the process of care shapes the product, which opens up the prospect of a feminist reconstruction of rationality itself as a responsible rationality – responsible to people and to nature alike.

3

Feminism and the Academy: Success and Incorporation

> Existing between a social movement and the academy women's scholarship has a mistress and a master and guess which one pays the wages?
>
> Linda Gordon, 'What's New in Feminist History?'

A sociology of feminist knowledge?

Before entering more deeply into the debate within the feminist critique of science, which has so richly flowered over the past decade, as a set of rich, competing, borrowing and friendlily quarrelling ideas, I want to consider it as a debate which has taken place within specific historical and geographical contexts. Such a historicized relationship to theory requires that texts and contexts are interrogated together, not least those produced by feminism, and it resists an overly postmodern feminism, which in its strong focus on discourse theory and deconstructionism drives out the historical subject as surely as did structuralism.[1] The price of too strong an embrace of postmodernism by feminism is not inconsiderable, for it is only now, when feminism has massively delegitimized the hegemonic voice of the white bourgeois male and valorized the voices of oppressed women in all their diversity, that postmodernism declares the 'death of the subject'. As Nancy Hartsock has so succinctly commented, 'Why is it that just at the moment when so many of us who have been silenced begin to demand the right to name ourselves, to act as subjects rather than objects of history, that just then the concept of subjecthood becomes problematic?[2] Denise Riley's proposition that 'woman' is 'discursively constructed and always relatively to other categories which themselves change', though not necessarily framed in the language of deconstructionism, is basic to any feminist sociological and historical enquiry; if pressed to the point where woman is 'only' a 'fluctuating identity', then feminism itself (with its root in the Latin *femina*, 'woman') comes into question, not at least as a social movement of historically and geographically specific women demanding radical social change.

All accounts of knowledge – even of feminist knowledge – need to be 'externalist'; that is, conscious of the social and especially economic conditions of their own production.[3] The empirical test of 'internalism' – the theory that knowledge proceeds by its own internal coherence and logic – was ultimately tested to destruction by the reduction of funding for the British research system by Margaret Thatcher's government.[4] Not only is there a passionate campaign to 'Save British Science', but in 1992 it was reported that no less than 20 per cent of Fellows of the Royal Society have brain drained, and that the association of academic publishers claim that university libraries only buy one and a half books a year per student. As surely as when Virginia Woolf asked for £400 a year and a room of one's own,[5] feminist intellectual production requires material resources. Who provides them and why?

Feminism and modernization

So far there has been little exploration of the differing role of the state, the foundations and their relationship to the production of feminist knowledge in different national contexts, and national explorations as in the spate of books on 'French', 'German' and 'Italian' feminisms do not discuss the resourcing of feminist enquiry. This is an economic and political lacuna in a generally reflexive feminist discourse and one that, even without the kinds of detailed empirical study which would indicate who was putting how much into what kinds of feminist academic production, feminism needs to bring into self-conscious scrutiny. Otherwise we have slipped, perhaps most reprehensibly for those of us who are engaged in feminist science studies, into an implicit assumption that the state and/or the major foundations are neutral, or even irrelevant, when it comes to influencing the direction of feminist academic production. In a period when there is an everyday acceptance that knowledge is socially produced, and a widely known history of the role of Rockefeller in shaping biomedical knowledge,[6] feminism as a body of thought needs to be aware of its influences, if not determinants.

As Juliet Mitchell suggests in her 'reflections on twenty years of feminism',[7] feminism cannot afford to give up analysing material reality, for the material has such immense determining power. I want to argue that the dichotomous choice between either materialism or postmodernism, which theorists such as Joan Scott and Denise Riley propose, is not compulsory for feminism. Other feminist scholars such as Jane Flax[8] and Deborah Cameron[9] offer fruitful theoretical and political openings, where they pay a nuanced attention to discourse while locating it within the material constraints within which it is produced. To locate texts in contexts, or science in context, is not to explain them away, to suggest in some mechanical way that texts can be read off from material circumstances; rather it gives feminism the possibility of developing a sharper sense of what might or might not be achieved within specific historical circumstances. What I am suggesting should be sympathetic to the entirely laudable attempt by deconstructionists to remove the claim of 'innocence' from the task of building reliable knowledge. I see a continuing need for a feminist analysis of social structures and institutions, and I want to make a substantially political argument for methodological and theoretical pluralism.

In recent years feminism and feminists have entered the academy, manifestly unevenly not only between subjects but between countries; what has been striking in this entry has been the radical change in the nature of feminist theoretical production, from being a largely outsider knowledge to one that constantly speaks of itself as being both outside and inside, precariously balanced between the academy and the movement.[10] One of the problematic issues we have to consider is the form of the feminist movement over this period of intense social change.

This has been a quarter-century of major capitalist and patriarchal restructuring.[11] It has seen, in the old capitalist countries of the North, the death of heavy industry, the relocation of manufacturing employment to the South and the creation of new employment structures based predominantly on service industries and informatics. These changes have been accompanied by a language of crisis. The challenge of 1968 was a demand for the strengthening and democratization of civil society, which, in its critique of professional control, helped usher in some of the means through which the right was able to delegitimize the welfare state and establish new relations between state and civil society. The destabilization produced by 1968 and the new social movements called almost everything about social life and indeed our relationship to the environment into question; it also provided the space within which the old social formation could restructure and open a new phase of modernization in which women would play a new part. As Beatrice Campbell observed during the 1993 Charter 88 debate on the monarchy, this restructuring was given a 'feminist froth' by the twin presence of Margaret Thatcher as the first woman prime minister and Elizabeth Windsor on the throne.

For Britain this intervening period since the end of the long economic boom has seen the systematic manufacture of unemployment as a means of promoting change, ensuring that the burden has been borne overwhelmingly by working-class women and men. A failing government has been largely unable and unwilling to create the infrastructure to support the processes of structural change, and has instead turned to cutting welfare as a means of limiting state expenditure. At a personal level these structural changes have demanded – and in a very contradictory way the new social movements have themselves called into existence – new subjectivities. Women and the subjectivities of women have been at the frontiers of these changes. In consequence, feminism's entry into the academy is part of and, in attempting to build feminist theory, self-consciously reflects these processes.

An ebb tide or a change of direction?

Because many of the contributors to what Sandra Harding has called 'the science question in feminism' invoke the feminist movement, it is important to begin by recalling the profound changes that have taken place in the movement itself. As the women's liberation movement, it began as a commanding and dramatic presence on the streets and in daily life struggles within the home and within employment, whether located in residual welfare capitalist states like the US, advanced welfare states like the Nordic countries, or rather in-between ones like the former West Germany, France, the Netherlands, Italy and Britain. Women worked to negotiate the space between dreams of new becomings and existing realities. But this was not only a matter for those locked into the patriarchal societies of the West, as an increasingly global culture – whether that of California, Calais or Cairo – meant that feminist ideas arose and were shared, not always in simple accord, the world over. Even now, in the shift from state socialism to some form of the market, acute new contradictions are accompanied by theoretical and political innovations. Fragments of a new feminism are appearing within academic discourse, at least in those countries which have so far managed to avoid either the violent disintegration of the former Yugoslavia or of those parts of the former Soviet Union[12] where bloody civil war destroys the possibility of the growth of civil society. But even in the former Yugoslavia, women's peace groups still manage to organize to denounce killings and the mass rape of women by both the 'enemy' and their 'own' men.

That dramatic and optimistic period of the 1960s and 1970s, with its consciousness-raising groups, huge street demonstrations, painful domestic struggles and illegal abortion networks, brought into existence a feminist culture which contained both revolutionary and liberal reformist strands. Such was the self-confidence of the movement that even the 'modest' liberal demand for half the pie could be understood by the socialist feminist Zillah Eisenstein as speaking of 'the radical future of liberal feminism'. As the increasingly bleak and right-wing eighties closed in, the German marxist feminist Frigga Haug, addressing the European Socialist Feminist Forum, observed:

> After travelling through fifteen countries I can no longer shut my eyes to it: like water in a mountain stream the women's movement is drying up. It is true it has pushed its way into society's mainstream; it has produced changes in laws and created paid positions for a few; but there is undeniably less of a political movement, if by this we mean a political form.[13]

Yet despite Haug's cautious assessment, other feminists, while paying homage to the 'movement', spoke in tones of increasing confidence.[14] This confidence stems from interpreting the changes differently, seeing feminism's ability to take new forms and still sustain feminist projects as a capacity to move in new directions, and not merely as the ebb tide of a particular form of social movement. The rage against men's violence and beliefs that they can do what they like to women's bodies has spread through different layers of women, reaching and mobilizing women whose class, race and age location might have inhibited them at the height of the street-based movement. In the US, the Hill–Thomas hearings spoke to the experience of women harassed in everyday life by male colleagues at work in the office, factory and shop, and also mobilized the elite women on Capital Hill. My feeling is that this fundamental subversion of patriarchal privilege has been slower to extend among British women, particularly at the upper levels of political and bureaucratic power structures, but that the grumbling and muttering is continuing to spread. There begins to be a distinctively feminist view of war; mass rape in Bosnia is increasingly understood as a war crime of men, not just of some nationally defined and gender neutral 'enemy'.

To speak of feminist activists and feminist theorists is not to create an antagonism, but to acknowledge that over time, because of the shifts in the movement's structure and to some extent as the price of feminism's success in entering the academy, a division of labour has developed between feminists. As I suggested in the previous chapter, however, the political and theoretical initative has by no means been entirely lost by community-based feminism.

I do not want to romanticize the late sixties and seventies, but theorizing at that time did develop in close conjunction with, and frequently directly out of, the collective process of the consciousness-raising groups.[15] These group discussions connected everyday experience and social structures with electrifying energy. The

intellectuals of the first years were, like Gramsci's organic intellectuals, seamlessly woven into the movement and largely created the
new knowledge outside and in opposition to the academic institutions. Although the left distinction between reformist and revolutionary politics was not always shared by the new social movements, not
least because the cradle of the new culture had been outside the
dominant culture, the movement celebrated an autonomous political
form. The taken-for-granted opposition to joining the mainstream
was shown by the intense debate in the late seventies as to whether
feminist scholarship entering the academy as women's studies would
lead to co-option and political weakness.[16] Ironically, despite radical
feminism's greater commitment to an autonomous women's culture,
it was more frequently the left feminists who saw entering the
academy as offering ghettoization and containment.

The key theorists in the early days were themselves predominantly
young, or academically marginal, or both.[17] Like other outsider
groups, the movement had to create both its own oppositional culture
and its own cultural capital, which it did through a proliferation of
journals and pamphlets. Much was self-produced, for second-wave
feminism coincided with technological advances in printing, so that
photo-litho offset printing put magazine production literally into the
hands of women, the other 'fragments' of the new left and the black
community, who through their struggles were building new communities and new cultures of resistance.[18] But for feminism in particular
there was also a fast-developing relationship with the publishing
industry and the market. Feminist books, journals and magazines
were soon appreciated as highly marketable, as an immense new
readership came into social visibility. Even in the depths of the
nineties' recession, feminist lists have remained strong,[19] so that the
market remains a complex ally in the task of disseminating the new
ideas.[20] But whether through self- or commercial publication, the
movement fundamentally spoke and continues to speak primarily to
'itself'; that is, to those women who in some way have been reached
by the new ideas and want to continue exploring them. The outpouring of feminist and feminist-influenced literature, from advice
handbooks, and business management to poetry, is at a historically
unparalleled level, influencing, changing, becoming a different culture.

In the sixties, a patriarchal higher educational system was
indifferent to or contemptuous of any attempt to bring the social
relations between men and women to visibility through the development of concepts such as 'sex roles', 'the sex–gender system',
'patriarchy' or 'gender relations'. Even now its accommodations are
uneven, and the success of the pressure for women's studies in
Britain owes not a little to the changed financial setting of higher
education, imposed by a highly ideological right-wing government,
in which universities are penalized for failing to recruit sufficient
student numbers. In this new climate the ability of women's studies

courses to attract students makes university administrators relatively friendly towards the new area, and the science and engineering departments hope that a new-found conversion to equal opportunities will fill their far from overflowing teaching laboratories. Accepting feminists as scholars has been more grudging and although the expansion has seen many more women academics in absolute terms, which gives an illusion of gain, the statistics point to a not very diffferent proportion being clustered around the lower grades of permanent appointments and to women being over-represented among the casualized sector.[21]

Feminist struggle at the turn of the century took, after the vote, the issue of access to education and science as one of its central objectives, and for the usual complex reasons both struggles were successful. Second-wave feminism took place in the context of more or less continuous expansion of higher education, so that increasing the numerical presence of women was only part of a new objective of changing the knowledge system itself. The expansion did, however, provide the conditions in which feminist intellectuals, as an increasingly large and visible group, have been able to move from a weak position largely outside the publicly financed production and transmission system of knowledge to one where, in certain areas of knowledge (mostly the humanities and the social sciences), they are a visible and influential presence. Natural sciences and engineering have, particularly in Britain, remained relatively unscathed, though there are increasing signs that the feminist critique of science as gendered begins to enter the discussions of women scientists.

Something of the social processes of this advance can be seen by way of making an analogy with Bourdieu's study of *Homo Academicus* (Gallicus), where he demonstrates that most of the most influential theorists of both structuralism and poststructuralism had rather weak positions, if any at all, within the French academy, and achieved their fame precisely through refusing the rules of the academic game and playing for cultural power as outsiders.[22] Participation in key journals, and contributing regularly to cultural debate through journalistic activities, were crucial in this alternative trajectory.

But the analogy is limited. There are major differences, and his title flags a not unusual clue. For while Bourdieu analyses the different strategies open to French intellectuals (and includes women),[23] he is, in a fundamental way, concerned with the intellectual and academic world of men. Even where women do achieve a place in the 'intellectual hit parade' they are treated by Bourdieu as if they were the same as – that is, identical with – men. Thus, although Simone de Beauvoir is high in the 'hit parade', Bourdieu does not reflect that where feminist intellectuals (superstars and all) have achieved their recognition within a patriarchal reward system they also have to be understood as part of the historical project of feminism, as the legitimacy of feminist intellectuals is crucially bound up with a specific social movement. Further, Bourdieu's elite are all drawn from what we may loosely call the humanities; he cannot see natural science as culture nor scientists as producers of culture. The power of science to transform culture, economy and society is erased. It is as if the impersonal voice with which science speaks described a culture of no culture, created by no one. The circle of invisibility is complete and the sociologist of cultural reproduction cannot break through.[24]

Theoretical currents and national contexts

Second-wave feminism's cultural power base within the academy was primarily built up around women's studies.[25] Theoretical production during the eighties and into the nineties has become very much tied up with this development, and thus has been significantly shaped by the structure and policies of national higher education and research systems, and their potentiality for change of both organizational forms and also the substantive content of knowledge. While feminism has long discussed the different mix of theoretical currents in particular national contexts, there has been much less said about the differences and similarities between the research policy responses of different nation states to the demands of feminism. Here I want to look particularly at the situation in Britain, which I know best, in the US, because it is the world's richest research system, and in Scandinavia,[26] where I have spent a considerable amount of time in recent years. As I will show, these three symbolize very different ways in which the shifts in the modernization project of the last quarter-century have related to the demands of feminism.

Britain

It has become almost a part of of feminism's conventional self-accounting that in the US, liberal feminism has been a powerful current, and that of the revolutionary currents there, radical feminism has been the more influential and socialist and marxist feminism rather weaker. By contrast, the account continues, in Western Europe, liberal feminism has been relatively weak and the strongest theoretical current has been that of socialist feminism, with radical feminism preferred by activists. Yet within Britain the term 'socialist feminism' has long concealed as much as it reveals, particularly during the early eighties when there was a tremendous radical and popular revival at the constituency level of the Labour Party, in which the word 'socialist' was given new strength. This upsurge encouraged the Labour-controlled Greater London Council to experiment with a new women's committee which drew in, and celebrated, an immense diversity of women. It was as if the GLC, symbolically facing across the Thames to Westminster, was attempting to build a rainbow alliance as a practical and popular alternative to Thatcherism. This exhilarating period, which was repeated and extended elsewhere within the socialist cities, for a while changed the trajectory of Labourism. Pre-existing categories of left, socialist, marxist and labour were for a few creative years put to one side. It took the ruthless crushing of the miners' strike, followed by the systematic destruction of the GLC and that entire level of local government, to signal that the Conservatives would brook no opposition.

In the 1990s socialist feminism has shared in the difficulty currently experienced by socialism. Even though the theoretical and political departure point of the New Left was its hostility to contemporary forms of communist society, or what was called with irony 'actually existing socialism', 1989 has unquestionably had a negative impact. So far as the British Labour Party goes, the word 'socialism' itself (along with the associated category of the working class) was almost unused during the 1992 election. This shift against socialism is not simply about the manoeuvrings of party politicians to secure power on any terms, though it did not help foster an alternative social vision. Did 20,000 women surround Greenham merely so that in the run-up to the 1992 election Labour's defence spokesman could claim that a policy of three Tridents against the Conservatives' four was a sufficient reason for voting Labour?

Even those who theorized the role of the new social movements in social change underwent a not so subtle shift which effectively wrote class out of the analysis, so that where the seventies spoke of possible alliances between the old social movements and the new, by the eighties the language was only of alliances between the new social movements. The future was to be consciously constructed without reference to class-based movements.[27]

These political changes have had their reflections within feminism, as numbers of theoretically oriented feminists have shifted toward broadly poststructuralist positions, paying attention to subtleties of difference yet somehow not naming class as a major source of difference even at a time when class gaps have sharpened – not least in terms of who lives and who dies. For that reason social welfare feminism, much as it has been throughout the century, is still located between the materialists and the liberal reformist tradition. The continuities between early twentieth-century Fabian feminism and many of today's social welfare feminists is noticeable. The latter participate effectively in the technical discourse of state policy makers, but rarely take part directly in the social movements of welfare-dependent women.[28] Within such technicized discourse the non-relational concepts of 'women' and 'poverty' stand in for the relational concepts of 'gender' and 'class'.

The sense that feminist research is supported by any research policy objectives is quite difficult to feel in the context of Britain, where for fifteen years a highly ideological and philistine right-wing government, with increasingly dirigiste policies, has given rise to a feeling of the need for unremitting struggle to protect what we have, let alone make advances. It is a struggle (against, occasionally with and sometimes going around the existing structures) to secure enough teaching posts, materials, books, grants for students and recognition for courses, let alone to find time to secure research monies and get research done. Because women in higher education are often located within professional training such as teaching, social welfare and nursing courses, they have been particularly exposed to the increasing limitation of intellectual space resulting from the professional and bureaucratic definition of education and research.

Major changes in the funding of higher education during the 1990s are likely to have still further negative impact on feminist research, for these reforms have ended the dual system of funding whereby the universities, in recognition of some residual commitment to free enquiry in half the binary system, were allocated baseline resources for research as well as teaching. Research funds will now become increasingly competitively awarded, giving the state increased control over research monies. While ending the binary divide removes one class – and race and gender – division in higher education, in the future the bulk of research funding is likely to be allocated to what will be the research universities. Feminists are under-represented in the elite institutions and feminist research with its craft system of research production is unlikely to be among the big money getters; so unless the emergent research universities feel, for whatever reason, that they must take women's studies seriously, feminist research is more likely to he hindered than helped by the latest round of reform.

Far from the new dirigism of the British higher education system contesting the androcentricity or the uniculturalism of course content,[29] traditional disciplines remain firmly in the ascendant. Even while the research councils affirm the desirability of interdisciplinarity, the Research Assessment Exercise – the mechanism through which university departments are assessed and allocated research monies – remains firmly located within the old boundaries. Interdisciplinary fields like women's studies research are thus located outside the structures and funding mechanisms of research policy.[30] This peculiar resistance within British higher education is a function of two contradictory strands within contemporary Conservatism: on the one hand a liberal desire to free the market, which might acknowledge women as individuals, and on the other a conservative desire to restore the family, which certainly will not. While Conservative politicians in a range of activities from using prostitutes' services to having extramarital affairs, and the royal family in its proliferation of single-parent families, display the impossibility of maintaining even a charade of the bourgeois family form, this tension at the heart of the British New Right means the equality aspect of late twentieth-century modernization is very weakly supported.[31]

While such an ideological confusion is endemic elsewhere, for instance in the US during the Reagan/Bush administrations, the much less centralized nature of the US political and indeed research systems has produced an inconsistent set of policies which have been simultaneously *for* women's studies research and *against* women – especially working-class and welfare-dependent women.[32] By contrast in Britain, increasing centralization during the Thatcher years produced a more consistently woman-unfriendly politics and research policy.

In this situation the establishment of the UK National Women's Studies Network has been achieved entirely by self-activity – with no governmental or foundation grants, and no institutional benefactors to ease its path. There has been no research policy of establishing centres for research on women, or their equivalent; those that have developed have been constructed bottom-up using soft money, typically with a strong policy profile to secure what resources are available.[33] The duration of their existence has been determined by the scale of their grants. Inevitably many feminists are active in consultancy work, training managers and generally working for survival. Developing a long-term critical research programme is an almost unaffordable luxury.

For good reason British feminism, whether inside or outside the academy, sees itself as largely oppositional. The state is for the most part seen as both hostile and hard. There is a sense of consistency in the attacks on women's daily lives through the cuts in welfare service provisions, the erosion of employment rights, and the stronger but still precarious place of women's studies in the academy. The market, in that there are numbers of feminists who want to take women's studies courses and who continue to buy feminist books, has been a better ally.

Even the most modest suggestion of a research initiative in gender studies – seen by its proposers as safer (more academic) than the dangerously oppositional women's studies – was firmly rejected by the Economic and Social Research Council in 1992. The almost two decades of support given to women's studies research by most of the Scandinavian countries as part of their overall equality project finds no equivalent in the UK, except through the underfunded and politically nearly toothless Equal Opportunities Commission.[34] By contrast the strongest feminist research developments have been made within the state-defined policy fields of employment, health and welfare services. The achievement of academic feminism has been to enter this policy-defined terrain, turn the research programme to feminist-defined objectives and secure the resources. Even this gain has had to be made against research councils and their committees, on which women are under-represented, and which work without a clear commitment to secure equality objectives in the organization and content of research. Such research structures and practices indicate that the research councils believe that deciding about research is gender-, race- and class-neutral – except in so far as the Conservative government has increased the representation of industrialists.[35] To date, no major British foundation has a policy of support for research in women's studies, although there are signs that the patient pressuring of feminists inside and outside the foundation bureaucracies is making at least a number of them friendlier to proposals from feminist researchers. By and large, British foundations have interpreted their role as one of collaborating with the research councils, and their substantially overlapping committee membership means that few strike out radically distinct research programmes.[36]

Some of the professional societies, notably the British Sociological Association, have, however, offered their cultural capital to foster the new feminist enquiry. The first major initiative came in 1974 when the annual BSA conference was devoted to Sexual Divisions in Society, and several hundred women from many disciplines celebrated the possibility of their liberation from the canons.[37] Today, in the 1990s, women's committees and groups proliferate within many disciplines, supportive to women currently active, rewriting the history of women in their disciplines into the teaching of their students, and in some cases reconceptualizing the knowledges themselves.

The United States

By contrast with the British, proceeding largely through self-help, in the US Ford established a national fellowship programme in 1972 and two years later supported the establishment of Centers of Research on Women at two elite institutions: Stanford on the West Coast and Wellesley on the East. As the former is a leading research university and the latter a training place for the social elite, the move made sure that both elite systems were significantly entered. In 1974 the Carnegie Foundation had also sought to lay the foundations by supporting a nation-wide conference on the under-representation of women in higher education. Major foundations, including Rockefeller and Mellon, have continued systematically to support academic feminism, playing an influential part in the direction, organization and content of women's studies research. By the late eighties there were some forty campus-based research centres and a further thirty independent centres focusing on public policy, the arts etc. The activities of these were linked by the National Council for Research on Women, in which the Ford Foundation again played a significant role.[38]

Almost certainly the strong current of liberal feminism within the US movement contains part of the explanation, for many such women placed in positions of high office within the foundations and governmental agencies are deeply committed to feminism in a way that is rather rare in Europe. The enthusiasm expressed by a number of liberal feminists in the US at the election of Thatcher in 1979 was expressive of these different political and cultural traditions. As high-achieving women they identified with her success, and assumed that she like them would be committed to extending the influence of women. They were rapidly disabused of this by the British prime minister herself. But for US elite women that theme of inside and outside, echoed strongly in the US literature, has ensured that key highly placed women have been of importance in pushing resources towards women's studies.

At the same time as developing women's studies research, Ford and other foundations like Mellon were supporting 'curriculum integration', by which was meant reforming the content of the mainstream disciplines.[39] What is particular to the US situation is the strong push given by the foundations and government agencies towards changing the knowledge system and its work force at the apex of the research system, However, this account should not be read as suggesting that women's studies courses or indeed women's studies research as a whole have been adequately funded or recognized. There that familiar story once more appears, of women's studies being developed through the commitment of feminists on the staff of regular departments, particularly in the state university system, where women are generally better represented. Here, other than in the top institutions such as Berkeley or Minnesota, there are higher teaching loads and little research support. Where additional assistance is needed it comes less from new tenure track positions than from part-time and temporary teaching posts. But as part of a movement conscious of the history of the ghettoization of women in home economics, where new teaching appointments are secured, staff are typically attached both to a conventional department and to the women's studies programme.

Some crucial threshold has been crossed in the US feminist research effort, so that it becomes difficult to think of a research area where there is not considerable impact. Nowhere is this more true than in the attention paid by women's studies to the natural sciences. Research on science, like research in science by women, is very much a minority activity, even in the US, but there has been financial support for studies of women in science, for theory-driven research on the content of science, and for governmental research agencies to redirect biomedical research to take account both of women's different health problems and of their historic erasure within biomedical research on the 'human' (in actuality the male) body.[40] This redirection of research has been largely secured by the pressure generated by the women's health movement and public feminism, and was powerfully fostered by Bernadine Healy during her period of office under the Bush administration as Director of the National Institutes of Health (NIH). The related pressure to establish a nation-wide health care system has been less fruitful, and feminism is one of the several groups currently looking to Hillary Clinton to achieve this.

Because feminist knowledge production is small-scale craft produc-tion, time, particularly in the overcrowded lives of women, is a key element. Time, which perhaps was simply generated out of the abundance of the collective energy of the seventies and the fewer responsibilities of that younger age group at the movement's centre, has now to be either found after the double day or funded institutionally. While women's studies as degree programmes rose initially in the state universities, the foundations chose to direct their financial support for research primarily to the private universities and the top researching public universities. These, like the elite univer-sities in the UK, were reluctant to admit women's studies courses, and the foundations substantially bought a space for women's studies research. An alliance of femocrats (to use the Australian term for feminists working in government) and academic feminists has leveraged access to a relatively small part of the immense wealth of the US research system. It is this toe-hold on wealth which has substantially been responsible for the strength of US academic feminist research – not least in the feminist studies of science.

This wealth of foundation support is very clear in the key theoretical texts in the feminist critique of science. If I look at the acknowledgements in the books and papers by my desk, I find Evelyn Fox Keller[41] thanking the Exxon Foundation and a Mina Shaughnessy Award for providing time to work on her biography of McClintock; Sandra Harding thanking the National Foundation for the Human-ities, the National Science Foundation, a Mina Shaughnessy award, and a Mellon Foundation award to work on *The Science Question in Feminism*; Donna Haraway thanking the Alpha Fund of the Institute for Advanced Study at Princeton, and the Wenner Gren Foundation for Anthropology; Helen Longino thanking the National Science Foundation for two grants and the Mellon Foundation for a third. These theorists, even where they have to piece together small grants, have been able to secure that commodity that exists only in the lives of a minority of women: time to think.

But even in the US this precious commodity has mainly been allocated to the scholarship of white feminists. For the scholarship of black feminists there has been a bleaker story. bell hooks writes of working part-time to keep herself while writing her path-breaking *Ain't I a Woman?*, and of the conflict of her feelings when the white feminist historian Gerda Lerner was funded to do research on black women's history, whereas black women like herself could not get research support. However, the criticism has been to some extent heard and there are now increased – if not adequate – funding sources for minority scholars. Meanwhile the situation in the UK is typified by Lynda Birke's acknowledgements in *Women, Feminism and Biology* to her membership of the Brighton Women and Science Group, the support of women's studies colleagues and individual friends. Wendy Hollway's *Subjectivity and Method in Psychology* thanks friends; at best, Janet Sayers's *Biological Politics* thanks her university for a year's study leave.[42] My own time debts are primarily to the Swedish Collegium for the Advanced Study of the Social Sciences and to the University of Minnesota for a visiting professorship. In the UK, very few women, white or black, get funded by UK sources for science theory, however well they wrap it up within other concerns.

This grant-supported knowledge-production system in the US particularly describes those who have moved into the theoretical aspect of the feminist studies of science. By contrast the feminist biologists who critically fought sociobiology have had to take time out of their laboratories, in some cases using their tenured status to leave them more or less permanently, in order to write. Neither Ruth Hubbard, Marian Lowe nor Ethel Tobach acknowledges foundation support, although Ruth Bleier acknowledges time and a development grant for one modest semester to begin her *Science and Gender*.

While unquestionably this support has been given in part due to the pressure from an immensely well-organized, diverse and powerful US women's movement, feminism has not explored the motivation of either these big foundations or the state. My hunch is that this responsiveness on the part of the state and the foundations to the demands of academic feminism can best be understood both as an expression of the confidence of what is still the richest and most powerful liberal democracy in the world, and as part of an immense project of modernization, to maintain that pre-eminence over the last decades of the twentieth century.

Possibly too, emphasizing middle-class, highly educated women as a modernizing project has been politically convenient, and has served to turn public attention from other areas of evident political domestic failure, not least the failure to find a solution for the growing homelessness and poverty within the cities.[43] Ideologically and culturally this modernization project has secured substantial gains, but within the liberal democratic structure of the US, so that by the beginning of the nineties what had been unthinkable in the seventies seems entirely possible. Modernization has also been accompanied by a 'new class war', initiated by Reagan and extended by Bush, which has had immense and very negative effects on the lives of both black and white working-class women. This negativity has generated such adverse criticism that in the run-up to the 1992 elections even Bush had to announce a programme to tackle the infant mortality figures of the cities – the worst in the industrial world. Politically, focusing on the issues of middle-class and predominantly white women, however just in itself, has served to distract from the injuries of class and race borne by other women.

Scandinavia

Over a broadly parallel period the Nordic countries have, as 'friendly states' (not a conception of the state shared by other Europeans or North Americans), seen women's studies research as part of a general 'equality' project, which, although located within welfare corporate capitalism, bears some comparisons with the US. The radical difference, however, is that because the project is framed within a social democratic perspective, issues of class have consistently been addressed.[44] The civil research programmes in the Nordic countries are generous compared with that of Britain; indeed the UK and the Swedish social science budgets are similar in size even though they serve very differently sized populations. Unlike Britain and the US, the Nordic countries do not spend approximately half their public research budget on military research, so it is possible for them to pursue social objectives through earmarking research money for what is typically spoken of as 'equality research'. However, even the Nordic research system has come under pressure as the recession bites first Denmark, then Sweden, then the hitherto buoyant Finnish economy, for a while leaving only oil-rich Norway looking reasonably comfortable.

To overcome the problem of a knowledge-production system controlled by the professoriat (almost entirely the preserve of men) and the history of the older and more powerful universities as often bastions of conservatism,[45] the Scandinavian governments, with the energetic support of women parliamentarians, have initiated a series of women's studies centres to introduce both teaching and research. Something of the way that this new speciality is understood is contained in the androcentric science policy literature, which indicates that 'women's oppression' is to be regarded as 'a social problem along with alcohol and drug abuse, research on working conditions, the mass media etc. – that is one for which government and the local authorities deem it important to put in special efforts'.[46] Thus for the Scandinavian countries, with Sweden as the archetype, equality research has been integral to sectoral policy in which the state unequivocally determined the objectives. Women in this construction have to be understood as a 'sector'. On a number of occasions during the eighties, women parliamentarians headed off attacks on the women's studies research budget. For example, in Sweden 'equality research' found effective political support against the budget-cutting intentions of the financial department. Pressure from the women's research centres led the Swedish Parliament in 1990 to ask the research council for a research programne on 'female approaches to science and technology'. In 1993, even with a change of government, there is hope that the resulting programme, which includes teaching and research posts and an earmarked research budget, will secure the necessary parliamentary support. As the rightist move of the nineties continues, this hitherto successful stance has become a little more uncertain; earmarked funding has been removed in Denmark but the anxieties that this might spread throughout the Nordic research system have so far not been realized.

It is not by chance that in the highly corporate welfare capitalism of the social democracies, as against the liberal democracy of the US, there has been a double focus on women in both reproduction and production. Informing the research strategy is a state commitment to a restructuring of gender and gender relations within both these locations of work. While unquestionably these societies are public patriarchies and a long way from the dreams of feminism, they are also the most civilized public patriarchies in the world. Nordic feminism, as I pointed out in the previous chapter, has a nice appreciation of this duality, but this has not inhibited self-criticism of the excessive orientation towards policy, which has led to an underdevelopment of feminist theory-driven work.[47]

In the field of the feminist studies of science and technology this preoccupation with societal management has meant that while there is extensive research on technology, especially workplace-related technology, there has been rather little on sciences.[48] The link between science and technology, above all as the new technosciences that a number of feminists want to foreground, is cut away by this tight boundary. Instead policy-oriented research and the emphasis on soft money and contract research has not only led to the theoretical development perhaps being less than its potential; it has also made it easier for the higher education system to keep feminist scholarship at a distance. Recently science policy analysts have drawn attention to the weaknesses of the sectoral system, and mainstream researchers have striven to distance themselves from the production of sectorally driven knowledge – where women's studies is structurally located. Mainstream social sciences and humanities have become increasingly theory-driven, with abstraction, itself associated with a particular construction of scientific masculinity, commanding higher status than applied research. Feminists with a record of productive but policy-led research find that their cultural capital is devalued when it comes to competing for academic posts.

Scientific and technological Europe

Despite the differing trajectories of research in the Scandinavian countries and the UK, over the past decade they have been increasingly embraced within the wider grouping which is becoming known as Scientific and Technological Europe, encompassing both the countries of the European Community – now the European Union (EU) – and those in the process of joining it. The EU has rapidly become a substantive player on the research scene, with an expanding budget (allocated by national governments) for distribution via Brussels, and with a clear sense of goals. Thus it was the European modernization project, given clout by EU directives which, for example, pushed British research towards gendered labour market studies, presently renewed especially for science and technology as governments contemplate the shortfall of young people into the labour market, and so once more look to women as a hidden reserve.

This so-called Scientific and Technological Europe is unblushingly unidimensional in its macho objective of technoeconomism, which seeks to harness research to 'catching up' with Japan and the US. Despite the environmental concerns implied by arguments concerning the limits to growth and the need to define social objectives for science, or even signing up at Rio, 'catching up' is constructed almost entirely in terms of technological innovation and the will-o'-the-wisp of economic growth. This drive has been apparent from the earliest days; behind the glitz of acronym research projects (EURATOM, EUREKA, BRITE, FAST) lies a programme of meshing science more and more closely to the innovative needs of European capital.

The needs of the European peoples in all their diversity (even the needs of those regarded as citizens within Fortress Europe, let alone those of the excluded)[49] were not seen as part of the research or policy problematic. Instead the European modernization project has sought to secure flexibility of the labour force as crucial to facilitating capitalist development. This has had two components; the first, harmonizing training and skills across the European countries, and the second, overcoming what are spoken of as 'traditional'[50] rigidities, which include gender segregation, within the labour market. Overcoming some of the national rigidities within the production of knowledge has been supported at the educational and training level by both ERASMUS (for students) and the Human Capital and Mobility Programme for post-doctoral training.

Little energy, at least until now, has been directed towards the rigidities associated with gender and 'race' in the employment of scientists and technologists. Overcoming gender divisions has been seen only in terms of socialization and training. Thus considerable amounts of money have gone from the Social Fund into training women in non-traditional occupations while the objectives of the research system and indeed its labour force composition remain unscrutinized. By contrast the Norwegian feminist Harriet Holter[51] and her colleagues have recently called for 'half the kingdom of research'. This modest demand for equality, which has all the subversive and infectious feel of early second-wave feminism's fair and impossible demands, is a useful starting point for what might be feminism's reconstruction of the political and research objectives of Scientific and Technological Europe.

Success and incorporation

Thus, whether scraping in by its fingernails or relatively graciously welcomed, over the past decade the making of feminist knowledge has secured an academic address in most of the old patriarchal and capitalist countries. As I have tried to indicate, this represents both success and incorporation: success because it has sustained a multistranded critique of androcentric knowledge; incorporation for two reasons. Firstly, the critical knowledges of feminism have facilitated the modernization project of capitalist patriarchy. The US, with its liberal democratic tradition and its immense wealth, has found it easier to accommodate the entry of women and women's research interests not merely in the lower ranks of academia but even in the tiny elite of research decision makers. The social democratic Nordic states have been formally responsive to the demands for greater gender justice even whilst the academic system has shown an extraordinary resistance to the entry of women to senior university positions (it is easier for women to become members of Scandinavian parliaments and even governments than to become professors). The destruction of many democratic and civil rights in the UK over the 1980s, along with the profound weakening of the research system under four successive Conservative administrations, has instead fostered self-help, leaving only technoeconomic, policy-oriented research relatively well supported – a trend strengthened by the directive influence of EU funding.

But secondly and perhaps more importantly, in entering the academy, feminism and feminists have themselves not remained uninfluenced. Professionalism, as Nancy Cott[52] observed for an earlier generation of feminists, generates its own discourses, its own research problematics, which serve to separate the feminist academicians from the movement which fostered them. Outsider knowledge adapts to insider knowledge; the prerequisite for participation in the feminist debates becomes an exhaustive knowledge of androcentric theorization, so that gradually the strong relationship with the social movement, which meantime has so profoundly changed its form, becomes increasingly attenuated. The theoretical difficulty, even the embarrassment, of early gynocentric discourse yields to a highly professional discourse in which all the subtleties of difference are acknowledged, where subjectivity is explored with sophistication, but where the raw interests of 'actually existing women' are with difficulty constructed so as to demand political attention. Instead a subtle linguistic battle is engaged today, a battle between scholars, where every nuance of the wordplay is to be admired, but where almost everyone other than the scholar/word-person is reduced to audience. Such a move from 'outsider' to something akin to 'insider' knowledge has privileged the voices of feminist academics over and above the voices from grass-roots feminism. It may be that this process is part of a developing technical division of labour between feminists rather than a social division involving hierarchical relations. I do not want to argue that the brilliant academic feminists whose work can give intense aesthetic pleasure should somehow subdue their brilliance and stop producing demanding texts, challenging pictures and all the rest, and waste time bewailing their class and probably 'race' privileges; but much as we welcome the work of feminist teachers, poets, doctors and scientists, it is worth remembering how many more women whose lives are presently impoverished could also be creative and demonstrate their brilliance in a less flawed society.

Others share these concerns. The French feminist philosopher Michelle le Doeuff in a witty, provocative letter to her British counterparts speaks against what she sees as a tendency towards academicism developing within women's studies, in which rules are established about who is and who is not an 'acceptable author'. Instead she advocates 'being a Renaissance person hating enclosures and restrictions in reading and loathing anything that could be the authoritarian limits of any School'.[53] She goes on to denounce unequivocally that 'main feature of academicism which is to turn one's attention away from the situation of the oppressed and the vile results of social conflicts'. For le Doeuff it is precisely the socially based agonies of women which make it important to have women philosophers, and I would want to add lawyers, scientists and all the rest. Despite such concerns le Doeuff herself, like so many academic feminists the world over, gives off an immense vitality and political commitment which every day resists any drift to academicism and any abandonment of women's socially based agonies.

While my next chapter turns to recent debates within the feminist theory of science, I want to draw attention to the double frame which surrounds us. The feminist science theorists are marked by a strong political commitment, first to feminism – to those socially based agonies of women – and second to a political sensibility which acknowledges that those that pay the piper have an influence even if they do not unilaterally call the tune.

4

Listening to Each Other: Feminist Voices in the Theory of Scientific Knowledge

What becomes very clear, however, is that feminists have now entered the debates on the nature and power of scientific knowledge with authority: we do have something to say. The only remaining problem is what, and here we are speaking in many voices.

Donna Haraway, 'In the Beginning was the Word'

Embodied politics: embodied knowledge

Earlier I suggested that the feminist critique of science only got under way during the second part of the seventies. Then the invisible college of feminists working in and on science was relatively small, and while the movement, not least because of the attack from the biological determinists, was in principle supportive of these efforts, feminism was slow to become interested in science.

The one area of scientific knowledge that second-wave feminism has been passionately interested in was what science spoke of as the biology of human reproduction. Women had been made to feel simultaneously that their bodies were somehow shameful and also that they had been kept in a state of childish ignorance and dependence. But understanding our bodies was not equated in any one-to-one way with the biology of human reproduction, and the agenda of what was to become the women's self-help health movement was always as concerned with the lived experience of the body as with biomedical science.[1] In consequence these early groups talked about how women's bodies looked, how they smelt and felt, and tried to link this shared, subjective experience with the abstract (and not infrequently blatantly sexist) accounts in the medical textbooks.

Although the feminist projects of understanding both technology and biomedical science were accompanied by a mass of leaflets and self-help groups, there was a clear and taken-for-granted difference between knowing about how a car worked and how one's own body works. While learning to change a washer or mend a fuse was easier with friendly support, the dimension of personal experience was not seen as absolutely intrinsic to understanding. Nor did looking at the inside of a tap involve quite the same breaking of a taboo as that required by self-inspection using a speculum 'Down There' (to cite one wonderful pamphlet title of the period). Body politics were a primary and passionate concern of the women's movement from its earliest days: winning back the control and knowledge of our own bodies was a political objective. 'With my speculum I am free' was a cheerful and mobilizing slogan of the period.

The group, for example, which produced *Our Bodies Ourselves* in the early seventies had originally come together as the result of a women's meeting in Boston in 1969 on Women and their Bodies. After an initial photocopy version, the first edition was published in 1971 as *Our Bodies Ourselves* went on to become a worldwide best seller. Published in no fewer than fourteen languages, including Braille, by 1992 it had sold 3.5 million copies. An icon of the women's health movement, it was also a primer of a liberatory knowledge for women, and as such stands for the myriads of similar texts which were, and continue to be, produced, as women bring together a critical reading of biomedical science with the complexity of living in a female body within a patriarchal, racist and profit-driven society.

The point I want to make is that in this success story of millions of women sharing and developing a new and emancipatory understanding of their own bodies, the word 'science' as such more or less disappears, carrying no special weight within the developing feminist discourse.[2] Often the idea of 'hard facts' or 'reliable information' needed for managing everyday life represented what those trained as scientists spoke of as 'science'.[3] The word itself, 'science', seemed an irrelevant or even alien name for this new, transformative knowledge which had brought together a shared subjectivity and a critical reading of old objectivity. It was only later that a number of feminists began to interpret this fusion as the new feminist science in the making,[4] offering a better, more truthful, account.[5]

Claiming the science debate

In this situation of an innovatory knowledge without a name, it was all the more important that *Signs: Journal of Women in Culture and Society* and its editor Catherine Simpson used the journal's prestige to put 'science' onto the general feminist agenda as a matter of both politics and also scholarship. This was the move that confirmed what feminists in the invisible college were already convinced of; science was far too important to be left to masculinism. Then and now within the natural sciences the field to be contested was biology, for it was biological science which claimed to reveal women's destiny, and new biomedical technologies which – for her own good[6] – saw women's bodily functioning as both distressingly 'natural' and undercapitalized. A feminist reading of the modern biological revolution could echo the chapter in *Capital* on the nineteenth-century story of the machine entering industry, and see the drama of the late twentieth century of fast changing biomedical technologies entering the physiological life cycle of both sexes, but above all the bodies of women. The rate of technological change over the period of second-wave feminism has been intense; ultrasound screening, IVF, egg donation, gene probes and genetic manipulation are the stuff not of science fiction but of everyday reality for many women in the advanced industrial societies. Overcoming gynaecological mystique with self-examination was a fine slogan and activity for the early seventies, but by the eighties there were few easy slogans.[7]

The *Signs* papers, published in 1978, were for a number of years probably the single best guide to the range and direction of US feminist work on science.[8] They were divided between those which examined the social organization of science and those which were concerned with the content of scientific knowledge. Thus, while the former discussed the gender structure of science and what forces controlled women's access to particular fields within science, the latter explored the masculinism of scientific knowledge, and reviewed past and presented new feminist attempts to oppose it.[9] Many of the authors who contributed to that early attempt to set the theoretical and research agenda continued to play a central part within the debates of the eighties.

While poststructuralism/postmodernism as an influential current emanated from within France, the eighties' debate between post-modernism and standpoint theory or realism within the feminist critique of science was most clearly articulated in the US. Because many took part in these debates, I am faced with all the intellectual and personal discomforts of selection, so my hope must be that the texts I discuss below demonstrate the changing emphases, the sharpest moments of conflict, whilst now, in the 1990s, offering a substantial measure of agreement. Both postmodernism and realism, in the sense that I am using the terms here, stand in for a set of theoretical currents, so that postmodernism includes poststructuralism, social constructionism and deconstructionism, while realism, by which more precisely I mean critical realism,[10] includes standpoint theory and some, but by no means all, of what has been called feminist empiricism.[11]

Reclaiming reality

The publication of *Discovering Reality* in 1983 signalled that feminist theorists had now moved confidently into the discussion of the 'epistemology, metaphysics, methodology and philosophy of science' and that feminist claims to provide a better, more truthful account of reality had now been issued. Sandra Harding and Merrill Hintikka, as philosophers and editors, drew on their familiarity with the technical language of the history and philosophy of science to enable feminism both to appropriate and make over powerful tools hitherto marked for the discourse of mastery, and also to set out the grounds on which feminism's own knowledge stood.[12] They asked themselves and their potential contributors the fundamental question: whether there were distinctively masculinist perspectives in the prevailing theories of knowledge and in the metaphysics that supported them.

The book's title flags the terms of the debate which was to occupy the rest of the eighties.[13] It both echoes the 'land ho' quality of the sciences' claims to 'discover' knowledge, and also stakes feminist political perspectives. While the authors generally adopt a realist perspective, three key chapters from Jane Flax, Sandra Harding and in particular Nancy Hartsock set out the claims of 'standpoint' epistemology. This is not to say that there were not other perspectives, simply that the strong articulation of standpoint theory set the terms of the eighties' debate.[14]

A number of the contributors drew on Nancy Chodorow's feminist object relations theory,[15] as at the time this seemed to offer a way of both getting at the profundity of the gendering process and neither abandoning the body nor collapsing into biological determinism.[16] The theory kept US socialist feminist theoretical debate focused on the relations between women and men, on the sex–gender system as well as on capitalism, whereas at the time the British debate was absorbed by the place of women within capitalism, a debate which all too frequently enabled men to escape mention or confrontation[17] (US marxist feminism, due not a little to the debt that the US women's movement owed to the experience of the profound struggles for the civil and social rights of black Americans which took place during the sixties, had better understood that race was deeply entwined with class).

In *Discovering Reality* the political theorist and therapist Jane Flax, the political theorist Nancy Hartsock and the mathematician physicist Evelyn Fox Keller thus all utilize object relations theory, but for rather different purposes. Keller uses it to bring out the gendered character of science and points to the possibility of a non-gendered science.[18] Hartsock by contrast is concerned to develop a specifically feminist epistemology through feminist struggle, and draws on object relations theory as sympathetic to her thesis, but it is by no means central in the way that it is for Flax and for Keller.

Flax is concerned to demonstrate how psychoanalytic theory and in particular object relations theory are crucial tools for feminist philosophy to move into the issues of epistemology and ontology. They represent a systematic attempt to understand human nature as the product of social relations in interaction with biology and offer a means of understanding how the processes of denial and repression of early infantile experience influence political theory. In an ambitious piece she takes on Plato, Descartes, Hobbes and Rousseau to illuminate the patriarchal unconscious and its implications for epistemology. She begins her discussion of object relations theory with an insistence, derived from anthropology, on the commonalities of women's experience. 'While there is considerable variation in men's participation in child care, to my knowledge there is no known society in which men assume the primary responsibility for children under six.'[19]

Because within object relations theory psychological birth is distinguished from physiological birth, the development of the infant is seen as an extended process from the early symbiotic closeness with the care giver – usually the mother – to a gradual process of individualization and separation. 'By the end of the third year a "core identity" or a distorted one will have been established', writes Flax,[20] a position she supports by citing other psychological research with very different theoretical perspectives. As Nancy Chodorow had already pointed out, this primary care giver is almost always a woman, and these very deep feelings and struggles of love, rejection and identity for both female and male infants are played out against an experience of only one gender. Hence what needs to be considered is the negative consequences of this arrrangement on the formation of gender, which is not derived from biological necessity, but is both constituted by, and in its turn constitutes, patriarchy.

Flax's analysis of the return of the repressed within the philosophers begins with Plato's Republic as the meritocracy of reason. She recalls his constant distinction between mind and body, true love and sexuality, love of knowledge and the love of boys or women, always celebrating the 'higher' and eschewing the 'lower'. Women are seen as inherently dangerous, so capable of stirring up the passions, so associated with the low, that they must be excluded from public office until they are over forty (as 'women of gold', which I take to be signifying women who are post-menopausal, so no longer sexually dangerous). The philosopher's state as the expression of social justice requires its citizens to live in the higher realm of disembodied reason, with the flesh and all its desires left to lower beings. Flax poignantly asks whether this bleak message of repression, this denial of sexuality and the body is the only way of achieving justice.

We begin to see what Hegel's 'unhappy consciousness' behind most forms of knowledge means in terms of the denied and repressed self. Descartes's passion for control of nature, including bodily nature, his belief that the only thing he could be certain of was 'my essence' and 'my thought', his conviction that any knowledge not built on mathematics is worthless, and his splitting of the mind and body were not only significant for Western philosophy and particularly for science; they also render him vulnerable to Flax's psychoanalytic reading.[21] Within object relations theory, a response in which the outside world is purely a creation and an object for oneself can only be understood as profoundly narcissistic. 'This frozen posture is one of the social roots of the subject–object dichotomy and its persistence within modern philosophy.'[22] She concludes that the dichotomy cannot be resolved from within philosophy alone, for what it speaks of is the problem of psychological development within patriarchy.[23]

Flax's proposals for a feminist epistemology are located firmly within her commitment to feminist psychoanalytic theory, for this provides the means for disentangling that 'unhappy consciousness'. It is important to add that this commitment by no means prescribes lying on the couch, and substituting individual therapy for feminist politics; it prescribes active engagement in the therapeutic and

existing forms of rationality and consciousness have been historically produced – not least within a sex–gender system, so that men are the embodiments of reason and women of passions – she urges that feminism needs to re-examine the epistemology of all bodies of knowledge which claim to be emancipatory, including marxism and psychoanalyis. Flax's subsequent move to embrace postmodernism is perhaps less surprising when we read that she proposes that (1) all concepts should be relational and contextual, (2) knowledge must be self-reflective and self-critical, (3) knowing should be understood as activity, as dialectics; and (4) women's experience is not in itself a ground for theory; it must be incorporated and transcended through consciousness raising.[24]

Standpoint theories

Nancy Hartsock's rallying call, 'The Feminist Standpoint: Developing the Ground for a Specifically Feminist Historical Materialism', has been a central text of standpoint theory and sturdily resists the postmodernist appeal to dismantle truth.[25] She opens with marxist meta-theory, with ontology and epistemology rather than the critique of capitalism. She proposes: (1) that material life (Marx's class position) structures and sets limits to understanding; (2) if material life is structured in opposing ways the vision of each will be an inversion of the other; (3) the vision of the ruling group structures the material relations for everyone and therefore cannot simply be dismissed as false; (4) and in consequence the vision of the oppressed group must be struggled for, as it requires both science to see beneath the surface of social relations, and the learning processes of struggle itself. As an engaged and subjugated vision the adoption of a standpoint theory both expresses the most accurate account of relations between human beings as inhuman and offers a historically liberatory practice.

Her approach is rooted in the sexual division of labour, and because she insists on keeping 'corporeal reality' firmly in her account she deliberately uses the most bodily concept of 'sex' rather than 'gender'. She follows Sara Ruddick's[26] argument that although both men and women can parent, which allows for many of the messages from object relations theory to be incorporated, only women, barring scientific developments not likely to occur imme-diately, give birth. This entirely self-conscious 'essentialist' move is characteristic of the standpoint theorists[27] – or feminist realists – as they ground their epistemology in the corporeal as well as in socially produced material reality. Thus they embrace both a generously defined reproductive labour and women's constant sense of living in a body which is not seen as fully human.[28] Hartsock points to the gendered character of the two worlds, one abstract and of high status, one concrete and of low status. And this gendered antagonism lies at the heart of a series of powerful dualisms of mind/body, reason/ passion, culture/nature, abstract/concrete; dualisms echoed and reinforced by a social order.

As a political theorist Hartsock is acutely conscious of how profoundly these dualisms resonate within philosophical and polit-ical theory, and argues that 'Abstract masculinity . . . can be seen to have structured Western social relations and the modes of thought to which these relations give rise at least since the founding of the polis.'[29] Like Flax she understands this abstraction as not only partial but perverse, because the price of the abstract masculinity which is equated with the fully human is a masculine association of sexuality with violence and death. Women's place in these social relations is to perform that reproductive work which is seen as less than fully human and which systematically harms and degrades those who perform it. Not for Hartsock is there any 'feminine' celebration of women's experience and ways of knowing[30] – within her standpoint perspective there is little romance with individual knowing and a strong commitment to engaged political struggle.[31]

While object relations is central to Flax's standpoint epistemology, and a useful but not crucial adjunct to Hartsock's, within Sandra Harding's position of qualified support it no longer appears. Harding begins with a different question which asks 'why the sex–gender system has only now become visible'. Arguing that it is an organic social variable, not an effect of other more primary causes such as the class system, she points to the immensity of its social dimension. Protecting her arguments from the charge of over-universalizing she cites the anthropological evidence for the existence of male domin-ance as an organic feature of most recorded social life. Harding then considers the kinds of epistemological claim that are made and might be made by feminists.

She distinguishes three epistemological stances and effectively dismisses the first two. The first, 'empiricism', explicitly holds that 'historical social relations distort our natural transhistoric abilities to arrive at true beliefs', and conversely that different social relations will not enable these abilities to provide better truth-claims about reality. At its core lies the conviction that it is only social influences which produce distortions, and that, if these can be removed, then our faculties will enable us to produce better knowledge. She suggests that a number of feminist researchers and theorists have taken such a stance. Feminism removes the blinkers from women's eyes, and they see more clearly and generate more truthful knowledge as a result. Very gently she indicates this is a less than adequate epistemology.[32]

Her second category of functionalist and relativist epistemologies is concerned with the new social constructionists, particularly the 'strong programme' associated with David Bloor and Barry Barnes.[33] While epistemological relativism enables its holders to show how appeals to objectivity function as a resource in science, its holders cannot go beyond that point. They cannot, for example, explain why particular scientific theories may grasp the regularities of either the natural or the social world better than other theories. Thus they cannot offer any reason for overthrowing a weaker theory in favour of a better (a position I had earlier called the new hyper-reflexivity)[34] in that such philosophical relativism lacks any ground on which to stand in order to make its own claims.[35] Harding having dismissed men's social constructionism, which I take to be one of the several strands of mainstream postmodernism, simultaneously leaves the door open to feminist postmodernist currents, as she sees these as a powerful defence against false and almost certainly Eurocentric universalism.

The third position Harding explores is that of feminist marxism. Although this accepts the legitimacy of class distinctions, it also insists on the commonalities of the gender division of labour across those distinctions.[36] This division in the context of explicitly feminist struggle offers the possibility of a distinctively feminist standpoint. Such a perspective must be sensitive to the differences between women through class, race and culture, but it can offer a more complete knowledge than the partial gains offered by the old proletarian standpoint.[37]

Despite her evident sympathy for standpoint theory, Harding observes that it still cannot answer the question of why the discovery of the sex–gender system occurs at this time in history. What conflict, she asks, 'objectively outside us' is in us 'the reflex of thought'?[38] Despite the confidence of the title there is a hesitation in confronting the theoretical issues, a hesitation which enables her to return fruitfully to explore these issues further in *Whose Science? Whose Knowledge?*[39]

The linguistic turn

But in the eighties, as we know, standpoint theory was not unchallenged. The commitment to a feminist marxism which took material activity (as a radically enriched concept of labour) and bodily existence as its central categories and which worked within the framework of a feminist materialism was being transmuted into a preoccupation not simply with language as a means of deconstructing androcentric science, but with language itself.[40] Within the feminist critique of science this movement can be seen very clearly in the work of the historian of science Donna Haraway. In 1978, envisaging the tasks of socialist feminism, she spoke of 'accepting our responsibility to rebuild the life sciences'[41] and continued:

> I understand Marxist humanism to mean that the fundamental position of the human being in the world is the dialectical relation with the surrounding world involved in the satisfaction of our needs and thus in creation of use values. The labour process constitutes the fundamental human condition. Through labour, we make ourselves individually and collectively in a constant interaction with all that has not yet been humanised. Neither our personal bodies nor our social bodies may be seen as natural, in the sense of existing outside the self-creating process of so called human labour. What we experience and theorize as nature and as culture are transformed by our work.'[42]

By 1981 Haraway had embraced the world of texts, or, as she was to put it slightly later, of story telling, and saw feminists as faced with two choices in how they might respond to the crucial challenge posed by the life sciences: they can either retell the original story in a way that is favourable to women, or they can tell an entirely new story. She marks this with an essay called 'In the Beginning was the Word'.[43] Through the discussion of two widely read books in whose production the feminist biologist Ruth Hubbard has been a central figure, *Genes and Gender* and *Women Look at Biology Looking at Women*,[44]

Haraway points to what she sees as 'repeated unexamined contradictions'. These contradictions result from Hubbard's method of
exposing 'bad science', revealing its 'fictive character' and then
proposing the 'real feminist facts'. Through a close reading of her
influential article 'Have only Men Evolved?'[45] Haraway notes
Hubbard's deconstructionist criticism of theories of representation and
ideologies of objectivity and draws attention to her discussion of the
centrality of language in this process. She approvingly quotes
Hubbard:

> For humans, language plays a major role in generating reality . . .
> However, all acts of naming happen against a backdrop of what socially
> is accepted as real. The question is *who* has a social sanction to define
> the larger reality into which one's everyday experiences must fit in
> order that one be reckoned sane and responsible . . . at present science
> is the most respectable generator of new realities.[46]

But Haraway reads Hubbard as if she had written, and only written,
'language generates reality.' Hubbard qualifies this statement, referring to 'a major role', and saying that 'science is the most respectable
generator of new realities'. Without the qualifier Haraway is able to
conclude that '(language) does not *stand for* or *point to* a knowable
world hiding somewhere outside the ever-receding boundaries of
particular social historical enquiries.' She thus reads a contradiction
between this and Hubbard's longing for a science which is 'more than
a reflection of various aspects of ourselves and our social arrangements'. As an example of this contradiction, she goes on to commend
Hubbard's nuanced reading of the male-engendered stories of human
evolution. Then she draws attention to a sentence in the middle of
Hubbard's deconstructionist account, which, without any sense of an
epistemological problem, asserts a fact. (The actual sentence is about
the palaeontological finds that led to the conclusion that the main
features in human evolution were upright stance, brain size and
reduced teeth size, conclusions that had themselves been the subject
of dispute – though not around gender.)
    What Haraway is quarrelling with here is the possibility, if
'language generates reality', of making any 'true' claims about human
evolution. Now arguably I have made more of Hubbard's 'a major
role' than is entirely reasonable, but I want to suggest that using
deconstructionist techniques, or adopting a social constructionist
perspective, does not of itself remove the truth-claims of the science
in question. There is no reason why Hubbard cannot simply stop
deconstructing; certainly deconstructionist others, not least Derrida,
have a nicely tuned sense of when to stop. Hubbard's use of
deconstructionist techniques to criticize masculinist science has to be
seen as connected with her practice both as a biologist and as a
contributor to the work of the Boston Women's Health Collective in

producing the provisionally reliable and liberatory knowledge of *Our Bodies Ourselves*, with which I began this chapter.

What has been called the 'linguistic turn' is a good reason for being grateful to postmodernism, which has indeed been feminism's ally in sharpening our ears to hear the construction of knowledge and its coupling with power, but gratitude does not carry with it any necessary commitment to abandon truth claims. While a historian can read natural science as stories, leaving the scientists with their problems of truth-claims subverted but not resolved, a natural scientist and/or a feminist engaged in health struggles has to be a realist, has to care about 'hard facts'.[47] As I show later, the nearer feminist historians come to studying the work of living feminist scientists in something like the same time and cultural space, the more their claims are transformed to those of a re-visioned realism.

While self-accounts are not the last word, we also have the later statement from Hubbard of what she sees as her position, for she returns briefly to this debate in the introduction to her book *The Politics of Women's Biology*, in which she makes it plain that 'In the current debate within feminist science criticism I stand with those who argue that the political insights feminism provides can lead us to more accurate, hence truer accounts of nature than we now have.'[48] Hubbard also makes it clear that she is arguing for cultural not philosophical relativism, citing James Fleck to argue that 'In science, just as in life, only that which is true to culture is true to nature.'[49]

Haraway is at this point arguing that feminists cannot both have a deconstructionist account and claim realism for their own. Yet as her own work unfolds over the eighties, while rather carefully re-visioning her concept of objectivity, realism's strongest word, she is not willing to abandon it.[50] The feminist theory of science during the eighties (and this article of Haraway's is a clear example) constitutes the watershed of the 'either/or' position between a postmodernist project of dismantling truth and that of standpoint theory or critical realism. As the decade wore on, the feminist critique of science developed, not without difficulty, a position of 'both/and'[51] and in doing so helped feminism avoid the more fruitless 'either/or'. While personally I began closer to the realist side of the new 'both/and' position, what I think most of the participants were conscious of and spoke about was the extraordinary seriousness and feminist good practice with which the debate was carried out.[52] 'Listening to each other' was a distinctive mark of this. Unquestionably the stakes were high and understood to be high, for the issues were and are central to feminism. So at this point I want to break off my discussion of Haraway and Hubbard and turn to Sandra Harding, as it is the intervention of her 1986 *The Science Question in Feminism* which prepared the ground for the precarious strengths of 'unstable categories' and which generated perhaps the most influential of Haraway's papers in response.[53]

Claiming both/and

Harding's book appeared at the high tide of deconstructionism in science, both in the mainstream and increasingly among feminists, and while she criticizes the hegemonic claims of a 'feminist successor science' as too much part of an Enlightment tradition, she never entirely abandons the standpoint perspective. However, she does not repeat the critique of the throughly relativistic epistemologies she made in *Discovering Reality*, and she gives generous space to the contribution of the concept of difference offered through feminist postmodernism. My reading is not uncontentious, for Janet Sayers,[54] from a marxist feminist perspective, sees Harding as simply giving up on realism, dispensing first with science because its legitimacy and authority are fatally compromised by its androcentricity, and then going on to do 'much the same for women'. Reading Harding in much the same way but to rather different effect, Sarah Franklin and Maureen O'Neil[55] enthusiastically welcome her as a notable convert to the full deconstructionist programme. I believe that both misread Harding in terms of their anxieties and desires, and while it is a truism that we all do, I do not think that texts are infinitely plastic. Harding herself says quite unequivocally, 'It should not need to be said – but probably does – that I do not wish to be understood as recommending that we throw out the baby with the bathwater'.[56] And a little later she continues, 'I am not proposing that human kind would benefit from renouncing attempts to describe, explain, and understand the regularities, underlying causal tendencies, and meanings of the natural worlds just because the sciences we have are androcentric. I am seeking an end to androcentrism not to systematic inquiry.'[57]

To save truth-claims while accommodating the new postmodernism, Harding draws extensively on Quine's critique of empiricism, in which he observed that in practice theory choice draws on a coherence criterion located within a framework of belief. For Quine meanings and facts cannot be entirely disentangled, and he sees physics and metaphysics as rather closer than in the conventional view from a positivistic philosophy of science. His conception of total science as a field of force whose boundary conditions are experience leads him to suggest that any conflicts at the periphery produce changes and adjustments elsewhere. Indeed, that Quine's model makes space for anomalies and sees them as problematic also provides a way of maintaining truth-claims[58] and avoiding relativism.[59] While criticizing Quine as too behaviourist, even reductionist, in his attempt to analyse science scientifically, Harding draws on him to arrive at a position quite close to the 'science in social content' position as it both locates knowledge and protects truth-claims. But of course exposing dogmas is not synonymous with ending them, and

Harding was absolutely right that 'excessively empiricist beliefs still haunt most of the feminist critics of science', even if now there is also a strong move towards deconstructionism.[60]

Harding's favoured 'model' for the natural sciences is that of the social sciences. Like most feminist critics of science she rejects the assumption that physics has some permanent pre-eminence (even as a model for physics itself), and in consequence she recruits Dorothy Smith's project of writing a feminist sociology. Within mainstream history and philosophy of science this is still something of a heresy, not least because of the long tradition – from say the liberal Thomas Kuhn to the radical and anti-sexist Brian Easlea – of physicists moving into the social studies of science. In consequence the theory of science is peppered with references to the history and to their personal experience of doing physics. Despite the siege which has been laid to the physicists' claim that physics *is* science, the belief is distressingly alive, and even given renewed vitality, in the extreme reductionism of the new genetics and molecular biology.[61] While I am sure that between doing some biological and some social sciences there are a number of ways in which the methodologies draw close, I am unhappy with the notion that any field of enquiry should become 'the model' for the rest, whether physics, social science or, as I sometimes feel nowadays, even feminist literary theory, as the belief in an ideal model fails to acknowledge the diversity and relative autonomy of the knowledges.

There are substantial differences between even biology and the social sciences. Social science does not share the cumulative nature of an experimental biology whose potent relationship with technology means that it is simultaneously about generating changing accounts of nature *and* creating new biotechnologies, which themselves impact on nature, including human bodily nature. Growth in the social sciences, while not cumulative as is the case with an experimental discipline, or at least only relatively briefly and within a specific paradigm, does deepen over time, but through the extensive reading of its practitioners. Thus the kind of theory of representation feminists might share in common across the knowledges has to be sensitive to the different fields, as well as enabling feminists to take a stand for or against specific accounts within them. Without such a theory and without political engagement, then the present powerful current of postmodernism beckons, which, unhinged to any political project, simply presides over nihilistic wise-cracking.

Dorothy Smith, in a number of books developed since the seventies, has been clearing the way for a distinctively feminist sociology 'for women'.[62] In our deeply gendered times she is manifestly right to specify the gender she wishes her science to serve. We remember that Galileo's universalistic claim that the new knowledge would be a 'science for the people' worked out in practice so that science came to serve a largely ruling class. We remember too that the would-be corrective of the 1960s' radical science movement's determination to re-vision the project, so that a renewed 'science for the people' would truly serve the working classes, was unable to see the social relations of gender and was only partially successful in theorizing 'race'. But Smith does not precisely speak of the 'feminist standpoint' but of the distinctive 'standpoint of women'.

Harding reads her as saying that the forms of alienation experienced by women enquirers make it possible to carry out successor science and postmodernist projects simultaneously and without contradiction. She quotes from Smith:

> Here I am concerned with the problem of methods of thinking which will realise the project of a sociology for women; that is, a sociology which does not transform those it studies into objects but preserves in its analytic procedures the presence of the subject as actor and experiencer. Subject then is that knower whose grasp of the world may be enlarged by the work of the sociologist.[63]

Here Harding seems to be imposing the categories of the science debate onto an older problem within sociology which has little to do with today's dismantling of truth at the hands of postmodernism. Because Smith brings together the voice of the enquirer and the voice of the 'subject' of the enquiry, putting them on the same epistemological level, Harding suggests that Smith is fusing hitherto 'incompatible tendencies towards interpretation, explanation and critical theory in the philosophy of the social sciences'.[64] Smith argues that this kind of science is objective, not because it uses the impersonal third voice but because it uses the more complete and less distorting categories of historically located subjugated experience.[65]

Smith's theoretical stance fuses feminist marxism, which stresses women's material activity, with a commitment to embodiment and to ethnomethodology. The latter has given new life to the tradition of phenomenological social enquiry which has stood in antagonism to positivism, emphasizing meaning without giving up on its own claims to provide a truer account. Where ethnomethodology in its initial form seemed to dissolve the world into the microsociology of interpersonal practices, and in consequence offered no way of getting a grip on those larger matters of structure, whether they are called patriarchy, capitalism or whatever, the fusion gives a way of being faithful to both structure and the intentional practical actor.

Thus my reading of Smith's achievement is rather different, since ethnomethodology has for almost two decades entered both mainstream and feminist sociological enquiry as offering a means of dealing adequately with both agency and structure.[66] The mainstream adoption of ethnomethodology has been extensively discussed;[67] by contrast, feminist sociology has given less time to elaborating its epistemology and more to providing truer/better accounts of women's lives. The latter has justified its concern with the intersubjectivity of social science, for the most part not from an explicit theoretical attempt to develop a hermeneutics of a feminist social science, but from political and cultural values which are committed to extending women's subjectivity as agents.[68] Smith's contribution has been to provide a powerful theoretical apologia for what I see as a widespread research practice among feminist sociologists and social psychologists.[69]

The innovative epistemological stance implicit in much of feminist sociology is unacknowledged by the mainstream theoreticians, and their citation practices suggest that feminist literature, even where they claim feminist sympathies, is either unread or unnoticed by them. They seem unaware that within specific fields of, for example, the sociology of health, social policy and stratification theory, the discourse has been changed and gender as a concept has come from the margins towards the centre. By contrast, Dorothy Smith's feminist sociological theorizing is acknowledged by numbers of mainstream theorists. This is no small cultural as well as political move from margin to centre within general social theory. She has made a powerful and subversive move supporting embodied knowledge within social science, which as part of science Hartsock so rightly denounced as 'abstract masculinity'.

Certainly Smith, like many sociologists outside the numerically small group involved in the sociology of scientific knowledge, is simply unimpressed by the seductive charms of postmodernism. Indeed she makes this clear in a rejoinder to Harding in which she indicates that she has no wish to go down the road of 'the repudiation of the very possibility of a master narrative, of knowledge, and its replacement by multiple partial knowledges derived from multiple sites, none prevailing, each equally valid'.[70] For her such ontological tolerance is incompatible with enquiry itself.

Radical deconstructionism and situated knowledges

By 1990, not least with the publication of Haraway's brilliant *Primate Visions*, the methodological debate had shifted.[71] No one makes this clearer than Haraway, for this is a scholar for whom the thought of an innocent text is unacceptable. But before turning to Haraway's theoretical approaches, it is important to say something about her field of enquiry, for primatology and the study of human origins are more made for deconstruction than almost any other area of science that I can think of, and it is to Haraway's credit that she was alert to this. Thus while experimental biologists talk about 'god's organism', by which they mean the one organism which for their purpose best speaks for nature, primatology and human origins could well be described as the 'goddess's disciplines' for a feminist deconstructionist.[72]

Primatology and human origins are surely disciplines which will speak with clarity to the feminist and anti-racist enquirer. The issues surrounding the origin stories of the human species are always bound to be rather more obviously culturally and politically located, not to say ideologically suffused, than, say, the biochemistry of liver function,[73] not least because the evidence is confined to deductions made from reconstructions of rather dodgy fossil remains combined with present-day anthropology, and the history of both are rooted in imperialism and the 'white man's burden'. What else might we expect but that the origin stories offered by palaeo-anthropology would be tailored to fit the myth of the moment and contemporary science fashion (from the innate aggression of man the hunter to optimal foraging strategy)? Primatology, by contrast, seems to be better able to accumulate data and more bound by its own evidential base – and indeed in the scond half of the book, where she turns from the one to the other, Haraway comes very close to acknowledging this.

However, there is one way in which primatology is much closer to social science than most areas of biology, for it is an observational, not an experimental, science. Once the camera replaced the gun in 'shooting' the subject of enquiry, primatology became a non-violent discipline, and so more compatible with the construction of late twentieth-century – albeit adventurous and comfortably off – feminine and feminist women and the defence of 'nature'.[74]

Observational and experimental sciences

Before continuing my exploration of *Primate Visions*, I want to say something about the way in which the observational sciences like primatology are a minor, although publicly highly visible,[75] current within the life sciences. It is the experimental sciences which are linked more closely to biotechnology that occupy most of the budget and have been the major focus of feminist criticism of contemporary technoscience. Both experimental and observational sciences are immensely powerful as shaping culture.

Physiology was the pioneering experimental science within the life sciences, and required and requires harming and killing animals – 'murdering to dissect' as the opponents of physiological reductionism described it. Such an experimental science has very different implications for the gendering of the science and how that has been read by feminists. It is worth making a brief detour into physiology and its reception by feminism, so as to highlight the contrast between an experimental – a deliberately interventionist – science and an observational one.

The cruelty without limits of pioneering nineteenth-century physiology, when animals were experimented on without anaesthetic or any other controls, has been widely documented. The founding father of physiology, Claude Bernard, had a fundamentally Cartesian conception of animals which could not admit their having sentient capacity. In consequence the removal of a limb was in his understanding more or less analogous to removing the leg from a chair. The justification for the disassembling of these animal mechanisms was that only experimental science gave the possibility of true knowledge – all the rest was mere nature study.[76] Even Burdon Sanderson, who worked in London rather than, like Bernard, in Paris, and so was arguably less influenced by the mechanistic metaphors of Descartes, was seen by many of his scientific contemporaries as incapable of acknowledging his experimental animals' capacity to experience pain. Even now the struggle to secure better controls over the use and care of laboratory animals is still waged, and the more radical option suggested by the geneticist J. B. S. Haldane almost half a century ago, of developing a 'non-violent biology' which did not use animals, is strongly pressed by the contemporary animal rights movement.

While there is a widespread hostility to the misuse and even use of animals in research, the links between women and experimental animals are less strongly drawn within contemporary feminism than in the late nineteenth century. Then, particularly inspired by the writings and campaigning of Frances Power Cobbe,[77] sections of the feminist movement made a direct link between the cruelty of men as scientists to animals within the laboratory and the cruelty of men as husbands to women within the home. The women's movement then as now was torn on the issue. Even the pioneering women doctors were divided: Elizabeth Blackwell was opposed to vivisection, and to gynaecological examinations where women were strapped and bound, as both brutalizing and degrading, while Elizabeth Garrett Anderson, by contrast, thought that women doctors should study physiology. None the less the metaphor of rape, and the need to resist it, is strongly present within the feminist anti-vivisection literature from the 1880s onwards.

Those persuaded by Cobbe might well feel that it was not by chance that Burdon Sanderson was estranged from his wife, and that as his widow she left her estate to animal homes. Nor do I think that this connection with a particular construction of masculinity and experimental science has entirely retreated. At a recent Cambridge seminar on the history of the laboratory in medicine one of the men historians made the observation, 'In the laboratory and in the whore house you can do what you like.'[78] The point is not the sexual offensiveness of the observation (and we do not need to be very sophisticated about the analysis of language to know the gender of that 'you'), but that the Freudian lapse spoke of the connection between masculinity and the construction of women and animals into the common objects of men's desires.[79] There is no love in that construction of power and knowledge.[80]

By using the example of physiology, which drew such strong feminist criticism, I want to underline the controlled violence required by most of the life sciences (ironically receding in some areas, notably in the new industrial approaches of molecular biology) and not to let the hard task of making such sciences feminist be glossed over, as we consider the less violent (and more re-visionable) observational sciences.

Primatology: the goddess's discipline?

Before I return to Haraway's theoretical positions, let me deal with an initial problem the book poses, recognized indeed by Haraway herself, whose suspicion that it is hard going leads her to say encouragingly that each chapter can stand alone. But the sheer difficulty of her prose, stemming from her commitment to decon-structionism, is both real and more than worth working around. Gayatri Chakravorty Spivak is surely right when she observes, 'Deconstructionism has taught us that taking contingency into account entails the labour of forging a style that seems only to bewilder.'[81] Despite Haraway's democratic desire to make the book 'interesting for many audiences', the heteroglossia of her prose can at times not only be 'pleasurable and disturbing'[82] but leave the reader shut out and longing for a less bewildering style.[83] Yet always, for Haraway as for all the feminists who work on and within natural sciences (at least I cannot think of any radical subjectivists in the study of the natural sciences),[84] there is a sense of the real 'out there' which is simply not shared by many of those who discuss fiction, history, cinema and psychoanalysis.[85]

Haraway describes her work as set within the four theoretical approaches ('temptations', as she speaks of them). The first is social constructionism and her exemplars are Bruno Latour and Steve Woolgar, who reject all forms of epistemological realism and are thoroughly social and constructionist. What attracts her is their insight that science is a fresh form of power in the social material world – and that scientists invest their political ability in the heart of doing science.[86] What she does not say is how Latour and Woolgar's literary and Machiavellian account of science can explain why one scientific account is seen as better than another. (They have, essentially, an E. H. Carr view of history: who wins is right).[87] When resources are equal (always a serious matter for anyone other than a head-in-the-clouds internalist), is the outcome of the struggle over facts always determined by 'political ability' – and what precisely is that? The argument of political ability gets very close to the position that skill (a historically acquired and acknowledged attribute) and power are right. If we were to try to explain the growth of science on the basis of political ability, as against restricting this, as Latour and Woolgar do, to an anthropological account of 'laboratory life' within a single laboratory and of the work of producing facts within it, then interpreting 'scientific scandals' becomes peculiarly difficult, particularly when associated with immensely powerful scientists, as in the ongoing saga of the 'Baltimore' or 'Gallo' cases.[88] My critique of strong social constructionism is old but serious; such total relativism gives us no way of judging between competing and plausible accounts.

Haraway's second 'temptation' is the historical materialist tradition, which claims the epistemic privilege of subjugated knowledges as accounting for the greater adequacy of some ways of knowing. This tradition claims to show that race, sex and class shape the most intimate details of knowledge and practice, above all where the appearance of that knowledge and practice is of neutrality and universality. While this approach rejects the relativism of the social studies of science, Haraway suggests that there are difficulties in specifying how it relates to detailed scientific knowledge, not least that of primatology. By contrast, I suggest below that her account in the second half of the book comes very close to doing just this. Haraway suggests that the deficiencies of historical materialism are overcome by the use of the concept of science as a labour process, originally advanced during the 1970s by the historian R. M. Young, a key figure within the collective producing *Radical Science Journal*.[89]

As with Latour and Woolgar's position, the argument of science as a labour process leaves no space for a realist or positivistic epistemology; every aspect of scientific practice can be described in the concept of mediation, making other social relations appear as derivative. Yet Haraway wants to insist, which is of great importance for primate studies, that childbearing and childrearing, despite an extended concept of mediation, cannot be contained by the category of labour. Given the wealth of British and Scandinavian literature, especially Norwegian (discussed in chapter 2), which explores the 'labour of love' and its associated rationality, I think Haraway gives up on the labour processes of caring as a means of understanding childrearing too easily, not least because here feminists have radically re-visioned the concept of labour.

But, as I indicated in chapter 1, I was among those who were highly critical of Young's approach at this period, as I believed that in abandoning realism he also abandoned the radical scientists as potential political allies.[90] Within the stance of 'science is social relations', only historians who understand mediations can speak; mere natural scientists with their commitment to reality are reduced to objects of historical study, their claims to create socialist, anti-racist or for that matter feminist knowledges of nature an epistemic absurdity.[91] (And while it is true that scientists have had too much to say in the past, and indeed present, construction of reality, the utopian project of the construction of knowledge which I want to see come into existence seeks to include the voices of all the relevant actors.)

The third theoretical approach which attracts Haraway is the claim of the scientists themselves that they are not simply concerned with power and control, but that their knowledge 'somehow translates the active voice of their subjects', and that they 'get at' the world. The power of this temptation, the seduction that what scientists do is really to describe nature, is conveyed in Haraway's prose as she has to insist almost against herself that, 'What does seem resolved, however, is that science grows from and enables concrete ways of life, including particular constructions of love, knowledge and power. That is the core of its instrumentalism and the limit to its universalism.'[92]

Her last temptation is simply story telling. While Haraway accepts that scientists have to have a commitment to the real, she suggests that humanists can live with story telling which lacks the power to claim unique readings. By contrast, modern science claims to convince the entire world with its one true story. Natural sciences are the other to the humanities, which are specific and local. Social and psychological needs, however, are served by the divided knowledges, by the division of labour and the production of the authority in the discourses. But – and here she begins to show how she weaves together her four approaches – she does not want to reduce the natural sciences to the cynical relativism that we would both agree lies in the social constructionism of much of the social studies of science.[93] Her argument does not claim that there is no world, no referent in the signs and meanings, no progress in building better accounts. Almost in the spirit of magical realism Haraway invites us to read her own text through the lens of science fiction, as an unfinished story, remapping the borderlands of nature and culture.

Primate Visions itself moves from telling stories about primatology and primatologists in the early years of the twentieth century (indeed, the thoughts and lives of dead white males apostrophized by feminists) to discussing current primatology and current primatologists – a number of whom are significant feminists. Haraway treats media and museum representations/stories and the accounts of primatology itself rather differently, which is particularly evident as the history becomes more recent. When Haraway is retelling the stories of the early days of primatology, deconstructionism is set to work energetically on the media representations, above all the museums, the science and indeed the scientists. She shows us that primate studies before World War II were an integral part of the system of unequal exchange of extractive colonialism, that the great

apes were crucial in the 'great chain of being', and that in
consequence during the interwar period the major European powers
had a mix of simian research colonies abroad and at home.

Haraway's account of the early days of primatology approaches the
anti-militarists' version of army recruitment advertising: 'Join up so
that you can travel to exotic parts of the world, wonder at the exotic
people, and kill them.' Initially, she tells us, the important thing to do
was to find the most beautiful and biggest gorillas, kill them, stuff
them and exhibit them in the museums of the West. Thus, to take her
first story of the 'Teddy Bear Patriarchy': Carl Akeley, artist, nature
lover, hunter and scientist (who, we are told in an aside which
borders on the surreal, had developed his taxidermic skills on the
deceased Jumbo, the showman Barnum's famous elephant), Akeley's
first and second wives, President Theodore Roosevelt, and the Giant
of Karasimbi, a lone silver-back gorilla, are the central actors in the
saga. Held together by the imperial, racist and violent hunting
enthusiasms and language of the time, Akeley's dioramas in New
York's Africa Hall powerfully convey that sense of Darkest Africa in
which the gorilla stood as the other to Western man. Loving, killing
and photographing this other was for Akeley, and indeed for his
patron and friend Roosevelt, the embodiment of white masculinity.
The Africa Hall, where these relationships were represented through
dramatic dioramas, was supported by leading capitalist philanthrop-
ists for whom its scientific story of the supremacy of the White Race
facilitated the developing convergence of conservation and eugenicism.

But after these early analyses, Haraway increasingly draws a
distinction between the account from science and the account from
the media. On the one hand there is a deconstructionist account of
the way the media – notably *National Geographic* but also a number
of more recent television programmes – have handled the stories of
human origins and 'man the hunter', on the other there is an almost
Whiggish history of the triumph of feminist primatology over
masculinist sexist or sex-blind accounts. Haraway addresses the
scientific papers of contemporary scientists less, focusing instead on
their popular writing or their media representations. Thus, despite
her best efforts, she erects an invisible wall between the 'science' and
the 'media stories about science' and avoids confronting current
science head on.

One of the clearest examples of this comes in her treatment of Jane
Goodall. Apart from the comment that in her early papers Goodall
fails to acknowledge her black co-workers except by their first names,
Haraway separates the 'real' Goodall, whose book *The Chimpanzees of
Gombe* she praises, from the media construction of Goodall – the real
Goodall is a co-operative worker surrounded by PhD students,
husband, mother etc.; the *National Geographic* Goodall is alone in the
jungle (later plus child, Grub). For the magazine construction the

numerous black assistants simply do not count. Similarly Haraway's discussion of Jeanne Altmann's attempt to eliminate observer bias inherent in the previously less systematic attempts to study primates in the wild, by the use of time-budgeting methods, treats this as something very close to a paradigm of good feminist science. Altmann's technique enables her to divert attention from showy activities like sex and domination towards the day-to-day business of 'juggling' both making a living and caring for the kids. Haraway thoroughly approves of this power move by Altmann which goes beyond critiquing sexist primatology to providing a different better account. Yet surely what Altmann does is of a piece with Hubbard's actions. Both delicately deconstruct and reveal existing concepts as 'bad science', whether of man's evolution or the concept of 'rape'. Each then uses the resources of power available to her, in Hubbard's case the position of being a tenured biology professor at a major university, in Altmann's that of being the US editor of the influential journal *Animal Ethology*, to persuade scientists to replace a bad account or concept with a better.

There is too more than an echo of a hermeneutics of the biological sciences in Haraway's notion that, 'The animals are material-semiotic actors in the apparatus of bodily production. They are not "prediscursive bodies" just waiting to validate or invalidate some discursive practice, nor are they blank screens waiting for people's cultural projections. The animals are active participants in the constitution of what may count as scientific knowledge.'[94] But the metaphor through which the advance is made is one which interprets the activity of female monkeys as analogous to the labour process of women in contemporary Western society. The trafficking between the standpoint theorists and the feminist radical deconstructionist seems increasingly complex.

Only in her treatment of the sociobiologist Sarah Hrdy does Haraway's distaste for sociobiology as the method of 'twentieth-century mutant liberalism' emerge (a distaste I more than share). Thus having told the stories of the primatologists Jeanne Altmann, Susan Fedigan and Adrienne Zihlman as accounts which could equally be read as good feminist science driving out bad sexist science, when it comes to Hrdy she is much more critical, and distances herself to allow Hrdy's scientific claims 'to be dealt with by the canons of their own discipline'.

In this rich, contradictory book – and in a highly contradictory world that description is by no means automatically a criticism – where do I think Haraway has moved to? First and foremost I think that Haraway's own close observation/participation of and in this outstanding group of feminist scientists has led her to spell out her recognition of their necessary commitment to realism. Their feminism means that this is always a *critical* realism, and their accounts of nature reflect their understanding that the stories they seek to overthrow are framed by the assumptions of their producers. Acknowledging their realism means that their stories, like those of the primates in their own studies, have to become part of the historian's account. This has set up an intense tension in her work between seeking to mark out and defend a feminist definition of realism and a 'usable definition of objectivity', combined with a radical constructivism, against a conception of realism and objectivity as merely part of a master narrative. Her alternative to relativism is 'partial critical knowledges located in multiple embodied sites, sharing the web of connections called solidarity in politics and shared conversations in epistemology'.[95]

Beyond either/or

Whatever is happening elsewhere in feminist theoretical debates, so that in literary theory and psychoanalysis the choice between postmodernism or realism is sharply put and postmodernism is in the ascendant, within the feminist critique of the sciences the debates are differently posed. In this field there is currently a robust attempt to refuse this either/or choice and instead to re-vision a defensible feminist concept of objectivity. This split beween feminists concerned with the sciences and those with the arts sometimes seems like a bathetic return of the (wrong even when first proposed) thesis of the two cultures.

Thus, having drawn on postmodernism's concern with difference to explore 'other others' and at times having been seen as coming close to the postmodernist fold, in her most recent book, *Whose Science? Whose Knowledge?*, Sandra Harding comes back to a renewed statement of standpoint theory together with a concept of 'strong' objectivity. 'Strong' in this context is taken with a pleasant sense of irony from the 'strong programme' of the Edinburgh relativists, and fused to objectivity in order to create a space for a distinctively feminist conception of such objectivity.

Harding criticizes the traditional concept of objectivity as 'weak' in that it only applies to the context of justification, that is the research methods, and not to the context of discovery, where problems are defined and hypotheses fashioned. This latter is seen as unexaminable by the rational methods of science even though it is here that the dominant values enter. It has been the context of discovery which feminist, left, environmentalist and anti-racist science critics have politically challenged as ignored within the myth of the neutrality of sciences.[96] Naming the construction of an objectivity which ignores or denies the context of discovery as 'weak' is a significant move, as it calls to attention the fragile underpinnings for the rhetorical claims of the objectivity with which those who wish to criticize science are continually confronted.

Thus it is entirely unsurprising that, for instance, the science of human difference, carried out from within in an imperial power with an interest in keeping a colonial people repressed, will tend, to put it no more strongly, to represent colonial peoples as naturally subordinate. It is only when there is a powerful social struggle being waged against such domination that those scientists who long to bring back an old, untroubled order feel the need to fix their results.[97] When hegemony prevails, scientists will effortlessly arrive at conclusions entirely acceptable within the rules of the context of justification. As Harding observes, it is precisely this weak concept of objectivity which has made science so eminently available as a resource to dominant groups.

This implication of this argument for a broader theory of feminist knowledge has been addressed by the African-American feminist Patricia Hill Collins in the specific context of the social sciences.[98] Collins develops an important statement of black women's standpoint theory which sustains a realist project without subordinating all other oppressions to those of black women. Perhaps it is the sharp sense of their own historicity that excludes the possibility of black feminist theorists giving a shred of support to any 'master' narrative. Thus the resources Collins uses to develop her concept of a standpoint theory, which sustains the entry of black feminist thought while avoiding both the relativism of postmodernism and the dangers of producing a new master narrative, are of key importance.

Beginning with the understanding that subjugated knowledges develop in the cultural contexts controlled by oppressed groups, she argues, like Fanon,[99] that the dominators always seek to permeate those contexts and control them from within. She combines a deep sensitivity to the need for these sometimes quite small autonomous spaces with an awareness of the distinctive systems of oppression of class, race, gender, age and sexuality.[100] She speaks of these systems of oppression not as unitary theory nor as dualism, but as a 'matrix of domination', and insists that no one system can be prioritized – not even the analysis of racism – so as to generate a master narrative. Only the specificities of the context can speak of what hurts most at a given moment. She shows simply and decisively how a master narrative which prioritizes one system subordinates others. Thus she is dismissive of a Euroradical left which can still argue, 'If only people of color and women would see their true class differences . . . class solidarity would eliminate racism and sexism.' Such a master narrative has at times been created within white Western feminism, not so much when gender as a pioneering relational concept was being torn out of grammar and into feminist analysis, but when the differences of race, class and nationality have been systematically subordinated to those of gender. This has been particularly the case in essentialist radical feminism, which has real difficulty in doing other than always prioritizing the divisions of sex.[101]

By contrast, within Collins's concept of a matrix of domination there are few pure victims and few pure oppressors, as each is located within the matrix. The boundary between postmodernism and standpoint theory again softens and a certain traffic between the two blurs the line. My only difficulty with the metaphor of a matrix is that it contains more than a hint of equality within the mathematical mesh, which tends to homogenize the power of the different systems. Yet Collins never lets this apparent equality remove a scrutinized historical context from her analysis. Steadily resisting biologizing thought, she refuses to reduce black feminist thought to a knowledge which can only be produced by black females, while at the same time acknowledging that historical experience means that black women are likely to be the majority creators of the new knowledge.[102]

What comes across from Collins's writing is a deep feeling of both responsibility and respect for all the actors located within the structures, from the mothers who understood laundry work and the right questions to ask, to past and present black feminist intellectuals. This powerful sense of community, not identity, which nineteenth- and early twentieth-century black women struggling for education spoke of as 'lifting the race while we climb', was forged in the history of African-American and African-Caribbean women's resistance to slavery. In claiming the privilege of this collective subjugated standpoint, Collins points to the vision of a black feminist standpoint more powerful than that offered by the standpoint of the isolated white, and often middle-class, women in Dorothy Smith's account. Yet like Smith, with her notion of 'bifurcated consciousness', Collins understands the critical value of being socially at the margins, or as she puts it 'the outsider within'.[103]

The powerful accounts of such black feminist theorists[104] reach out from theory into the everyday lives of black women and from everyday life into theory.[105] These are organic intellectuals who write from within their community of resistance. Such a linkage is politically strengthened by an explicit commitment to writing accessibly – though not less rigorously – thus making sure that as many as possible of the actors can check the story out.[106] For Collins, making theory is intimately and inescapably linked with empower- ment and with the strengthening of the community of resistance among black women. As an embodied and empowering knowledge it is the antithesis of 'abstract masculinity'.

I began this chapter by quoting Haraway's observation concerning feminism's entry into the debates on the nature and power of scientific authority, and her confidence that feminism had important things to say. At that moment, at the beginning of the eighties, it was easy to share her conclusion that there was a problem about what the multiplicity of voices were saying. Having myself been among the standpoint theorists who wanted to make over for feminism the epistemological claim that the perspective of the subjugated was

necessarily more reliable, for this could empower women both in the
production of feminist knowledge and in everyday life, the initial idea
of a plurality of accounts was disturbing. Standpoint theory had
sought to provide theoretical justification of why we should take the
views of the oppressed more seriously than those of the oppressor; in
very practical terms, of why women's accounts of sexual assault or
violence were likely to be more truthful than the accounts of the
alleged perpetrators. What was at stake in the 'postmodern turn' was
whether there was now such a celebration of difference and diversity
that there was no means of granting epistemic privilege to the
perspectives of the oppressed, whether such an awareness of
language meant that we lost the power to claim reality. I did not want
to lose standpoint theory and the possibility of going 'beyond
masculinist realities'[107] but felt that the combination of the theoretical
arguments and the politics of difference was compelling. Put in the
terms which have informed the feminist debate, for me it has been
the new fusion of standpoint theory with situated knowledge which
has both extended and re-visioned feminism's truth-claims.

5

Gender at Work in the Production System of Science

It isn't really necessary to live such a peculiar life as I have, you see I so love science.

Marie Curie, *Eve Curie, Madame Cu...*

Gender in the global production system of science

That women are under-represented and men over-represent[ed] science appears to be a universally recognized phenomenon, at in conventional Anglophone accounts. As a phenomenon it requ essentialist explanation either as lying in the nature of wo[men] themselves or, by a focus on the exclusionary processes of transmission of scientific knowledge, as embedded in the educatio[nal] system. Arguably this second proposition is itself a version of essentialism which, rather than naturalizing women, naturalizes science – or rather modern, Western science – rendering it above and outside history; gender-, class- and race- free. Yet the sciences, as organized bodies of knowledge about the world, as methods of acquiring that knowledge, and as the institutions which sustain the acquisition, preservation and transmission of that knowledge, are not unique to Western society. Forms of science exist in all societies and throughout such human history as we have access to. African, Chinese, Indian and Native American sciences all predate the emergence of Western science, and the formation and relation of gender within and to these sciences cannot be simply assumed to match that of Western science. Arguments about whether science is gendered inexorably become entangled with the issue of which 'sex' produces science.[1] But even posing the question that way erases many women in science, for it renders invisible that great army of under-labourers, the technicians, secretaries and cleaners, who inhabit the laboratories but are excluded from the definition of who constitutes the scientific workforce by the elision between science and the construction of masculinity. Class, and in the case of the life sciences gender, play a strong part.

For that matter the boundary lines between 'science' and 'non-science', which are frequently portrayed as so strongly framed as to be immutable, as in C. P. Snow's influential and yet fundamentally flawed concept of 'the two cultures', are themselves constantly under negotiation.[2] As Michael Young observed, 'What "does" and "does not" count as "science" depends on the social meaning given to science, which will vary not only historically and cross culturally, but within societies and situationally.'[3]

Parallel with feminism's criticism of Western science has been the development of the ethnosciences, which have sought to root systematic knowledge of the world in specific cultural contexts, and to establish alternative ways of knowing. These have taken two forms: one a recuperation of past sciences erased by Western imperialism; the second the attempt to integrate the specific knowledges of Western science into the cultural and philosophical forms of non-Western traditions. Examples of the first include the recovery of the mathematical reasoning embedded within Mayan or African cosmography, or Van Sertima's account of the sciences of Benin,[4] so that people from Latin American or African backgrounds looking at Mayan calenders, complex weaving patterns or Egyptian scarabs can see in these the autonomous development of sciences other than those presented by the dominant culture as the one true science.[5] Examples from the second include the work of Islamic scholars who, emphasizing that different civilizations have produced different sciences, are engaged in claiming a distinctively modern Muslim science.[6] The attempts to develop 'new sciences' by feminism and by those working within the ethnoscience movement[7] have for the most part followed parallel tracks, but where feminism[8] increasingly sees the ethnoscience movement as a potential ally, there are rather fewer signs of ethnoscience being other than an androcentric development. The signal exception to this lies in the work of Vandana Shiva, who fuses the forestry knowledge of the Chipko women with her own training as a physicist to make a compelling critique of contemporary development approaches, with their destructive implications for people and forest alike.[9]

Where Western science has differed from the earlier traditions was in its peculiarly interventive, reductive approach to the natural world.

As I have already mentioned, this interventionism, so central to modern science, was written into the foundational texts of the new masculine knowledge: 'There was but one course left, therefore, to try the whole thing anew upon a better plan and to commence a total reconstruction of the sciences, arts and all human knowledge raised upon the proper foundations.'[10]

Europe was the centre of the production system of this new science; first Italy in the Renaissance, then seventeenth-century England, later, by the eighteenth and nineteenth centuries, France and Germany. German hegemony was to be broken only by the arrival of Nazism, and the subsequent shift to the US and, more recently, Japan, began. The 1917 Bolshevik revolution ushered in a period in which Soviet – or, more precisely, Russian – science, from its initial French and German tutelage, developed its own powerful ideologically autonomous tradition, only destroyed with 1989. As I will discuss below, for women the historical construction of scientific production in the Soviet Union and Eastern Europe was both like and unlike that of the capitalist West.

Those public knowledge-production systems which originated in the twin birth of capitalism and science in the seventeenth century were from their inception deeply and self-consciously gendered. Yet although from the moment of birth the Western science-production system was closely linked with military technology and economic objectives, in its early days the pursuit of science was not professionalized; its practitioners were largely gentlemen amateurs for whom science was a cultural activity.[11] This inclusion of science within the sphere of acceptable amateur culture made the home laboratory not merely the possible but the probable site for experimental work, a tradition which persisted at least in biology until the time of Darwin, whose research, for decades after his return from his *Beagle* voyage, was based at his home, Down House in Kent.[12] By the end of the nineteenth century this was sufficiently rare that the Haldanes, with their north Oxford domestic laboratory, were exceptional[13] (though, as discussed in the next chapter, the physicist Hertha Ayrton had a laboratory in her own house at around the same time). Elsewhere, the practice continued amongst the wealthy, as in the case of Rita Levi-Montalcini discussed in chapter 7.[14] Today such privacy is seen as eccentric or the mark of opposition on the part of wealthy men scientists to the scientific establishment. The Nobel Prize-winning biochemist Peter Mitchell left Cambridge in the 1960s to found a private laboratory on Bodmin Moor in Cornwall; the formulator of the Gaia hypothesis, James Lovelock, also moved to a private laboratory a few years later.

Despite the formal exclusion of women from the developing institutions of science, because it was largely produced within the domestic space the boundary lines around scientific production were relatively permeable. The interested wife or clever daughter could be recruited, or recruit herself, into the practice of science. During the seventeenth, eighteenth and even nineteenth centuries, numbers of economically privileged women, through participation in the private cultured spaces afforded by membership of particular families or through their location in the salon milieu where scientific and literary matters were discussed, were part of both the making and the discourse of science. Sometimes acknowledged within their peer group, such elite women, who contributed as scientific ladies, were at the same time only marginal participants in a gentlemanly scientific culture. Such participation is revealed in the intense pictures of Joseph Wright of Derby, where entire bourgeois families are depicted at home, absorbed in watching the outcome of experiments in physics, or in the fashionable success of the Royal Institution in London, whose scientific lectures at the end of the eighteenth and beginning of the nineteenth century became highly regarded social occasions for women and men of society.

Working-class men might assist (as what would later be called technicians) in the production of knowledge, but unlike bourgeois men they could be neither its creators nor, like bourgeois or aristocratic women, its witnesses. As for those women who none the less did contribute to science, the problem of their marginality, not to say social non-existence as 'cultivators of science',[15] was commonly managed by getting men connected by kin – like William Herschel for Caroline Herschel – to represent their work at scientific meetings. Scientific women, like literary women, needed to stand behind a masculine identity in order to publish their work, but the former operated under even tighter constraints. Literary women could simply invent a mythic male pseudonym, but scientific women had to have a real and supportive brother, father or husband (unrelated mentorships like that of Charles Babbage and Ada Lovelace, nineteenth-century mathematicians and inventors of a forerunner of modern computers, were highly problematic and perhaps only possible where the woman was aristocratic).

It was with the professionalization and industrialization of the sciences, and the steady transfer of the production of scientific knowledge from within to outside the home – first chemistry in nineteenth-century Germany, later physics and most recently biology – that women came to be systematically and sequentially excluded from the new occupational structures which, at their apex, were linked to new forms of economic and social power. Centrally part of the contested stratification system of a class society, such profession-alized and industrialized science was to become an intensely meritocratic means of achieving and confirming class power for men, and, as I shall show, for very few women. Women were increasingly confined to those areas of more contemplative, less interventive science which were not yet industrialized (such as botany or nature study in the nineteenth century) or to newly developing areas (biochemistry in the 1930s, crystallography in the 1940s and 1950s, computing in the 1960s and 1970s). Women were able to gain a foothold in these new fields of science, which lacked fully elaborated career structures.

Although it is my primary purpose to examine the processes by which women are able to study the natural rather than the social world, it is true that much of social science and historical research, even in the most industrialized countries, is still a craft activity and includes many distinguished women social scientists and historians. Thus both primatology and its historical analysis are craft activities, and hence places where women can excel.[16] But these craft areas are now the exception in modern knowledge production, particularly in the natural sciences, as these have become largely industrialized over the course of this century. Descriptions of work in the new industrialized science increasingly parallel the factory, a process of intensely skilled, closely focused teamwork with a marked division of labour and hierarchization, which is in turn rapidly replaced by smart technology, the deskilling and even the robotization of scientific production. The contempt with which leading molecular biologists, such as James Watson and Sydney Brenner, speak of the 'science' involved in the $3-billion Human Genome Project, discussed in chapter 8 ('even a team of monkeys could do it') speaks to this process in the most recently industrialized of the sciences. It is interesting that the selfsame men whose scientific eminence was achieved largely through craft science both have helped father the new industrialized science and at the same time seek to privilege their own activities and strategies as somehow representing 'real science'.

National differences

In many countries the late sixties began a period of explosive growth in higher education. Looking at the university teachers, as a key group who both transmit and, in a number of countries, also produce knowledge, UNESCO figures report that the overall numbers in the world surged during the sixties from 1.4 to rather over two million by the end of the decade, giving a large annual growth rate of 8.5 per cent.[17] During the following decade this slowed to an average of 5.5 per cent, ranging between Latin America, where there was massive growth of 14.3 per cent per annum, to the more modest levels of 4.2 per cent in Europe and 3.6 per cent in North America. The differential is explained by the fact that some countries started from a much lower base and expanded rapidly to catch up, in contrast to the old industrial countries which already had strongly developed university systems and therefore expanded more slowly. Even so, by the end of 1980 there were 3.5 million university teachers in the world, and of these almost a quarter worked in the USA.[18]

What slice did women get in this expanding higher educational pie? Certainly the enrolment of women students increased, so that by 1980 women constituted 40 per cent of the world's higher education student population, with 44 per cent in developed countries and 33 per cent in developing countries. North America with 47 per cent and Eastern Europe with 49 per cent were the two regions with, for very different reasons, the highest proportion of women. However, the proportion of women university teachers by no means paralleled this growth, whether we are discussing the Anglophone countries, in which teaching and research are both carried out substantially within the university system, or those countries which follow the French and German model – particularly in science – where research institutes carry out the bulk of the research. Yet the presence of women as tertiary-level educators in the comparative statistics of UNESCO is sometimes the best approximation available to indicate the presence of women in research. Indeed as Ann Cacoullos observes, 'women-in-science qua problem is intimately linked to women-in-education and to women-in-politics. This,' she continues, 'is not a new idea. The problem of "why so few?" in science and technology is co-extensive with the same query in politics: it has to do with the absence of women in the centres of power.'[19]

The very absence of such gender statistics in the scientific labour force tells its own sad story. Even though there has been an extensive discussion of their relationship to economic development, science and techology are not understood either as entailing social develop-ment or, except rhetorically, as needing women in the labour force. Thus, despite the intermittent enthusiams of, for example, the European Union or the Organization for Economic Cooperation and Development (OECD), to increase the number of girls and women studying 'non-traditional' subjects such as science and engineering, there has been a failure to follow through at the level of gathering even the most basic scientific labour-market statistics. The main provider of comparative statistics thus remains UNESCO with its broader and more cultural brief. Of course the fact that UNESCO has had such a brief brought it into conflict with the highly conservative administrations of Reagan and Thatcher, who therefore withdrew US and UK funding during the 1980s, a cut which neither Bush nor Major restored. The kind of data collection exercise which ought to have been developed by the European Union simply has not been pushed, a lack which reflects the extent to which those who seek purely a common market have been more successful than those who are trying to build a political community. The 'democratic deficit' within Europe also means that neither the Parliament, nor the energetic women's group within it, has anything other than minor powers of scrutiny over the working of the different Brussels directorates, and there is no supervision even of the employment policies of the immense research budget handed out by the

Commission. No data is gathered on the gender and ethnic composition of the scientific labour force hired with EU money[20] (giving rise to the comment that the collective noun for data about women in the labour market is 'an absence').

Thus any discussion of where the women are in the production of science today has to be constructed with very varied material. While the data on the US is immense and highly detailed as the social composition of the scientific labour market is seen as a policy issue, and that on the UK has the strength of a relatively long time series of university statistics, the comparative data is fragile. Consequently it is only the broad pattern rather than the individual statistics (for either subject or nation) which is revealing. But even this does not give much cause for comfort.

To some extent the tremendous expansion, which has meant so many more women in absolute numbers researching and teaching in tertiary education, has masked the very modest slice women have been permitted of the scientific and academic labour market. The fact that many feminist academics owe their present occupations to this expansion, and have a clear sense of the personal privilege between generations (for instance of being a university teacher and a researcher as I am, rather than say teaching in a primary school as my mother did), perhaps contributes to masking the modesty – to put it no more strongly – of apparent gains. As usual a personal sense of generational mobility is misleading. In a period when higher education takes on an increasingly important role in training and educating the skilled labour force, then both old occupations which were trained outside universities, such as nursing and schoolteaching, and entirely new occupations, such as those of information technology and biotechnology, become university based. It would be surprising if the numbers of women in higher education, whether as teachers or students, did not expand. Indeed it is virtually only those countries which are both 'developing' and Islamic where there are still less than 10 per cent women as university students.

But even in the long-industrialized European countries, the story has not been one of automatic growth and progress. Thus while it is customary in women's studies circles to regard positively the situation in the Netherlands, with its strong women's studies centres and chairs, there too the situation for women academics has deteriorated over the past two decades. Where in 1970 there were 2.7 per cent women professors, by 1980 this was down to 2.2 per cent and by 1988 to 2.1 per cent. But the Netherlands is a small country and the raw figures are more graphic: thus where there were 65 women professors in 1970, by 1988 there were only 50. The middle rank of associate professors from whom the professoriat is recruited had also shrunk, in this case from 312 to 105 women. Women are still concentrated at the assistant level: in 1970 there were 571, in 1988 there were 927; that is, 14.7 per cent of all women employed as university teachers are appointed at the bottom level. Getting hold of this double story of achievement and loss is important for feminism.[21]

Western European feminists also look with some envy at the state feminism of the Scandinavian countries, with women occupying a third of the parliamentary seats, where there are strong social programmes which support everyday life.[22] None the less Scandinavian academic institutions cling to their patriarchal power and have been rather successful at resisting structural reform.[23] In Sweden, where university teachers are part of the civil service, and which, along with Denmark, has the most extensive public child-care provision in the Nordic countries, there are still only 4 per cent women full professors, pretty much level-pegging with Britain despite the latter's exceedingly conservative government and conspicuously weak commitment to gender equality. Even Norway, which has recognized the persistent under-appointment of women at the professorial level, has only managed to move its percentage from 4 per cent in 1985 to 6 per cent by 1990. Science and academia in Scandinavia seem to have a demonstrable and defensible autonomy from the state which enables them to resist the claims of women to share even approximately equally in the forms of social and economic power of the knowledge classes.[24] Yet this resistance within the Scandinavian academy to appointing women at the professorial level, while it affects the structure and direction of research, is not reflected in the pattern of student recruitment or in the composition of employment, particularly in engineering, where, compared with the situation in Western Europe generally, women are well represented.[25] Thus by the end of the eighties women formed between 19 per cent and 21 per cent of the engineering students in Sweden, Norway and Denmark.[26] The proportions for Finland[27] and tiny Iceland were lower, but were to some extent compensated for by the high proportion of women in maths.

While admission to higher education and to the subject is the first step, there are also significant differences in the scientific labour markets. Uneven development has meant that although there is one dominant and global production system of science, it is still produced from within different economic and social contexts. These differences in context are associated with very different sexual divisions of labour within scientific production. For example, from 1917 in the Soviet Union, and from the 1940s in Eastern European countries with communist governments, women increasingly entered all branches of science and university teaching.[28] At the end of the Second World War, women comprised about half the research staff of the prestigious Ioffe Physical-Technical Institute in Leningrad (now St Petersburg). Until the shift to the market, over half the research positions in Warsaw's leading biological institute, the Nencki, were occupied by women. By contrast with the situation in the US and Western Europe, women participated in the military science of the former Soviet Union in very considerable numbers.[29] Such examples can be multiplied many times. What is clear is that these women seem not to suffer from those 'inborn deficiencies' which seem to make their appearance, particularly within Anglophone cultures, every time that women seek entry to study, teaching or research, particularly in the natural sciences and technology.

However, despite the ideological commitment to equality in Eastern Europe and the former Soviet Union, the overall proportion of women teachers reached only the middle to upper thirty percentiles (until 1989, when higher education and science like the rest of the economy began shedding female labour even faster than male labour). The current rise of religion and religious values celebrating domesticity for women in the former Soviet Union give cause for concern that ideological opposition to women's participation in teaching and research will exacerbate these economic and political trends. While the Western focus has been on the rise of Islam in the FSU, the erstwhile Christian areas such as Russia and the Ukraine are also experiencing a strong rise in 'family values', so that women are being urged by political and academic leaders to return home, develop their spirituality and nurture the moral renaissance of their country.[30]

A certain Anglocentrism and methodological individualism[31] in searching for explanations as to why women do not do science has masked significant differences even within Western Europe.[32] In an important paper, Beatrice Ruizo[33] has pointed to the relationship between the participation of women in academic production and the level of technological and economic development. She suggests that considerable numbers of elite women (37 per cent in 1978) take part in science in semi-industrialized contexts, such as her own country, Portugal, precisely because scientific production is still part of cultural production and is not yet fully locked into technological and economic growth.[34] This high level of participation – between a fifth and rather over half of all researchers – is to be found in semi-industrialized countries as scattered as Mexico (21 per cent), Singapore (24 per cent), Finland (27 per cent), the Philippines (38 per cent) and Argentina (57 per cent), where elite women simultaneously take part in what is a global production of science and also practise science as an acceptable cultural activity within a national context, much in the way that women from similarly privileged backgrounds participated in the past in the now old industrialized societies of northern Europe and the US. The difference is that today's researchers share in the rhetoric of modernity and equality, so that they are paid for their scientific work.

By contrast in developing countries, women are under ten per cent of the scientific labour force – ranging from 2.5 per cent in Madagascar to 9 per cent in Togo[36] – and also very few in absolute terms, for the number of scientists as a proportion of the total labour force is also very small. Economic and technological development of the nation state on a Western model, like the industrialization of science itself, works to exclude women. Exceptionally, the very high present-day representation of women in Turkish higher education (32 per cent) has to be understood as substantially a residue from Ataturk's modernization project in the 1920s, which crucially turned on the weakening of Islam. Ataturk and that tradition which followed him would far rather see bourgeois women than working-class men admitted to the academy.[37] However, the rise of religious fundamentalism in Turkey in recent years, and its promise of class mobility for men at the expense of women, offers to change these quite exceptional statistics.

Why so few? The question within Anglophone feminism

Despite this differentiated picture of the place of women in the global science-production system, within Anglophone societies the picture is one of exclusion. However, it was not until the 1970s that second-wave feminism woke up to the realization that instead of the gains of the heroic days of the 1870s – when women had more or less successfully fought to secure access into higher education and were starting on the second struggle to gain admission to the research training that would enable them to secure university posts – being consolidated, the situation had actually deteriorated. The new generation of feminists expressed their anger in no uncertain terms. 'A woman in Britain today has more chance of going to a mental hospital than going to university' wrote Phillipa Ingram in 1972,[38] arguing that quotas represented the only way to secure equal opportunities. Even access to higher education showed the structural resistance of class to educational reform, especially when class was linked to gender. Westergaard and Resler[39] noted that despite the battery of educational reforms which were supposed to have opened the educational system, the numbers of women students securing places in universities whose fathers had unskilled working-class jobs was the same minute proportion in the sixties as before the 1939–45 war. It was clear that the massive overall growth of higher education – and complacency – had masked women's lost ground.

The evidence of this deterioration was particularly marked in the US; between 1870 and 1970 the proportion of women on the academic staff of higher educational institutions fell from 33 per cent to 25 per cent.[40] In 1940, 28 per cent (40,000) of the total academic and professional staff in higher education institutions were women. By 1964 the percentage had dropped to 22 per cent of the academic staff, even though in absolute terms the numbers of women had increased, along with the growth of higher education, to 110,000. A similar situation prevailed in the award of doctorates: in 1920, 90 women were awarded doctorates, comprising some 16 per cent of the total; fifty years later, in 1970, 3,980 women were awarded doctorates, but these made up only 13 per cent of the total.[41] (Even this figure represented a recovery from the nadir of around 9 per cent during the fifties.)[42] At the undergraduate level the situation was also slipping. In 1920, 47 per cent of admissions were women; in 1970, 40 per cent were, and there was, until the women's movement challenged it through the legal action of Bernice Sadler,[43] a straightforward quota system working to exclude women.[44] For teaching staff, vertical segregation has been and still is particularly strongly marked in the US, as in other Anglophone countries. In 1969, national surveys showed that 21 per cent of women academic staff were in the humanities, 16 per cent in education, 8 per cent in social sciences, and only 5 per cent in the natural sciences.

The situation has not greatly improved in the subsequent two decades. Thus Cornell, having admitted women in 1870, well ahead of most of the eight Ivy League colleges, in 1988 had only 5.7 per cent women full professors and was at the bottom of the League. Columbia at 10.8 per cent was top of the mixed institutions (the women's college Barnard, Columbia's stepsister, did dramatically better), but this performance among mixed institutions was exceptional; the average was 7.8 per cent. The best of the 'big ten' state universities were roughly parallel with Columbia with an average of 8.9 per cent.[45] Even the difference between these figures and those for the Ivy League indicates the manner in which the proportion of women increases, and their chances of promotion become greater the further they are from the centres of educational prestige and power. Women academic staff are better represented in the smaller universities, which award only undergraduate and not research degrees.[46] Nor, despite the fact that by the year 2000 one third of the US population will be 'minority', and in some areas, such as parts of the West Coast, the 'minority' will be the majority, are these demographic figures reflected in academic posts.[47] Of the few minority people who have academic staff positions, those with tenurable or tenured posts tend to be men. Minority women are mainly to be found in temporary posts and often overloaded with administrative responsibilities so as to demonstrate that their institution is doing a good job in meeting 'diversity' objectives.[48]

It is the award of the doctorate that gives the crucial entrée to university teaching and research, and here the pressure on the US system to change continues, so that there has been a doubling of women PhDs since 1972.[49] In social sciences and the humanities, 45 per cent of the doctorates are now women, and in education, for long a preserve of women, there are now 54 per cent. In physical sciences the proportion has moved from 7 per cent to 16 per cent, in the life sciences from 7 per cent to a dramatic 34 per cent, while engineering has moved from under 1 per cent to 7 per cent.[50]

In Britain, despite recent expansionist moves, a much smaller percentage of the population goes on to university than in the US, and for many years the predominantly women's professions of schoolteaching, social work and nursing were excluded from the higher education statistics, as these 'semi' professions had no associated degree and were substantially based on an apprenticeship learning model.[51] However, there was an improvement in the entry of women into the universities over the sixties and by the end of the decade some 30 per cent of the intake were women. As in the US, women undergraduates, postgraduates and academic staff were both very bunched into particular disciplines and subjected to a remarkable thinning out process as they sought to rise through the academic hierarchy. Thus only 16 per cent of the students who went on to postgraduate work, and only 9 per cent of the university staff, were women. In 1970 women were a small minority within the applied sciences; only in the humanities did they form the majority of the students, although even there they only formed 14 per cent of the teaching staff. Women professors were extremely rare, there were no women university vice-chancellors or polytechnic directors, but there were some women principals of women's colleges (and institutions which had orginally been women's colleges).

The eighties saw some modest changes. In 1981–2 almost 40 per cent of the university students were women, but women staff still only constituted 14 per cent and slightly over 2 per cent of professors (119 women). Women students were still being primarily recruited by the humanities and the social sciences; in 1982, French had 80 per cent women students, English was close behind with 66 per cent, sociology had 64 per cent, biology recruited 46 per cent women, and medicine, once the quota system had been removed during the seventies, moved up towards the 40 per cent level. Mathematics at 28 per cent, chemistry at 24 per cent, physics at 13 per cent and electrical engineering at 4 per cent recorded the persistence of a strong vertical segregation between disciplines. The sexualized language of hard and soft subjects, and the exclusionary cultural practices of the science departments, did little to soften the boundaries.[52] In addition to horizontal sexual segregation dividing the subjects, vertical segregation marks the difference between men and women teachers in higher education. By 1987–8, women constituted 3 per cent of the professoriat, 7 per cent of the readers and senior lecturers, and 16 per cent of the lecturers. But other changes had taken place in the structure of the academic labour market over the course of the eighties which both expanded and casualized the lower levels. Increasingly, lecturers and research staff were part-time, short-term contract workers appointed on soft money, and women were comparatively over-represented among them.[53]

Despite a great deal of publicity advocating the merits for women of a career in engineering and the sciences, aided by government ministries,[54] the Equal Opportunities Commission and the professional engineering and scientific institutions, there has been extremely slow growth in the percentage of women in science and engineering courses. The Engineering Council reports that for the whole of higher education for a five-year period beginning in 1982–3, the combined figure was just under 24 per cent, while at the end of the period it was just under 25 per cent. This masks the much lower proportion of women students in engineering than in science. Whereas over a third of science students are now women (a figure substantially pushed up by biology), engineering in 1986–7 still only recruited 11 per cent women in the university sector and 9 per cent in the former polytechnics. Despite feminist resistance, computing studies increasingly lost women, the figure going from 24 per cent in 1980 to 10 per cent in 1987, and by the end of the eighties some degree courses had none.[55]

By the early nineties there were still no women vice-chancellors, but there were two women polytechnic directors, and among the London colleges a woman, Baroness Blackstone, was appointed as Master of Birkbeck (which as the title indicates was hitherto patrilineal), but the merged Royal Holloway and Bedford New College – which on the Bedford side of the merger had hitherto always had a woman principal – saw the appointment of a man, Norman Gowar. The new RHB now joined the mixed colleges, obscuring its origins as a women's college and as a locus of feminist scholarship.[56] The senior management of both universities and polytechnics remained overwhelmingly masculine. Arguably the shifts in the management structure of universities over the eighties towards a more Fordist managerial style and away from a collegial model facilitated the preservation of the male managers.[57] It was not until the Hansard Report in 1990, which was scathing about the under-representation of women in senior positions within the universities, that there was any public recognition of how poor the British situation was, and some modest acknowledgement on the part of the dominant gender of the need to make some corrective moves. At the beginning of the nineties most higher educational institutions had a formal commitment to an equal opportunities policy, but very varied levels of effective implementation.[58] Certainly there has been no sudden and visible rush to promote women, or bring their pay to the level of their male peers.[59] More optimistically, the recent ending of the binary system of universities and polytechnics may help weaken the resistance of the universities to reform, as elite groupings within education, such as the Committee of Vice Chancellors and Principals, which were solely the province of men, have to learn to work as groups of women and men.

Explaining the difference

Education within the UK reform tradition has long been seen as the means both of providing the appropriately skilled labour force needed for the economy and the policy (during the long years of the welfare state consensus) of increasing social solidarity, and of diminishing if not overcoming the antagonisms of class, and, in an uneven and usually lesser way, the divisions of gender[60] and race.[61] As against the naked transmission of privilege through privately purchased education, primarily for the sons of the ruling class, the new state-sector schools were to offer what was called equality of opportunity for all. Co-educational and comprehensive secondary schooling was to be developed on the grounds that it would help open educational opportunities for working-class children and girls from all classes. But apart from the Robbins Report which preceded the massive higher education expansion in the sixties, policy emphasis and theoretically informed educational research has long focused on schooling, and within that chiefly on the schooling of boys.[62]

There were, broadly speaking, two approaches to explaining why a meritocracy failed to come into existence despite the successive reforms promoting equality of opportunity: the first adopted a cultural deprivation perspective, focused on the individual pupil, and tried to work out how he or she could be enabled to perform more effectively; and the second took a structural perspective, arguing that far from education being an open meritocratic system it was integral to the reproduction of social class inequality over the generations.[63] Some of the new theorists, particularly the US marxists Bowles and Gintis,[64] were highly deterministic, their subjects made into stereo-typical examples of what sociologist Dennis Wrong had, in the 1950s, so presciently called 'the oversocialised conception of man'. Working-class boys were sent to school to be taught appropriate skills, but even more importantly to absorb the passivity and obedience which would guarantee the preservation of the class formation. This theoretical prioritizing of structure left no space for agency, so that its protagonists were left vaguely hoping that the passive working-class recipients of ruling-class ideology might somehow find an alternative space through which to develop a critique.

Something of this monolithic determinism and hopeful voluntar-ism found its expression in feminist 1970s' criticism of education and indeed of pre-school socialization, as 'conditioning'.[65] Such was the context of optimism that this 'conditioning' was seen as a phenom-enon which could be denounced by the women's liberation move-ment and simply overcome by voluntary action. The recognition that 'agency' and 'subjectivity' were rather more complex, and that the canonical knowledge itself was gendered, came only later when this early, utopian optimism had failed to achieve instant, let alone lasting, results.

By contrast, the cultural deprivation approach, embodied for example in the 1967 Plowden Report, saw working-class children as 'little deficit systems' who needed topping up with suitable middle-class values in order that they might benefit from education. This lack of middle-class values was laid at the working-class parents' – and by implication the mother's – door; the task of policy makers was to make good the deficit. In its strongest form this deprivation thesis was articulated by the Conservative Secretary of State for Education Keith Joseph, who saw children failing in the educational system as victims caught in a 'cycle of deprivation' whereby shiftless values were transmitted through the generations. Because (apart from Michael Young's pioneering work in the 1970s) there had been a failure to develop a specific sociology of science education which took the content of the knowledge itself into account, the cultural deprivation thesis, with all its individualizing, victim-blaming power, was ready waiting to explain why girls and women did not take science at school or college or enter scientific research.

Such cultural deprivation theories have been applied to girls and women to explain not why they fail to perform and secure entry to secondary and tertiary education, for that manifestly is not at issue, but why they choose 'feminine' subjects and therefore do not have open to them many well-paid occupations, not least those of science and engineering. Both the Girls into Science and Technology project (GIST)[66] and the Equal Opportunities Commission project of Women into Science and Engineering (WISE) capitulated to a view of girls and women as inferior and the subjects done by men as necessarily intellectually superior.[67] The projects' compensatory tasks were therefore to 'top up' girls and women until they could compete in the superior world of boys and men – in short, until they could do science.

Unsurprisingly, neither project, though they have passed into the well-publicized annals of feminist approaches to science education, was particularly successful. Yet although both projects have been justly criticized within feminist circles, there was a suggestion within GIST of a 'girl-friendly science' which hinted at links to the more theoretically informed critique of the science/gender system, and which perhaps could have been fostered. Possibly the positivist framework of the evaluation of GIST inhibited the potentiality of 'girl-friendly science' from being explored and linked with US work such as Sue Rosser's feminist pedagogy for science,[68] or Jan Harding's in the UK.[69] Certainly Alison Kelly, as the sociologist within the GIST study, subsequently wrote an autocritique of her 1982 paper in which she largely accepted the structuralist view of education and the feminist critique of the science/gender system. She wrote, 'I would put more emphasis on the role of the schools in dissuading girls from science, and less on the girls' internal states. The article suggests that it is necessary to change the image of science. I now think that it is necessary to change *science*.'[70] Kelly's conversion marked the closure among feminist researchers in the sociology of science education of the deficit view of girls and women; subsequently most made some sort of acknowledgement to science as gendered. The trouble was that there was still a long way to go between acknowledging the problem and developing a feminist sociology of science education. Even Kim Thomas's examination of the powerfully gendered social organization of university English and physics departments and the difficulty they posed for women[71] was unable to enter the discourse of physics itself.

Meanwhile the other main strand of explanation, which had been preoccupied with structures which 'determined' the educational and occupational lives of boys and men, began to change, and a more nuanced account (still entirely in the world of working-class boys) was offered by Paul Willis's *Learning to Labour*.[72] Willis's boys were no passive recipients of class ideology; instead they actively resisted the school and its values, even though their resistance meant that they failed to acquire those educational skills which would have enabled them to achieve some measure of class mobility. This new tension between structure and agency was appropriated and reworked by marxist feminist sociologists of education, who saw education as an important means of transmitting class and gender relations, but as one in which the recipients may partially reject or filter knowledge, or even use it to their own advantage.[73] However, unlike radical feminists, notably Dale Spender,[74] socialist feminists working in the sociology of education did not question the gendering of knowledge itself, even though within the feminist critique of science (whether produced by radical or socialist feminists) science was seen as deeply gendered.

Potentially the new poststructuralist feminist pedagogy, with its Foucaultian commitment to the power/knowledge couple, offers a way to explore the gendering of science education, but so far the discussion of the context of science itself has yet to begin. One of the problems is that entering the power/knowledge couple in science education means entering scientific knowledge, and all too often feminists as gendered beings have themselves been shut out of that knowledge. Thus the possibility of developing a feminist sociology of science education requires that more feminists understand science, or the field stays at the level of making a research agenda rather than carrying it through.

Nothing less than half the labs?

Some years ago I wrote an essay called 'Nothing Less Than Half the Labs',[75] with the aim of drawing attention to the extraordinarily radical implication of the claim for equality of a serious commitment to liberal feminism. Liberal feminism has rightly drawn attention to the fact that science and engineering are among those reasonably interesting and well-paid forms of employment predominantly carried out by men. The point I wanted to emphasize, both in the earlier article and in more detail in this chapter, is the very modest progress to equality in the scientific labour market which has been made since the admission of women to higher education more or less at the turn of the twentieth century. While it would be possible make projections to indicate at what time in which country women might achieve the goal of half the laboratories; certainly at the present rate of progress (and in some subjects reversals) such a project is sufficiently distant to be utopian.

By contrast, radical and socialist feminisms, while rarely directly opposing such a project, have been rather more cautious in actively advocating that women should enter an activity whose ends have been variously described as 'the death of nature' or 'the end of everyday life'. Instead they have concentrated on criticizing (and, as I indicate in chapter 8, at times rejecting) patriarchal science and technology. Yet because effective criticism requires a willingness to understand and so enter the knowledge of both science and technology, the differences between the different strands are perhaps fewer than the commonalities.[76]

Looking back over the twentieth century, particularly the last quarter-century, what is all too visible is the patriarchal resistance, not least in the developed capitalist societies of the West, to the numerically equal presence of men and women within the labs, and, as the next two chapters explore, a specific refusal to admit women to the positions of greatest eminence and cultural power within science. Such intractability hints at a not so elusive connection between the liberal and the more radical strands within feminism. Securing the liberal goal of 50 per cent women in the laboratories may well have more than a passing linkage with the radical objective of producing a different science and technology. Getting more girls and women into science, in the context of a feminist movement, is likely to be achieved through changing both the social organization of science and its content, for the power/knowledge couple, whether that of Foucault or of Bacon, sits at the heart of science.

Yet the two agendas cannot be separated, for tackling the social organization of science so as to make it more hospitable to girls and women is also part of the task of changing knowledge, since it means admitting that it is possible to see and become responsible for hitherto invisible practices of exclusion, from discrimination in hiring, paying and promoting to sexual harassment. Bringing invisible practices into visibility is part of that long process of changing the knowledge itself. In my optimistic moments I believe that working for the equality project of liberal feminism, or making science feminist, as in the project of socialist and radical feminism, or de- (and thereby re-?) constructing the power/knowledge couple of feminist postmodernism look at least as much like potentially complementary political projects as like competing ones. In this situation the goal of 'nothing less than half the labs' should be seriously pursued.

However, while the different feminist approaches to science and science education may have good reasons for developing a politics of solidarity, this is not to underestimate the extent of resistance. For to Evelyn Fox Keller's concept of the science/gender system, I would want to add an emphasis on the intimate relationship of that system to other forms of power: cultural, economic and military. Science is locked as much into these overlapping systems of power as it is to gender, and through examining the history of extraordinarily talented women at the apex of science we can see the practices of resistance by the powerful. For women to be admitted to these institutions is to be admitted to the complex and multilayered forms of power to which modern Western science provides access. I therefore turn in my next chapter to the account of the manoeuvrings of the men of the Royal Society when, after three hundred years of exclusion, they realized that they finally had to admit women to this, the oldest of prestigious scientific societies.

6

Joining the Procession: 'Man'aging the Entry of Women into the Royal Society

Do we wish to join that procession? On what terms shall we join that procession? Where is it leading us to, the procession of educated men?

Virginia Woolf, *Three Guineas*

The history of the Royal Society, that archetypal account of the move from the invisible to the very visible college of science, is told by the mainstream history of science as a Whiggish tale of the democratic institutionalization of scientific eminence, with the period during the eighteenth and early nineteenth centuries in which it was little other than a fashionable aristocratic club merely a shocking episode of degeneracy and decline. For feminists, this episode is by contrast a mere blip on the historical screen; instead, the interesting (and really shocking) questions concern how, for almost three centuries, male exclusivity operated and how this came to an end. Was the Society's foundational metaphor of the 'new masculine knowledge' part of the exclusionary discourse against the talented Margaret Cavendish[1] in the 1660s? Or today, when passing Mary Somerville's bust on the stairs in Carlton Terrace, do the Fellows reminisce about how in the 1830s they had managed to treat this brilliant scientist so badly?[2] How did they manage to keep women out despite the tremendous pressure from the feminist movement in the early years of the twentieth century, not least during the post-1914–18 war reforms? And why, although women were finally admitted in 1945, did it take another forty-six years before a woman became an officer of the Society?[3]

Social scientists are discouraged from leaping to conspiracy theories of history to explain events, but, in reasonably sophisticated circles, they are also encouraged not to be so fearfully high-minded that they entirely exclude the possibility of conspiracy. The story of the Royal Society's failure to admit – or even consider – women as Fellows, and in particular of the long years between 1922, when the legal advice the Society was given made it clear that such exclusion was no longer tenable, and 1945, when the first two women candidates were finally elected, would seem on prima facie grounds to require a conspiracy of silence as explanation. For while it is no surprise that the Society, as a bastion of self-proclaimed 'masculine knowledge', had been able to resist women since its foundation in 1660, those final years take rather more explaining.

The relevant internal documents have only recently come into view, as elite scientific institutions are secretive even by comparison with the practices of national governments. The Royal Society has a forty-year rule on access to archives, ten years longer than that of the British government (to be fair, the Society seems not to be weeding its files with the same energy as the government). Thus, when feminist historians of science wish to unravel the complex accounts of the treatment of particular women scientists, a Griselda-like level of patience is required. But the archives chronicling the last stages of the protracted struggle to secure the admission of women scientists to election as Fellows of the Royal Society are now available, and the account of the internal, largely procedural, manoeuvring provides wonderful illumination as to the rhetorical and practical mechanisms by which a powerful scientific institution changes its commitment from androcentricity to an accommodation with the pressure for equality. The maintenance of an outward demeanour of serenity and the skilful management of potential internal turbulence provide a rare insight into the process of patriarchal reform.

For scientists, election to the Royal Society is a signally valued honour, but it is important also to understand the institutional power of the Society. Historically it is through the officers, as an elite within the elite, that the Society has been close to government. Presidents of the Royal Society are the dining companions of prime ministers. Indeed, using an exceptional clause, the Society has the power to elect to the Fellowship non-scientists who have made some special contribution to science. Thus, to the anger of many scientists facing cut research budgets and an underfunded education system, Prime Minister Margaret Thatcher[4] was so honoured in the 1980s. Something of the social prestige of the Society and the alchemy of the letters 'FRS' was later reflected in the press release of her resignation as prime minister, in which she was described as the Right Honourable Margaret Thatcher FRS, a point missed by few interested in the politics of science. Through its foreign membership the Society has a supranational role in science, and because of its location at the hub of the British empire and of science, it has over its history played a signal role in advising on international affairs.[5]

An illustration of how both men and women scientists felt about the Society and how skilled it was at excluding women is provided by crystallographer Dorothy Crowfoot Hodgkin, who writes:

> I remember sitting on the steps of the Royal Society waiting for some one – talking to Bernal[16] and I was telling him that we had solved the structure of penicillin. He said, 'you will get the Nobel Prize for this'. I said 'I would far rather be elected a Fellow of the Royal Society', and he said, 'that's more difficult', which just shows how they were viewing elections at that moment.[7]

Hodgkin was elected to the Society in 1947 at the age of 36 – young by any standard and even more impressive if we remember that this was still only the third year after women had been admitted. In 1964 she was awarded the Nobel Prize for her work on the structure of penicillin. She is also the first woman since Florence Nightingale appointed to the exclusive Order of Merit.[8] But unpicking Bernal's 'that's more difficult' requires an understanding of both the historical context and the minutiae of the internal political process, that rhetoric by which the men scientists of the Royal Society convinced themselves that the time to change had come. The rhetoric of the texts, however, can only be understood by paying attention to changes in context. Thus over the first decades of this century the men of the Society were confronted by a choice similar to that which Prime Minister Baldwin, when faced with class rather than gender upheaval, was to describe to his colleagues as 'to reform or be reformed'.

The story of this reform falls into three phases: first, the Society's refusal, after taking legal advice in 1902, to admit an exceptionally well-qualified candidate, Hertha Ayrton; second, the changed legal context of the post-1914–18 period in which the Society received clear advice that it could no longer refuse women candidates, and yet managed to hold out until the pressures of a second world war made the position untenable; and third, the period of intense manoeuvring between 1943 and 1945 during which a new president, Sir Henry Dale, recognized inevitability and forced the Society to face it, culminating in the election of the first two women in 1945.

The first phase: Hertha Ayrton

In 1902 the Council of the Royal Society was for the first time in its history obliged, in response to the challenge posed by the submission of the candidacy papers[9] of a well-qualified woman scientist, to seek legal opinion as to the admissibility of women. That opinion, drawn up by W. O. Danckewerts and R. J. Parker, turned out to be very much a last-ditch essay in Blackstone-ism[10] against the rising demands of feminism. It sought to close down any possibilities that a woman might have an independent, let alone scientifically creative, existence, and thus be eligible for election. Their view was that it was 'very doubtful' that the Charter of the Society would permit the admission of single women, while it was unequivocal that 'couverture' precluded the possibility of married women being admitted. Single women would become disqualified by marriage.[11] On the specific issue of the candidacy of the married 'lady' in question, Danckewerts and Parker's opinion concluded that she could not be nominated, and the certificate could be neither registered nor read.

The scientist whose candidature was thus rendered non-existent was Hertha Ayrton. She was a strong candidate, supported by eight Fellows including the engineer and physicist John Perry, who worked with her husband William Ayrton, also an FRS. Her exceptional talent is indicated by the fact that she was the first (and for a very long time the only) woman to be accepted as a Fellow by the Institution of Electrical Engineers. Even after her nomination had been refused by the Royal Society in 1902 for consideration as a Fellow, her work on the electric arc and on ripples in sand was recognized in 1906 by the award of the Society's Hughes Medal. Royal Society medals have great prestige within the scientific community and such an award to a woman was quite exceptional. Joan Mason's meticulous account of these events reports a private and wry letter from one Fellow, Lord Huggins, which makes it clear that it was only because he was indisposed that he was unable to be present at the medal committee so as to block her award.[12] In the letter, he explained that recognition of Ayrton would give joy to the suffrage leaders currently held in Holloway Prison, and ironically imagined orgies of celebration at the Cambridge women's colleges of Newnham and Girton. Clearly, while some scientists were prepared to acknowledge scientific merit separately from sexual politics, Lord Huggins (despite the fact that he himself published with his wife) was not. However, Lady Marguerite Huggins, notwithstanding her public image as dutiful wife, held rather different views and wrote warmly to Hertha Ayrton on the award of the medal, congratulating her on a well-deserved success.

Ayrton herself was unquestionably a woman of exceptional scientific talent and determination, who was also deeply committed to the cause of advancing women's rights. Born in 1854 into a talented but far from affluent Jewish family which had left Poland to escape the pogroms, she had the then not uncommon slow start of women and entered Cambridge at 23 to read mathematics with the support of the wealthy Barbara Bodichon,[13] who was also a significant benefactress of Girton. Given the name of Phoebe by her family, she had, anticipating a common move among second-wave feminists, changed her name to Hertha, after the heroine of Frederika Bremmer's feminist novel. Having adopted Hertha as an emblem of her feminism, she never reverted to Phoebe or relinquished her political views. In the 1870s women were still neither permitted to attend all university classes nor awarded degrees; in consequence Hertha's third-class marks were no indication of her future performance as a scientist.

Returning to London, she met her future husband William Ayrton through attending a scientific course he was giving at the Finsbury Technical College.[14] He was a pioneer electrical engineer and an FRS. When he met Hertha, William, some ten years older than she, was a widower who held progressive views about women's rights. These he shared with both his first wife Matilde, who died sadly young shortly after she had finally qualified in medicine at the Sorbonne,[15] and with Hertha. The personal relationship between the established and the prospective scientist developed, and they married within a year of meeting.

Institutionally marginal, her 'laboratory' was a room in their house; as a married woman she was a volunteer scientific worker and, like others of her generation, received no salary; none the less, intellectually and scientifically she was able to command growing respect in a distinguished scientific circle to which she was introduced primarily through her husband. Although the Ayrtons collaborated, both William and Hertha were conscious of the risk to the woman of joint husband/wife publications, and with his support she published alone. Her circle was wide, including scientists, the feminist activists of the Women's Social and Political Union with whom she worked, and also literary figures, not least the novelist George Eliot, whose interest in science, women and Jewish culture led her to base the figure of Mirah in *Daniel Deronda* on Ayrton. Eliot wrote this novel as part of that profound attempt made by a number of Victorian intellectuals, and most particularly by the novelists, to wean British culture away from its taken-for-granted anti-semitism and towards an appreciation of cultural diversity.[16] Ayrton herself was secular and like her daughter (who as Barbara Ayrton Gould was to become one of the earliest women Labour MPs) strongly feminist.

Thus, unlike Lord Huggins, Ayrton felt that her sex was irrelevant to the evaluation of her work as a scientist. 'Personally', she wrote, 'I do not agree with the idea of sex being brought into it at all. The idea of "women and science" is utterly irrelevant. Either a woman is a good scientist or she is not; in any case she should be given opportunities, and her work studied from the scientific not the sex point of view.' None the less although Ayrton wanted to be judged for her science she was always conscious that for her work to be studied from the scientific point of view meant a corresponding political effort. Ayrton's energy was formidable and she seems to have been as active in her politics as in her science, taking part for example in the great suffrage march of 1911, walking with the scientific section among the 800 women graduates in academic dress.[17] During the march some windows were broken and a number of women were arrested and put into Holloway Prison, including her daughter Barbara, a matter which occasioned some maternal pride.

Within that small group of women at the turn of the century who were somehow managing to do science there was a network of mutual support which extended beyond national boundaries. Behind the invisible colleges discussed by the masculinist history there were much more invisible colleges of women scientists. Hertha, for example, met and became friends with Marie Curie when the Curies visited London in 1903 and Pierre (alone) was invited to speak about their work at the Royal Institution. Subsequently, when Marie's work was attributed to Pierre in the *Westminster Gazette*, Hertha robustly refuted the lie. Still later, after Pierre's death, when Marie was being harassed by the press because of the allegations by his estranged wife of adultery with the physicist Paul Langevin (see chapter 7), she came, as Ayrton's guest, to take anonymous refuge in England. Nor was the support only one way, for Curie, at Ayrton's request, willingly signed the international letter in defence of the jailed suffragists. Signing was compatible both with Marie Curie's own feminist values[18] and with her recognition, as a Polish nationalist and socialist, of the hard battle against tyranny, and of the use by tyrants of prison to crush social and political struggle.

Second phase: the aftermath of the 1919 legislation

The Royal Society's rejection of Ayrton in 1902 was, it is reasonable to assume, aided by the choice to consult counsel who were known anti-feminists. Even then, legal opinion was divided on this matter and other lawyers could and did give other educational and scientific bodies different advice. None the less, the subsequent 1919 Act, on which admission was eventually to hinge, was won by the efforts of Ayrton and her sisters. The evidence of the crucial difference in the context – the change in the legal situation which meant that women scientists were eligible for election – is sparse but unequivocal. A Royal Society Council minute dated 8 December 1922 reports receipt of a letter from the Women's Engineering Society[19] (the secretary being Caroline Haslett) enquiring whether suitably qualified women scientists were eligible for election to the Society. The Council decided to consult Mr Barrington Ward, KC, and Mr Dighton Pollock, who gave, within the month, their opinion:

(1) We agree that it is very doubtful whether the Charters in themselves admit of the election of women (See Joint Opinion of 1902); but neither sex nor marriage, in view of the Sex Disqualification (Removal) Act 1919, can now be a bar to admission of women to a Society incorporated by Royal Charter. S.1. of the Act reads; 'A person shall not be disqualified by sex or marriage . . . for admission to any incorporated society (whether incorporated by Royal Charter or otherwise' . . .

In view of S1 of the Act we consider that both single and married women are now eligible for the fellowship under the Charters, and there is no need for a supplementary charter to admit women. The present Statutes only contemplate the admission of men. In order to make the position clear, we think that they should be amended . . . The exclusion of women thenceforth rests on the Fellows themselves who can reject women on voting for the election of Fellows from among the candidates.

This opinion concerning the legal implications of the 1919 Act was unquestionably a political bombshell for the Society, even though the last sentence can be read as a fraternal hint about how to stay this side of the law and still 'democratically' keep women out. It took time to decide how to manage the issue.[20] Indeed procrastination and keeping a low profile seem to have been the strategies of choice, for the matter was not reported back to the Council until 5 July the following year, and, with a touch of laxity surprising in any well-run committee, there was no indication of an appropriate letter reporting the outcome to the original enquirers. Nor did the Council feel the necessity to adopt the suggestion of the lawyers for the appropriate modification to the statutes.

Some three years later this failure to reply was noticed by the Women's Engineering Society, and Caroline Haslett wrote again in April 1925, this time securing a reply from the Society within twenty-four hours. This blandly said:

In reply to your letter of April 7th Council, as I stated in my letter of December 7th, 1922, took into consideration the question of the admission of women to the Society. There was a general opinion that women were already eligible, provided of course that their scientific attainments were of the requisite standard, but to make the matter quite sure we took Counsel's opinion, and were informed that women were, as we had thought, eligible under the Charter and present Statutes of the Society.[21]

The 1919 legislation, which in counsel's opinion unequivocally reversed the eligibility situation, has to be understood as part of the postwar settlement between the government of the day and the feminist movement, whose demands for the vote and for equality of access to higher education and the professions was pressed with particular forcefulness during the years leading up to 1914. Thus the 1922 opinion, and the eventual admission of women scientists, was Hertha Ayrton's and their victory. Although the reversal of 1922 was too late for Ayrton and her allies, for she died the following year, aged 69,[22] her courageous failure at the turn of the century was crucial to the subsequent success of the two first women to be elected, Kathleen Lonsdale and Marjory Stephenson, in the 1940s.

Two decades of collective amnesia in the Royal Society?

Yet despite the 1922 opinion, no women were elected to the Society until 1945 – and indeed, no certificates for women candidates were presented between 1922 and 1943. This extraordinary gap suggests at best a collective amnesia – or perhaps a repression of memory – within the Royal Society, in which the fact of legal eligibility and the political likelihood of success became conflated to become an unstated and legally false, but socially powerful, consensus that women were not admissible. Looking back, it seems inconceivable that those many women who contributed to the meetings and publications of the Society over the next twenty-two years did not include even one name that a male colleague felt worth submitting. Constance Elam, who collaborated with Geoffrey Taylor FRS on 'The Distortion of an Aluminium Crystal During a Tensile Test'[23] and who subsequently wrote what was to become the classic monograph on *The Distortion of Metal Crystals*, was perhaps such a possible candidate. Or, just to list the names of the contributors over one year, consider Edith Irving, Jean McMinn, Jane Sands, Dorothy Sands, Dorothy Wrinch, Elizabeth Scanaman and Ida Birkinshaw (who was the co-author of the Bakerian Lecture in 1925 with no less a person than the Secretary to the Society). Some, like Birkinshaw, were in due course 'lost' to science by the pressures of wifehood and motherhood, but it is difficult to think that there were no suitably qualified women scientists eligible for consideration on purely scientific grounds.[24]

But the pressure which had brought the reluctant admission that women were eligible had weakened. The interwar period, above all the depression and the Geddes Axe of 1932 which sent so many married women working in the public sector back to their kitchens, meant that the feminist organizations were functioning just at tick-over. Science itself was acutely underfunded and what positions there were went, through the family wage ideology, to the men. Twentieth-century science had ceased to be a cultured activity of gentlemen, and, as members of an ill-paid but interesting occupation, scientists were beginning to form an embryonic trade union in the form of the Association of Scientific Workers. In this situation, although women were gaining access to the laboratories in increasing numbers, it was on very weak terms.

The failure of the Society to consider women candidates during the interwar period was scientifically and even politically all the more questionable in that the presidency was held first by Frederick Gowland Hopkins (1930–5) and then by William Bragg (1935–40), in both of whose laboratories there were numbers of gifted women – including both the two first women to be elected, Kathleen Lonsdale and Marjory Stephenson. Yet the eligibility issue lay buried until 1943. Indeed as Dorothy Hodgkin observed, 'Marjory Stephenson was elderly, already very distinguished. She could have been elected twenty years before if anyone had thought to propose her.'[25]

Third phase: 'A much more dangerous breach than the election of a few women'

However, that no women's names had been submitted during that long period is precisely what the president, Sir Henry Dale, was able to claim in a most delicately worded communication circulated to all fellows subsequent to a meeting of the Council of the Royal Society on 16 December 1943. As Sir Henry put it, the issue of eligibility 'has now been raised in a practical form by the presentation of two certificates on behalf of women candidates'. The use of 'practical' enabled the Society to have inhabited a judiciously grey area in which, while it had not gone against the legal opinion it had received in 1922, none the less had not had to consider concretely the issue of the eligibility of women. This achievement has to be understood as part of a general conjuring trick within science through which even very distinguished women scientists were made to disappear. Margaret Rossiter[26] reports, as a reasonably common occurrence, occasions on which outstanding women scientists who had been invited to give the guest lecture to the annual meeting of a scientific society were not expected to attend the dinner that followed. The historian of science Charles Singer in his classic account of the history of biology manages simultaneously to dedicate the book to the botanist Agnes Arber (the third woman FRS) while not mentioning her or any other women's work within it.[27]

Dale's communication to the fellows explained that the legal position had been clarified by the officers with the assistance of the Society's legal advisers.[28] 'Clarified' referred to some rather considerable diplomatic work, as some members of Council had felt what the lawyer, C. R. Medley, tactfully referred to as a 'difficulty'. Medley informed the Society that 'as the Statute now stands it is obligatory to receive and register the certificate which we understand has been lodged in proper form, and that certificate must be suspended in accordance with the Statute.' He explains that he has had the opportunity of consulting Mr Sullivan, KC, who 'agrees with me that this is not a matter which admits of doubt'. Dale proposed that the Society vote on the two amending clauses proposed in the 1922 opinion:

> 1) To add at the end of Statute 1 the following words:
> 'Nothing herein contained shall render women ineligible as candidates'
> 11) To add a new Statute, to be entitled Chapter XXII, Statute 93, to read as follows:
> 'In the foregoing Statutes and in any Standing Orders of the Council and in any Rules and Regulations adopted by the Royal Society or by any joint Committee for the administration of Trusts to which the Royal Society is a party, words importing the male Gender shall include the female unless the context requires a contrary construction.'

This formulation avoided what Dale saw, in a briefing paper[29] he had previously set before his fellow officers, as two competing dangers: one, of the Society moving into a situation of illegality by refusing women as candidates; the other, of not being sensitive to the Society's internal democratic traditions (which had so successfully excluded women). Thus, while the fellows had to be consulted, it was important that they understood the nature of the consultation, namely that it was about the words of the proposed change, not the substance. It was important to handle the matter so tactfully that the fellows would gracefully accept the inevitable. To this end Dale accepted an ingenious suggestion from Sir Thomas Merton: to use the context offered by the war, and, invoking the War Emergency legislation, to conduct a postcard poll on the matter. This exceptional event would not 'raise trouble' by introducing the principle of a referendum which, as Dale saw it, constituted a 'much more dangerous breach with tradition than the eventual election of a few women to the Fellowship'.[30] Dale was a sophisticated strategist, and clearly, if women were to be admitted, it was to be properly managed. Indeed although the president and officers are required to conduct themselves with conspicuous impartiality, and not be associated with particular candidacies, Dale was to play an active part in the electoral process. He had direct knowledge of the candidate in question, Kathleen Lonsdale.

The Lonsdale saga raises a number of different issues, and is worth discussing in some detail. She held a research fellowship at the Royal Institution of which Dale was director. As the very young Kathleen Yardley she had originally come there to work as a research assistant to Sir William Bragg (senior) and had stayed on after his death in

1942, when Sir Henry Dale took over. Dale clearly thought very highly of her. One story that her biographer Dorothy Crowfoot Hodgkin (also both a crystallographer and an anti-militarist) tells is of Lonsdale's pacifism, which led, not without some effort on Lonsdale's part, to her being sent to Holloway Prison for a month in February 1943. Unable to carry out the heavy domestic work required from her as a prisoner, she collapsed, and it was Sir Henry who made arrangements for her to have both papers and instruments so that she could continue her scientific work. The friendly letter from the senior technician, H. E. Smith, which came with the material speaks to the warmth of her relations with her colleagues.[31] The prison experience led Lonsdale to become a prison visitor and a supporter of penal reform. A letter from her to the prison governor in which she points out that she had been able to put in seven hours of scientific work a day was a delicate but firm reminder that prisoners were (and are) appallingly underoccupied.

That Lonsdale was a woman was seen as a matter of considerable significance; however, there is no sign of anyone at the time batting an eyelid at her militant pacifism – even during 1943, a particularly grim period in the 1939–45 war: The Royal Society was politically (in a left/right sense) a good deal more tolerant in the 1930s and 1940s than seems to be the case in more recent years. Then the Society, like British science generally, included a number of active and committed communists, such as Bernal, Needham, Haldane and the physicist and socialist Patrick Blackett. The last, who later became President of the Society, was refused Foreign Office clearance to visit the Soviet Union on security grounds.[32] These socialist and marxist men, who were in principle committed to sexual equality, in practice were able to refrain from taking the initiative in proposing any particular woman candidate.

One of Lonsdale's fellow students and scientific collaborators at the Royal Institution was W. T. Astbury, who had himself been elected into the Fellowship in 1940. According to Hodgkin, Astbury had first suggested that Lonsdale should be considered shortly after his own election.[33] Bernal's obituary memoir, written for the Society, says that Astbury had a pronounced sense of fairness, which was able to recognize scientific merit in either sex. Certainly from correspondence between himself and Blackett, who then chaired the Society's listing committee, it is clear that Astbury gave Lonsdale his unqualified support. Blackett, while sympathetic to Lonsdale's case, sought to reassure himself that Astbury fully supported her.[34] Astbury first agreed with Bernal, then reminded Blackett that no less a person than Max Born had maintained that there is 'no doubt about her case'.[35] Astbury backed Lonsdale, not only against another possible candidate under discussion who also did X-ray crystallography, but as 'definitely one of the leading structural analysts of either sex'.

Thus the configuration of circumstances within the Society

favoured Lonsdale's chances. There were a number of left and liberal fellows politically committed to the equal treatment of men and women; crystallography was a rising discipline within the Society; and there was the presence of a personally committed colleague prepared to start the ball rolling. The larger context of an air-war in which the civilian population were being killed almost at the same level as the military, and in which women were active in many occupations previously exclusively occupied by men, is also likely to have played its part in softening – even if only for its duration – social and political attitudes to women's participation generally.

Having received a correctly drawn-up certificate for Kathleen Lonsdale, Dale next made sure that she was willing to be considered. Although in her young womanhood Lonsdale had been interested in feminism, while she was willing to face prison for her pacifism she was no militant for women's rights, and she replied, 'Not if it means dissension among the Fellowship.'[36] Such personal self-effacement combined with public scientific excellence provided Lonsdale with a means of managing her gender identity which was less likely to give rise to the antagonisms Ayrton had experienced.

Dale, however, aware that the Society had to register the certificate, took the view that more than one candidate was easier to manage politically. 'It won't do with one' was a constant refrain within his campaign, not least perhaps because the 'one' came from his own laboratory and had been in prison. The internal committee papers and documents of the Society reflect the politics of judicious acquiescence in which it seems that Dale's indubitable political skills were being exercised to protect the best interests of the Society. However, the Dale archives also point to a personal track record as a supporter of women's rights. He helped open the Physiology Society to women, subsequently served on the London University Committee which improved women's access to medical education, and, not least in the case of Lonsdale, was supportive to individual women colleagues. Perhaps the clearest expression of the warmth and imagination of his support is left in the record of a talk he gave in 1951 to the Pearse School at Cambridge (a fee-paying girls' school). Here Dale, speaking of the kinds of career opening up for women, described what a later generation was to call role models, and, unlike all too many scientists talking to young women today, sought to expand the horizons of the girls, not simply run a recruiting drive for science. Thus he encouraged them to think about becoming like Dorothy Russell, a London professor of pathology, like the economist Barbara Wootton, the biochemist Barbara Holmes, the archaeologist Jacquetta Hawkes, the poet Elizabeth Sewell, or the detective fiction writer Margery Allingham.

With this complex commitment to the Society, to science and to women scientists (in this order), Dale set about making sure of a second candidate. During mid-November (the closing date that certificates could be submitted being the end of that month), Dale wrote a number of letters to fellows whom he considered supportive of women's rights. Unusually in Dale's correspondence, these letters are marked 'strictiy private and confidential', though given that the president, who must stay above the election process, was busying himself very directly in the nomination process, such a warning to his recipients is perhaps less surprising. Among the letters, a key missive is that to the bacteriologist (later Sir) Paul Fildes:

> You have shown your interest in Miss Stephenson's claim, and there is still time, if you hustle to get a certificate signed and presented before November 30th. If you are hard up for signatures you might think of sending it to M. Dixon, and asking him to solicit support from Hopkins, Keilin and others at Cambridge who might be interested.

The letter concludes:

> You must understand that my hint to you must be strictly confidential. There must be no suggestion that I am advancing a particular certificate, or offering support for a claim in advance. Perhaps as you are going on the Council, it would be better not to sign yourself.[37]

Fildes was a shrewd choice, as there was a recognized complementarity between Gowland Hopkins's laboratory, which worked on bacterial metabolism and, during the war, pathogens, and Fildes's own laboratory at Porton (the biological warfare establishment), which worked on the nutrition of pathogenic organisms. Thus there was a delicate subtext balancing the martially commited Fildes[38] and his protegée with the pacificist Lonsdale proposed by Astbury and others. Fildes was also a known supporter of women's rights.[39]

Having committed himself to what was a dangerous course of action if his recipient gave him away, Sir Henry did not leave the matter to chance. He wrote on the same day to Haldane, known to be an outspoken supporter of women's rights. The secretary, A. V. Hill, had drawn Dale's attention to some correspondence he had been having with Haldane concerning his 'public allusion to the absence of women from the Royal Society'.[40] Dale used this both to inform Haldane that they now had a 'woman Candidate on the Physical side' and also to rope him in with an exceedingly broad hint to support Stephenson.[41] The next day Dale wrote a much longer letter to the now blind Gowland Hopkins (the letter was not marked 'Strictly private' – but then he was writing to an ex-President who would scarcely need reminding of the delicacy of the situation), mixing his evident regard for Hopkins's distressing situation with setting out the facts of the matter and making sure that Hopkins would support his colleague Stephenson.

Fildes's handwritten letter to Dale came by return, marked 'Private and confidential'. He reported that:

> The situation about Miss S is that without discussing it with me, Harington as a biochemist decided to pursue her and asked me to second before my name was mentioned in connection with the Council. It was obvious that the proposal should come from a biochemist and I agreed to second. Her certificate is now in Harington's hand and I have obtained the support of the bacteriologists who are all in the process of signing. I imagine H has dealt with the Cambridge fellows.[42]

He concludes, balancing his commitment to Stephenson's cause with a demonstration of his sensitivity to the political situation, with the added and underlined sentence: *'Your attitude understood.'*

Haldane wrote back less hurriedly but within the week, explaining that he had 'already signed Dr. Stephenson's certificate'. Haldane is the only scientist in this entire correspondence who acknowledges Stephenson's academic title; the rest follow the then common practice of speaking of even immensely distinguished women scientists by their marital status. But even he speaks of her as 'Miss' Stephenson within the same brief note. Not, however, known for his conciliatory style, Haldane tartly observes: 'I have no doubt at all that had she been a man she would have been elected to the Fellowship some time ago.'

Hopkins, although restricted by his disability, had signed the papers before writing back to Dale in early December. He writes:

> I have somehow been ignorant of the Council's former discussion concerning the legality of electing women to the fellowship, and I wondered just what had happened when I received Harington's letter concerning Marjory Stephenson. Their election will [the 'will' is crossed through and the more cautious 'would' is inserted] be revolutionary of course, but I think the time has come for the justification of the policy. We have heard much of equal pay for equal work, and we may well feel that this should apply to intellectual work and the honours accorded to it.

Hopkins's letter suggests that the collective amnesia concerning the 1922 opinion had been so pervasive that even as President, with many gifted women in his laboratory, including his own daughter, he did not 'know' of the legal eligibility of women.

The consensus among the bacteriologists and biochemists that Stephenson more than merited election is indicated by the unusually large number of names on her certificate. Usually eight to ten fellows sign the papers; Stephenson's had eighteen. Fifteen supported her candidature from 'personal' knowledge and a further three from 'general' knowledge.

Dale's judgement that he needed two candidates did not mean that he only actively canvassed for Stephenson; he also wrote to (later Lord) Adrian sounding him out about Honor Fell. While Fell was elected within a few years, there was no instant build-up of support as there was in Stephenson's case, and there is no sign of Adrian's reply in the archives. But there were good reasons for hurrying over Stephenson. First, Haldane was right; Stephenson, now 60, more than merited election. She also did not enjoy good health, so that if justice was to be done there was a special case for speed. And indeed she died in 1948, only three years after her election.

Ironically, in the crucial discussion within the Society, Astbury (according to Dorothy Hodgkin) had to argue passionately for his candidate, as the consensus for Stephenson had led to a proposal that Lonsdale's candidacy be deferred. Astbury insisted that she had already been de facto deferred and that she should be elected at the same time as Stephenson.[43] As we know, Astbury, aided by Dale's view that 'one won't do', won the argument.

Thus, by the time Dale convened the officers' meeting at which he proposed the plan of action, he had already ensured that all the pieces were in place for the subsequent meeting of Council: first, that there were two candidates, not one; second, that there would be a vote on words but not on substance; third, that there could be a non-precedent-setting referendum. He further reassured the Council of their own powers; they could legally proceed to accept the certificates, although he advised that the consideration be postponed until the vote had been taken. In all of this Dale showed himself to be a masterly general who could lead his troops into the historical defeat of three centuries of male exclusivity with such judicious grace that the reverse was made almost imperceptible to the participants, or even the observers. He made the final achievement of Bacon's new masculine knowledge its capacity to trascend the mere literal specifics of male production. Dale summed up:

'Altogether, I believe that we could in this way do what we wish to do without any suspicion of illegality, or danger to our traditions of indifference to any but scientific merits or demerits, which I regard as far more important to the Society than that of its purely male Fellowship.[44]

The council accepted Dale's advice and the postcard vote took place over a six-month period so as to ensure that all the Fellows would be able to respond without undue haste, allowing for the conditions of war. The strategy worked and by May 1944, 336 Fellows declared themselves in favour; only thirty-seven were against and a further three were in favour with qualifications. One of these, S. P. James, recognized the inevitable but continued to plead the men scientists' case: 'I record my vote for the proposals on condition that an increase of one is made to the total annual elections to the Fellowship . . . If women are made eligible then the percentage of men who succeed in the annual election will become even less than at present (about 14%).' James's plaint reminds us that in a situation where men as a single gender have all the cake, the most modest move towards gender equity means that men generally, and some in particular, will lose out.[45] A press release issued after the council meeting on 12 October 1944 made the news public. On 23 October the indefatigable Caroline Haslett wrote again from the Women's Engineering Society: 'I was delighted to see in *The Observer* that women are now eligible to become Fellows of the Royal Society. In 1922 and 1925 I had correspondence with the Secretary on this subject. It is good to see that the matter is now settled.'

Pioneering candidates and scientific oases: Lonsdale and Stephenson

Lonsdale and Stephenson were not only both exceptionally gifted scientists; they also came from exceptional laboratories, and in Lonsdale's case exceptional entire disciplines, where gifted women were welcome. Having an exceptional mentor is, as Harriet Zuckerman's study of US Nobel Laureates[46] suggests, a helpful condition for achievement and recognition. Both Lonsdale and Stephenson worked in the laboratories of Presidents of the Royal Society; however, in both cases the initiative of proposing them was made by peers, their mentors being signatories to the certificates.

As a scientist Lonsdale was probably one of the most extraordinarily gifted of her generation and hard to match as a candidate. Whereas almost all women scientists (and many men) at this period came from intellectually and economically privileged backgrounds, Kathleen Yardley was born in 1903 as the tenth child of a family so poor that four of the children died in infancy.[47] However, she was able to benefit from the educational reforms secured by an earlier generation, and after elementary school she won scholarships, first to the county high school where she was permitted to study science at the neighbouring boys' school, then to the London women's college, Bedford, where she secured admission to read mathematics when she was only 16. Hodgkin tells the story of how Kathleen's mother, who had accompanied her daughter to the interview, cried when the place was offered. When Kathleen took her BSc at 19, having transferred to physics, she came top of the University of London list. The crystallographer W. H. Bragg (Bragg senior) was one of the external examiners and was sufficiently free from prejudice to invite her to join his research team at University College on a studentship valued at £180 a year. This was wealth – as indeed it would have been to many working-class women or men at the time – and she was able to help out with the family expenses.

Crystallography in the interwar period was almost synonymous with the name Bragg, and in consequence to be recruited into an elite laboratory at quite such a young age with a background so utterly divorced from what Noel Annan has called the British intellectual aristocracy[48] made Lonsdale quite exceptional. The field had been divided with a certain simplicity by the two Braggs (joint Nobel Prize-winners when the son was only 25); the senior took the organic and the son the inorganic crystals. Bragg senior's laboratory, first at Leeds, then at University College London and later at the Royal Institution, was both an exciting scientific environment and rare for its welcome to gifted women and men alike.

Kathleen also met her future husband Thomas Lonsdale at the Royal Institution and they married in 1927 when she was still only 24. That year they moved to Leeds where Thomas had a research post working on silk. She considered giving up science, but 'Thomas would have none of it; he had not married he said, to get a free housekeeper.'[49] Reports of the highly organized Lonsdale housekeeping system touch a modern note: joint shopping once a week and Kathleen's thirty-minute meal, with Thomas and Kathleen both working on their research in the evenings. Although they did not live in Leeds for very long it was an important time for them both. They became Quakers; this was important especially for Kathleen and became the organizing principle of her life. Her earlier interest in feminism found a more congenial home in the egalitarian and socially concerned milieu of the Society of Friends. The discourse of pacifism provided the means of reconciling public conceptions of womanliness and science within a gender indentity. In her address as the first woman president of the British Association for the Advancement of Science in 1968, she tried to raise the larger issues of the world outside science and of the need for a scientist to be concerned with what she spoke of as 'the Good Life'. She also consistently worked against militarism and for pacificism with both the scientists' Pugwash association and the peace movement generally.

Kathleen was fortunate in her marriage; Thomas was always supportive of her scientific activities and shared her pacifist commitments throughout their life together. Kathleen wrote, 'For a woman and especially a married woman with children, to become an exceptional scientist, she must first choose or have chosen a good husband.'[50]

Scientifically Leeds was productive for Kathleen; she had been made welcome at the university with a part-time demonstratorship to eke out her Bedford College[51] research grant. Funds in the Lonsdale household were, however, exceedingly tight and the Royal Institution, at Sir William Bragg's suggestion, gave her £50 a year for someone to take care of her new baby while she did calculations on the structure of the benzene ring.[52] When they returned to London and Kathleen was working at home, Sir William wrote that 'Sir Robert

Mond is giving me £200 with which you are to get assistance at home to enable you to come and work here.' Kathleen did some sums and showed that in 1931, now with two children, it would cost £277 to replace her at home. Sir William found her £300, but although this speaks of the lengths that Bragg was willing to go for an exceptional woman scientist, it also speaks of the continuing grip of the family wage ideology.[53] She was on soft money until the end of the war when she secured her first academic post.

Maureen Julian, the feminist historian of crystallography,[54] has produced a genealogy of both Bragg laboratories which brings out their significance for the development of crystallography as – relatively speaking – a woman-friendly area in the physical sciences.[55] As an insider Julian knows well the derogatory language in which the discipline was spoken of – thus in some circles crystallography was derided as 'intellectual knitting'. The historian offers two explanations for the presence of women; one that the need for painstaking calculation – prior to the advent of computer methods in the 1960s – meant that there was stereotypically suitable work for women; and the other, particular to the Braggs, was that the laboratory style (of both father and son) was unaggressive and therefore welcoming to women. It is not necessary to choose between these explanations, as the one may have helped the other, and both were fostered by the newness of crystallography as a research area. What, however, was both unusual and true was that the Braggs were able to recognize the scientific contribution of women. Lonsdale for example kept a 1928 letter of Bragg senior's, in which he wrote, 'I think your new result is perfectly delightful: many compliments upon it! I like to see the benzene ring emerging.'[56]   Certainly no fewer than ten quite exceptional women worked in the Bragg laboratory (including Constance Elam, mentioned earlier as a possible candidate for the Fellowship). Of the more than seventy scientific workers and associates who trained with Bragg, no fewer than five were to become FRS, all of whom were to share and extend this open-minded tradition in their own laboratories. In addition to Lonsdale herself, the five included W. T. Astbury, who as her fellow student set the ball rolling for Lonsdale's election, and Desmond Bernal, who both gave shelter to Rosalind Franklin at Birkbeck after her conflict with Maurice Wilkins at King's College (see next chapter) and also was Dorothy Hodgkin's thesis supervisor and lifelong friend.

At the time of their joint election in 1945 Marjory Stephenson was 60 and Lonsdale 42. Stephenson's biography has to be understood as being of a piece with the slow career development characteristic of nineteenth-century women. Born in 1885 into a comfortably off Cambridgeshire farming family, she had a governess who was herself a strong-minded educationalist and sufficiently convinced of Marjory's talents to persuade the parents to send their daughter to a high school for girls. Her biographer, the protozoologist Muriel Robertson,[57] who was elected to the Royal Society two years later, tells us (and again it is worth remembering that these obituaries are written on the basis of notes provided by fellows, so it is a point Stephenson wanted us to know) that while her father had inspired her interest in science by explaining nitrogen fixation when walking in a clover field, it was her mother who made sure that both Marjory and her elder sister went to Newnham.[58] The relationship with the college was important and Stephenson gave Newnham her lifelong support. She read a Part I in the natural science tripos and although women were not at that point admitted to the classes in zoology and chemistry – two of the three subjects – she managed to gain a Class II mark.

Frustrated by shortage of funds in her original plan to read medicine, Stephenson studied and taught domestic science (that ghetto to which so many gifted women scientists were confined), but in 1911 was given the chance to join Dr R. A. Plimmer at University College London, working first on animal lactase and then on the metabolism of experimental diabetes. Despite the five-year gap spent teaching domestic science, Stephenson took well to research and in 1913, when she was 28, she was awarded a Beit Memorial Fellowship. With the outbreak of war she joined the Red Cross and worked as a nurse in France and Greece although there is no indication of her feelings about what for most of her generation was a devastating experience.[59] After the war she took up her fellowship and joined Hopkins's biochemistry laboratory in Cambridge, where she stayed for the next thirty years until her death in 1948. This, like the Bragg laboratory, was relatively an oasis for women scientists. Biochemistry, like crystallography, was a new area; in consequence there were few regular posts, i.e. posts suitable for men, as the subject was not yet part of the undergraduate teaching system. Whereas the Braggs divided crystallography, Hopkins's strategy was to foster new developments and new people, until Cambridge biochemistry, like Bragg crystallography, defined the field. Consequently there was a similar constant need for gifted scientists which made even 'irregulars' such as women welcome. There was also Hopkins's own egalitarian attitude toward women as scientists, for he had encouraged his daughter Barbara in her scientific interests, and she (later Barbara Holmes), together with Dorothy Jordan Lloyd, Dorothy Needham and Marjory Stephenson herself, were among the many talented women scientists who were trained in the Hopkins laboratories. Dorothy Needham's work on muscle chemistry subsequently secured her election as a Fellow of the Royal Society.

The progressive atmosphere of the laboratory meant that it also offered shelter for refugees escaping Nazi Germany, including a number of distinguished Jewish biochemists, such as Hans Krebs, Fritz Lipmann, David Keilin and his daughter Joan Keilin. Biochemistry at Cambridge was an oasis for both women and refugees. This was the golden age of biochemistry, and Cambridge biochemistry in particular; not for nothing was the departmental magazine called *Brighter Biochemistry*.

Marjory Stephenson went on from her early work on fat-soluble vitamins to develop bacterial chemistry as a distinctive field within biochemistry, consolidated by her book *Bacterial Metabolism*, published in 1929.[60] She saw that the bacterial cell was a crucial site in which to study the chemical activity of the living cell. Yet despite the high regard in which her work was held, she was employed on soft money until she had a secure position at 44. Her biographer, commenting on this, writes, 'It is a curious reflection of the difficulties in the path of women scientists and perhaps also a reflection of the distrust of a somewhat new subject that so original a worker should have been an annual grantee for so many years.'[61] The biographer might have added that it was also the novelty of the subject which played a significant part in opening the laboratory doors to women scientists. Finally, however, Stephenson's distinction was such that, despite the hostility to women in Cambridge, she was the obvious candidate to inherit Hopkins's chair, which she held until her untimely death in 1948.

The limits of patriarchal reform

While the actions of Dale and his colleagues have been read as working systematically to secure Stephenson and Lonsdale' election, they reveal both that these eminent men knew exactly which women should have been immediately elected and also an unblushing display of administrative flexibility to achieve the necessary electoral outcome. Such an insight into the workings of the British scientific elite indicates not just how much control lies in the hands of very few men but how they use it. But the subsequent history of how and whether the scientific men controlling this powerful institution have been willing to let women join the elite procession of educated men is hidden. Instead we have to rely on external indicators of change. These are not encouraging. Since the struggle, now some forty-eight years ago, when women were first admitted to the Society, only fifty-two women have been elected Fellows and a further five have been made foreign members. In 1990 there were thirty-one living women Fellows and 1,059 men.[62] Women Fellows form such a small proportion that what appeared to be an absurd suggestion in 1944, namely that there should be expansion by one Fellowship a year to compensate men for the impact of the election of women, about meets the situation, as the rate of admission for women has been a breath more than one.[63]

A dourer reading of the story of the admission of women and their current presence in the Society would be that, far from Sir Henry Dale rhetorically reassuring his fellow Fellows when he said that a referendum would be a 'much more dangerous breach with tradition than the eventual election of a few women to the Fellowship', he was setting out a plan for the minimum of reform compatible with the much larger object of preserving this powerful patriarchal institution. It seems that while 'his' sense of gender justice in science was present, it was a very modest affair in comparison with 'her' sense.

That passion for justice in which women would be judged for their science and not their gender, which inspired Margaret Cavendish to envisage her utopian scientific community Blazing-World nearly three and a half centuries ago, or Hertha Ayrton and the 800 other women scientific graduates who marched in the early years of this century, has yet to secure more than minimalist recognition in this bastion of masculinist and scientific power.

Nine Decades, Nine Women, Ten Nobel Prizes: Gender Politics at the Apex of Science

All I desire is fame . . . but I imagine I shall be censured by my own sex and men will cast a scorn upon my book, because they think thereby women encroach too much upon their prerogatives; for they hold books as their crown, and the sword as their sceptre, by which they rule and govern.

Margaret Cavendish,
The Description of the New World Called the Blazing-World

Gertrude Elion, Rita Levi-Montalcini, Barbara McClintock, Rosalyn Yalow, Dorothy Crowfoot Hodgkin, Maria Goeppert Mayer, Gerty Cori, Irène Joliot Curie and Marie Curie: nine women, ten Nobel Prizes for science (Marie Curie was awarded prizes both in physics – 1903 – and in chemistry – 1911), distributed over the eighty-five years between 1903 and 1988. They range in age, from Curie receiving her first prize at 36, to the three most recent, Elion, Levi-Montalcini and McClintock, being 71, 77 and 81 respectively. Apart from Irène Joliot Curie, who, at 38, emulated her mother in her youthfulness as well as her scientific talent, and was awarded a prize in 1935, the intermediate group of postwar prizewinners were all in their fifties: Gerty Cori, 53 (1947); Maria Goeppert Mayer, 57 (1963); Dorothy Crowfoot

Hodgkin, 54 (1964) and Rosalyn Yalow, 56 (1977). These nine women constitute some 2 per cent of the scientific Nobel Laureates.

Since Nobel Prizes are not awarded posthumously a number of commentators, including but not only feminists, viewing the three women honoured in the 1980s, have suggested that longevity is increasingly an additional criterion for women scientists to meet. It seems that the Nobel committee, in responding to the new pressure on it to recognize women scientists, feels safer in going back in history, to acknowledge those whose scientific eminence is unquestionable but who have been previously passed over. Perhaps men with the power to give public recognition suffer from an inability to recognize scientific merit in peer-group women, whereas they have no such problem with peer-aged or even younger men. However, in that a central rationale for awarding the cash-rich Nobel Prize was to free creative scientists from concerns about resources, then these most recently honoured women would seem to be ineligible, and certainly other older men scientists have been explicitly excluded on precisely these grounds.[1]

This anomaly, which in its repetition suggests a response to the increasing claims of gender justice and those of scientific merit, while possibly not at the level of conscious intentionality, is demonstrably effective as a means of constraining reform. The overdue recognition of these distinguished but now older women scientists limits the possibility of their exercising the usual powers of a Nobel Laureate. Their age means that, however brilliant, they are manifestly less likely to be in touch with younger up-and-coming scientists in their own field and less likely to be able to campaign for them. The move also diminishes the pressure to recognize those others, in their forties or fifties, who would be in a phase of their life and career cycle where they might best utilize the reward and the status. Even before the most recent awards, the time gap between their work and its formal recognition was already more strongly marked for women than men.[2] Nor is this unrecogized by the women scientists themselves, though perhaps it takes someone of Rita Levi-Montalcini's social and scientific confidence to reveal publicly her anger at the lapse of time and of the different treatment accorded to those she sees as in every way her peers. She notes that 'Two of my university colleagues and close friends, Salvador Luria and Renato Dulbecco, were to receive the Nobel Prize in Physiology and Medicine, respectively seventeen and eleven years before I would receive the same most prestigious award.'[3]

The prize as cultural capital

While scientific excellence has, with very rare exceptions, been successfully acknowledged by the Nobel Science committees (the

Literature and Peace Prizes have long had a more contentious record), the institutional and social origins of the Laureates have played a significant part. Just ten colleges, for example, produced 55 per cent of the 71 US Laureates studied by Harriet Zuckerman.[4] In a similar way the history of the nine women Laureates is in a number of ways a microcosm of the history of gender politics in science this century. The Nobel Prize sits at the apex of the status system of science. The Laureates are icons of the fusion of scientific knowledge and cultural power, so that where they are not already members of their national elite groupings of scientists, such as the Royal Society of London or the French or US National Academies, then it is customary that they are rather swiftly elected. Membership of institutionalized national and international scientific elites, as well as confirming such cultural and political power, also offers its bearer the prospect of participation in these institutionalized forms, and hence a close and uncritical relationship with the state. Members of this ultra-elite within science are invited to walk the corridors of power. Governments seeking scientific advice of a politically strategic nature frequently turn to their national academies or to specific disciplinary groups within them. The shadowy JASON group of leading Nobelist and near-Nobelist US physicists advising on US military strategy came into notoriety during the Vietnam war, but has continued ever since, today advising the US government on what is euphemistically termed national security and defence. Nor is the desire for scientific advice limited to powers temporal in the late twentieth century; the Pope, wishing to develop his thoughts on the environment, turned to the collective wisdom of Nobel Laureates, via the Vatican Academy, for advice.

This is a paradox at the heart of the Nobel system: scientific eminence is achieved through a small but innovatory piece of knowledge concerning a specific aspect of chemistry, physics, physiology or medicine, but winning the prize gives its bearers the ability to advise on global sociopolitical issues far outside their range of expertise. Feminists, in order to explain the systematic undervaluing of women within the labour market, have described women as 'inferior bearers of labour'; by contrast Nobel Prize-winners become 'superior bearers of thought', acquiring the power to speak and be listened to on topics where their competence is either at the same level as that of their fellow citizens, or even demonstrably less. Because this cultural power is rather concrete, few people are entirely consistent in their attitudes to its manifestation. Individual scientists have used their cultural capital to support their ideological and political commitments. Thus I have to admit that, like many anti-racists, I tend to point out that Nobel Laureate William Shockley received his prize for work on transistors and that he had no special competence to support his unquestionably hereditarian views on intelligence/IQ, but that when anxious to see nuclear power controlled I welcome seeing George Wald throw his political and scientific weight onto the socially critical side of the debate, and am less anxious to point out the modest connection between his Nobel Prize-winning work on receptor pigments in the eye and the scientific debate at issue. Many times have I welcomed the signatures of what seems to be a shrinking handful of anti-militarist British Laureates and FRSs, not least Dorothy Hodgkin and Maurice Wilkins, in protests against military aggression, without dwelling on the cultural power that their welcome presence reinforces.

It would also be ungenerous, particularly in periods when non-conformity with the state carries significant penalties, not to acknowledge the personal courage sometimes entailed. For leading non-Jewish German scientists to oppose the Nazis required an act of courage, as it did for leading US scientists to use their cultural capital to protest against witch-hunting during the height of the McCarthy era; it was much easier to deplore the excesses privately and subscribe to the politics of prudential acquiescence. The physicist Sakharov was rightly admired for his courage in using his cultural capital as the father of the Soviet H-bomb to play a leading role in the human rights movement. And although political persecution has not been a significant issue for British scientists, her anti-militarist activities and marriage to a communist meant that Nobel Laureate crystallographer Dorothy Hodgkin was proscribed from admission to the US except by a special CIA waiver until she was in her eighties. (Presumably the combination of her age and the collapse of the former Soviet Union led the CIA to think that she was not imminently about to engage in the violent overthrow of US liberal democracy.)

Thus I want both to salute individual Laureates and other eminent scientists for their sometimes quite concrete personal courage in the use of their cultural capital, in the face of sanctions which have ranged from exclusion and even death to various levels of social opprobrium, yet also to criticize a system which has amplified the cultural power of science, not least because of the extent to which, during the twentieth century, science has become incorporated and plays a predominantly socially conservative role.[5]

Ceremony and secrecy

As I indicated in the previous chapter, the archives of the Royal Society, itself one of the oldest scientific institutions, are accessible after forty years; but those of the Nobel Institute, created at the beginning of this century, are accessible only after fifty years; in consequence proportionately more of the Nobel's iceberg of secrecy is hidden. Such intense and prolonged secrecy about the affairs of the scientific elites, considerably longer even than that of the notoriously secretive British governmental tradition and an anachronism in

Sweden, where any citizen may have access to letters written by a minister, is in itself a matter of curiosity.[6] When it is remembered that these elites are choosing to honour creators of 'public knowledge' in science, not trade, military or diplomatic secrets, such secrecy speaks of the sense of cultural and political mystery with which Bacon's masculine knowledge has endowed itself.

The archives made it possible to go behind the public face of the Royal Society, with its discourse of the President being above the election of members, and scientific merit being the only criterion for election, and contrast this with the very particular ways, documented by committee minutes and correspondence, through which actual elite men scientists treated the claims of women scientists and finally came to the understanding that they could no longer exclude women. No such possibility, for other than the early years, exists in the case of the Nobel Prize archives. Lacking this account from the perspective of the powerful who manage such events, the story of the election of the women Laureates has for the greater part to be built from more outsider sources, including biographies, the rather rare autobiography, the occasional interview, and, as an important set of resources which have remained constant over time, the Laureate's Stockholm speech of acceptance, together with the biographical note and the photographic portrait which accompanies its publication.[7]

The occasion of the prize-giving is highly formal, and takes place in the presence of the monarch. It is the Swedish king himself, that symbol of a past military system of power, who awards the medals. The men attending the ceremony are required to wear white tie and tails. While for a number of recent men Laureates, perhaps particularly those from the US, who have rarely been known to wear anything except jeans and checked shirts, such dressing up is something of a novelty, it is also – as Virginia Woolf reminded us for the thirties – still very much part of the life of educated men. The academy has a passion not only for secrecy but for distinctive attire, a surrogate uniform on which medals signifying heroic performance on the field of truth may be displayed. The sharing of the military code and its honours is made all the easier in the Nobel ceremony because it is carried out at such a symbolic level; the constitutional monarch of a neutralist country is at once remote from the military and also the descendant of Gustavus Adolphus, the last Swedish king to die leading his troops in battle.[8]

Novelty and innovation are always central within the award of a Nobel Prize, even though the language has shifted over the decades from the 'land ho' quality of scientific 'discovery' in which the newly recognized phenomenon is equated with finding a new land (or at least new to the discoverers) to what the users doubtless see as a rather more nuanced language of a 'seminal contribution'. Women Laureates have to be innovators in an additional sense. Like the women who were first admitted to the Royal Society, they are likely to be entrants to new and therefore initially low-status areas of science where the discipline has not been fully formed, where there is no clear structure of employment and career, and hence where there is room for unpaid or badly paid pioneers whose passion is knowledge of the natural world.

Women who were recognized and women who were not

Marie Curie

Perhaps it was partly that the Nobel Prize was so new – not yet gelled in its prestige status – that made it possible in 1903 not only to invite Henri Becquerel and Pierre Curie to share the Physics Prize, but also to include Marie Curie at the astonishingly youthful age of 36.[9] (The terms of the Nobel award mean that it may be shared a maximum of three ways.) The introductory address on behalf of the committee spoke not only of the discoveries opening 'a new epoch in the history of physics' and of the close relationships of their producers, but of how:

> Les découvertes et les travaux de M. Becquerel et de M. et Mme. Curie sont en relations intimes les uns avec les autres: et les deux derniers ont travaillé en commun. Aussi L'Académie Royale des Sciences n'a-t-elle pas cru devoir séparer ces éminents savants, quand il s'est agi de récompenser par un prix Nobel la découverte de la radio-activité spontanée.[10]*

Equal producers the Curies may have been, but it was Pierre alone who was to give the Nobel address at Stockholm. There was perhaps some justification for this as he was eight years older than Marie, and had not been educated in a Warsaw lycée or transferred countries and languages before studying at the Faculty of Science in Paris. Nor did his father have the relatively modest occupation of a teacher in a Warsaw lycée, but was a French medical doctor. Marie and Pierre had met and researched together at the Ecole Physique and were married in 1895; in the same year he was appointed to a chair. (At the time of receiving the prize Marie had not yet defended her doctorate thesis.) Within two years of the marriage their first daughter, Irène, was born. Personal life and work thread Marie's notebooks; she describes her daughter's first steps, then speaks of the element she and Pierre have found which they propose to call radium; her next entry reports the

---

*[The discoveries and work of M. Becquerel and of M. and Mme Curie are closely related to one another: and the latter two worked together. Also the Royal Academy of the Sciences does not believe that it should separate these eminent scholars, when it is deciding to award a Nobel Prize for the discovery of spontaneous radioactivity.] Author's translation.

consolidation of Irène's walking. They shared a common commitment to socialism and to feminism, the last a matter of no small significance for the history of science.[11]

Because for the rest of the century this astonishing woman has been held up to all, and especially to all women scientists, as the example of what women are capable of achieving,[12] Elizabeth Crawford's[13] study of the early years of the Nobel Institution makes salutary reading. She reveals that the recognition of Marie's contribution to her and Pierre's achievement was not uncontentious; we suddenly find that we are back in an old story, recognizable all too often from our own lives. At the first hurdle, that of nominations, the French Academy had only put forward the names of Henri Becquerel and Pierre Curie. Marie, as a woman, was not seen as capable of producing scientific knowledge, and therefore was outside the committee's consideration either as a potential member or as a nominee.[14]

Within the politics of Swedish science things were a little better but still complicated. Ironically it was the monarchist 'right-wing' mathematician Gösta Mittag Leffler, a highly active figure in science politics, who, though outside the crucial committee structures, was more supportive of women than the liberal reformer and key Nobel commitee member Svante Arrhenius.[15] The Swedish mathematician had already shown his willingness to acknowledge women scientists in an earlier suggestion to Alfred Nobel that he establish a chair for the Russian mathematician Sophia Kovalevskaia. Nobel, incidentally, refused on the grounds that it was not necessary as 'Russia was less prejudiced', a comment which suggests that the founder saw himself as more open-minded to the claims of women than many of his compatriots. Thus when the nominations were being considered, it was Leffler[16] who became sufficiently concerned that Marie Curie might not be offered a share in the prize to write to Pierre Curie. Pierre replied: 'If it is true that one is thinking about me [for the prize] I very much wish to be considered together with Madame Curie with respect to our research on radioactivity.' The letter then goes on to suggest that giving the prize jointly will be 'artistically satisfying'.[17]

Curie's fame thus depends not simply on her work, and on the general processes through which scientists are recognized, but on the integrity and egalitarian values of two men: one a Swedish mathematician who shared his sister's feminism,[18] the other, her husband and collaborator who shared hers. This story of the recognition of Curie points to the peculiar dependency of a woman scientist, particularly if she is part of a wife-and-husband team, on her collaborator's unequivocal acknowledgement of her contribution. All too commonly the woman/wife's share of the work is only acknowledged by a dedication, and the crucial authorship/ownership is denied in a way that is rarer between men scientists. Without recognition by her husband/collaborator she stays in the private domain, for only he has the power to testify that she is a creative scientist, which will enable her to begin to enter the public world of science. Otherwise the two are one, and that 'one' is the man.

A dangerous combination of love and science   The recently recovered biography of Mileva Einstein Maríc[19] documents the dangerous combination of love and science for women, and its power to render women and their science invisible. After a painful beginning where she conceived a child by Albert Einstein out of wedlock and had the baby adopted, the marriage was initially happy and mutually appreciative. Einstein, for example, explained to a group of Zagreb intellectuals that he needed his wife as 'she solves all the mathematical problems for me'. Two key episodes document the process by which her work, if not actively appropriated, was certainly lost by her to him. In one episode Mileva, through the collaboration with a mutual friend, Paul Habicht, constructed an innovatory device for measuring electrical currents. Having built the device the two inventors left it to Einstein to describe and patent, as he was at that time working in the patent office. He alone signed the publication and patented the device under the name Einstein-Habicht. When asked why she had not given her own name of Einstein Maríc she asked, 'What for, we are both only "one stone" [*Ein stein*]?'. Later when the marriage had collapsed she found that the price of her selfless love and affectionate joke was that her work had become his. She also lost her personal health through trying to do the mathematical work to support his theorizing and simultanously take care of their children. One son suffered from schizophrenia and after the divorce Einstein was mean about keeping up with the alimony.

Troemel-Ploetz[20] points to the even more disturbing episode of the articles published in 1905 in the Leipzig *Annalen der Physik*. Of the five key papers, two of the originally submitted manuscripts were signed also by Mileva, but by the time of their publication, her name had been removed. These two articles, written in what was widely understood as Einstein's golden age, included the theory of special relativity which was to change the nature of physics, and for which he alone received the Nobel prize. Thus although the purpose of the biography was to restore Mileva's name as a distinguished and creative scientist, and not to denigrate Einstein, it inevitably raised the issue of his withholding recognition of Mileva's contribution to the achievement. A number of observers have also commented on the puzzle of Einstein's gift of the prize money to Mileva Maríc even though they were by then separated. This gift-giving was later emulated by George Hoyt Whipple, a Nobel Prize-winner in 1934. Although Whipple had the reputation of being very careful financially, he shared his prize money with Frieda Robsheit Robbins, his co-worker for many years, and with two other women colleagues. In Einstein's and Hoyt Whipple's circumstances, was the money meant to compensate for the system's, and perhaps their own, appropriation of their collaborators' work?

While Mileva's biographer is careful to indicate that Einstein was the creative thinker, she suggests that he could not have realized his theoretical insights without Mileva's mathematics. Between men scientists such a collaboration between theory and technique is rather difficult to ignore; between husband and wife scientists it was – and according to the context still is – rather easy. It was especially so at the turn of the century when bourgeois women, as wives, were only permitted to work as unpaid workers and when scientific work like housework and child care could be constructed – as they were by Mileva – as part of the labour of love. While Trbuhovic Gjuric's biography (not least because it was originally published in Serbian in 1969) has not had the impact of Ann Sayre's study of Rosalind Franklin, it has raised doubts in the physics community;[21] meanwhile feminists will recognize the pattern as characteristic, made possible by that early twentieth century scientific labour market in all its unbridled patriarchal power of appropriation.

Curie's second prize   Although Marie Curie's story is rather happier in the recognition given her by being awarded the Nobel Prize, together with what Crawford[22] speaks of as 'a watershed' of public interest in science aroused by the press reports of the immense effort required to produce radium,[23] its great commercial value, and the philanthropic selflessness of the Curies' attitude to their discovery, none the less the achievement did not give her a clear place in the French scientific establishment. The Academy refused to change its rules barring the admission of women and quite exceptionally for Laureates she was not admitted, although the debate was intense and she lost by only one vote. The Academy, in its profound androcentricity, only admitted women scientists in 1979. Yet the story of the Curies had produced for the 1900s a climate of sympathetic interest in science that would be hard to imagine in the context of the much less confident scientific establishment of the 1990s. The otherwise strait-laced newspaper *Le Figaro* described the Curies' story as a fairy tale, beginning its report with 'Once upon a time . . .', and *La Liberté* wrote, 'We do not know our scientists. Foreigners have to discover them for us.' Science, at least as done by the Curies, was popular, as evidenced by a large audience for Pierre Curie's address to the Royal Institution in London in 1903 and another to listen to Marie Curie defend her doctorate at the Sorbonne in the same year.[24]

But the pleasure from shared work and shared recognition was short-lived; Pierre was tragically killed in a traffic accident in 1906. Suddenly, as a widow and no longer a wife, Marie's scientific eminence was recognized by the University of Paris and she was appointed to the chair Pierre had held.[25] In 1911 she was invited once more to return to Stockholm, this time to receive the Chemistry Prize for the discovery of the elements radium and polonium. But even her apparently triumphant return to Stockholm was marked by gender and sexuality. Arrhenius, ever vigilant lest women should escape their special place, on learning that after Pierre's death Marie had become close to the gifted physicist Pierre Langevin (Langevin's estranged wife cited her in divorce proceedings), wrote to her urging that in order to protect the good name of science, the Nobel Institution[26] and so forth she should not come to Stockholm. With some courage Curie came, supported by her daughter Irène.[27]

On this occasion, Mme Curie's biographical notes as a Laureate extended to two pages, rather then the mere half-page of eight years before, and reported that, among other honours, in 1910 she had been made a member of the Swedish Academy of Sciences. Her portrait too had expanded from the matching small images of her and Pierre in grave impersonal profile with every inch except her neck and face covered with clothing; now the scientist, bare-armed and bare-necked, hand touching cheek, looks thoughtfully out. (See plates 1–3.)

But the Royal Society in London was still not minded to change its conventions. Although the physicists Rayleigh, Ramsey, J. J. Thomson and Rutherford were all both fellow Nobel Laureates and influential Fellows of the Royal Society, the Society felt no need to honour this prize-winning physicist any more than did the French National Academy. Indeed Rutherford was highly dismissive of Curie, persisting in seeing her as Pierre's underlabourer, the scientific and physical effort of extracting radium from pitchblende constructed as little more than an extension of housewifely skills. Given that seventeen (men) Laureates were to come from the Thomson and Rutherford stable, such views were decisive, at least within the British context.

Irène Curie

The right of the scientist widow, but not the scientist wife, to speak was reiterated in 1935 when the next woman to receive a Nobel Prize was also honoured jointly with her husband. Just one year after her mother's death, Irène Curie was awarded the Prize for Chemistry jointly with her husband Frédéric Joliot, for their work on the half-lives of 'artificial' radioactivity. Irène was three years older than Frédéric, and had completed her doctorate by 1925, while he completed his some five years later. Thus whatever the differential of skill between the older and more experienced male researcher and the younger and less experienced female which might be understood to have operated between Pierre and Marie, such a disparity was not operative here. Despite Irène's seniority within the Radium Institute, where the powerful presence of first both Curies, then Marie alone, ensured that the normal gendering of ordination was set to one side, in Stockholm it was considerably restored and only the man spoke. Once again the two Nobel portraits are matched in size but the silhouette conventions of 1903 have yielded to a more obviously gendered representation, in which the masculine gaze challenges the viewer while the feminine looks modestly away. Frédéric, formally dressed, looks directly at his viewer, whereas Irène, bare-necked, thoughtfully looks down, away from hers. (See plates 4–5.)

Although on the prize occasion it was Frédéric's speech which recorded their joint acceptance, in general the naming practices deployed in their publications show a considerable sensitivity to the task of ensuring the woman's ownership in joint scientific work. Thus in most scientific publications they used their given names of Irène Curie and Frédéric Joliot, while in later popular writing and in political life, Frédéric used the name Joliot Curie.[28] Irène also used this version in her Nobel biography. Like many French and British scientists in the thirties and forties the two were active in left and anti-fascist circles. Irène became Minister for Science (the only woman scientist Minister for Science)[29] in the brief Popular Front Government. After the war both, but particularly Frédéric, were active in the peace movement along with Pablo Picasso and Desmond Bernal.

The (scientifically) inexplicable exclusion: Lise Meitner

It is the story of Lise Meitner, as the woman physicist who was not awarded a share of the prize with her scientific collaborator Otto Hahn in 1944, which is probably the most disturbing in the Nobel history. This is not to say that there are not other injustices, as these are inevitable in any selection system, but the Lise Meitner case is outstanding in that it brings together issues of scientific, gender and 'racial' justice in acute form.

Meitner's early years in Austria getting a scientific education[30] show the typical difficulties of young women scientists at the time, but also point to her early recognition as one of the most outstanding physicists of her generation, particularly as an experimentalist. Her long collaboration with Hahn had led to a number of important advances, not least the concept of nuclear fission, proposed in a 1939 letter written with her nephew Otto (Robert) Frisch. But while her intellectual contribution was accepted, in everyday scientific production she was isolated on grounds of gender. Coming to Berlin from Vienna to join Max Planck in 1907 at the Kaiser Wilhelm Institute, she soon began to work with Otto Hahn because of their common interest in radioactivity. Hahn's autobiography reports that:

the Director did not accept women but made an exception in her case. With the condition that she was not to enter the laboratories where the male students were working she was permitted to work with me in the woodshop. In 1907 this was a really large concession . . . In time he also developed an attitude of fatherly friendship to Lise Meitner. But the rule that she had to stay in the woodshop (which was later extended to include another basement room) remained in force.[31]

It was only in 1914, when she was offered a post in Prague and there was a risk of losing this outstanding physicist, that the institute suddenly found itself able to find her a post. The Hahn–Meitner collaboration lasted until the consolidation of Nazi power in the thirties, when, although Meitner continued experimental work until the last minute, as a Jew she had to flee Germany for Stockholm in 1938.

Hahn was given the prize for discovering that barium was a product obtained by bombarding uranium with neutrons; the scientific debate is whether this was a project in which Meitner was an integral partner and therefore should have shared in the recognition, or whether this was a separable development from the joint work. Politically there is the question of what, in 1944, with the Nazi death camps still in operation, the recognition of Hahn meant, for although politics are not meant to play a significant part on the proceedings, it would be foolish to think they played no part. If the Nobel committee had offered the prize jointly to a German physicist and to a Jewish refugee, the Nazi regime would have forced Hahn to refuse. For that matter, recent historical work has suggested that Swedish 'neutrality' was rather more pro-Nazi than had been understood; in consequence, inviting Hahn and refusing Meitner might well have seemed acceptable. It is also a matter of record that Sweden (Meitner's first and far from happy place of refuge) was not free from active anti-semitism. While the government was sympathetic to the plight of the refugees, some sections of the Swedish academic community were openly hostile. In Uppsala, right-wing students, ostensibly on the ground of employment competition, secured a massive vote against admitting the refugees to academic posts.[32]

It is difficult to see just where Hahn stood during the crucial Nazi years in relation either to the Nazis or to the Jewish woman scientist; even the mainstream analyses of the role of the physicists, and Hahn in particular, conflict. Yet the festschrift edited by Otto Frisch for the joint eightieth birthday of Meitner, Hahn and Max Von Laue points to the power of the physicists' invisible college and Meitner's unquestioned place within it.[33] The political scientist Joseph Haberer,[34] for example, is critical of the lack of resistance on the part of the natural scientists to the growth of Nazism, and does not exclude Hahn from this, whereas the historian Alan Beyerchen[35] is less critical, suggesting that many did resist within their capacities, particularly Hahn. Beyerchen goes on to suggest that the physicists did not recognize the power which was to be given them by the role of physics in the 1939–45 war.[36] Yet this charitable view seems misplaced on two grounds. First, other European scientists, not least the refugee physicists, had a very clear view of the potentiality of nuclear physics and actively persuaded the politicians of the decisive power of nuclear weaponry, and there is no clear reason why others in this research community should not have shared this understanding. Second, when during the immediate aftermath of the war the German physicists were interned at Farm Hall in England, and their conversations secretly taped, several (but not Hahn) were to express their regret that the German developments had not been crowned with similar success as the Manhattan project.[37]

The double handicap of being a woman scientist and Jewish in such a desperate political period may well have been too great for Meitner to have been awarded a Nobel Prize, but her scientific standing was recognized in other ways. The Royal Society, in contrast to its earlier lack of enthusiasm for the Curie women, in 1955 elected Meitner to be the first of the five ever women foreign members.[38] When the 1944 archives of the Nobel Institute are opened, with the expiry of the fifty-year rule, there will be more than a passing interest in processes by which the physics and chemistry committees managed to ignore this extraordinarily talented scientist, and by which Hahn felt able to accept the prize alone.

Gerty Cori

Thus it was twelve years between Irène Curie winning the prize and the next woman Nobel Laureate, the biochemist Gerty Cori in 1947. She too was married to her scientific collaborator, Carl Cori, and they were joint recipients of the Nobel Prize for Medicine. Born in Prague, her class background ensured that she was educated at home, except for two years at the lycée. Her exceptional determination was soon evident in that she prepared herself for the university admission examinations. Entering the university at 18, within four years she had received her doctorate of medicine. During this period she met her fellow student Carl Cori. Carl's background, as the son of the director of the marine biology station at Trieste, and the grandson of a Prague physics professor, was strongly scientific; none the less he was an indifferent student until they found they shared a common interest in pre-clinical medicine – initially the biochemistry of blood serum. After working on serum together, they married and moved to the US in 1920, to enable Carl to take up a post as a biochemist at the Malignant Disease Research Institute in Buffalo, New York.

Gerty's biography says that she 'joined her husband at' the institute, which, decoded, meant that she was working as his unpaid assistant. They continued researching and co-publishing even though he was warned that to do so might be damaging to his career.[39] That it might be rather more than 'damaging' to hers was not an issue. Fortunately in 1931 Carl was appointed to a full chair at the University of Washington in St Louis, Missouri, and Gerty's employment status moves into visibility; she is now described as a research associate. The impersonal voice of the biography does not convey the achievement which the silences and tactful formulations hint at. But despite the years of unpaid research, scientifically this was an important period for their collaboration; together they worked on tissue extracts and eventually moved towards the isolation of enzymes. By 1936 they had successfully identified phosphorylase.

Cori's treatment aroused in Harriet Zuckerman a flash of feminist anger: 'she never had a regular academic appointment till she was 53, nine years after she had done her prize winning work.'[40] While they shared a number of honours and awards, mainly after the prize itself, Carl's Laureate biography reports that he was part of the institutions of the scientific establishment: a member of the elite US National Academy of Sciences, of the American Philosophical Society and of the American Academy of Arts and Sciences. By contrast, during the thirties Gerti does not even rate a star in 'American Men [sic] of Science' – the US scientific *Who's Who* – and she was elected to the National Academy long after receiving the Nobel. The fact that the Coris were working in biochemistry was probably key to their scientific success; biochemistry was a new research area and relatively open to women, not least because as a new area it had an inadequate number of permanent jobs attached to it. But while the man's achievement was early rewarded by his being allocated positions in the national network of scientific power, the woman's achievement was rewarded retrospectively and parsimoniously.

Their photographic and self-presentation at the prize-giving are full of the complexity of being a successful partnership and married. Carl's Nobel portrait reproduces the now familiar male gaze; he looks out sternly, seeking commanding eye contact; Gerty's is three-quarter profile – that favoured photographic angle of the forties and fifties which was to be read as feminine and flattering. She wears a jacket with the neck 'softened' by a pearl choker and she smiles pleasantly. (See plates 6–7.) After the bare gravitas of the Curies, mother and daughter, the Nobel Laureate women from henceforth are covered, but almost all offer the reassuring smile of femininity.

The Cori's Stockholm presentation broke new ground as a format, for it was made in three parts. The lecture, strongly technicist in style, was opened by Carl setting out the biochemistry of polysaccharide phosphorylase, carefully describing the procedures which had enabled them to isolate glucose-1-phosphate and thence to examine the mechanism of phosphorolysis. Gerty took over the middle section, moving the technical discussion further, then Carl concluded with the third and final section. What was happening in this theatre of presentation? The middle section is the longest but the lengths of the other two together are significantly greater. Why was this equal but unequal scientific partner only permitted to speak safely contained by the man's voice at the beginning and closing? What delicate compromise between an earlier wifely silence and a postwar construction of equality and difference was reached in this lecture division?

Appropriation and erasure: Rosalind Franklin

Between the awards to the biochemist Gerty Cori and the physicist Maria Goeppert Mayer was the triumphalist story of DNA and its soon-to-be revealed subtext of the appalling treatment of the X-ray crystallographer Rosalind Franklin. The account of the erasure of this outstanding woman scientist and the appropriation of her work was told to a wider audience in 1975 in the biography by Anne Sayre, who along with her crystallographer husband was a personal friend of Franklin's. Sayre's book made public the grave disquiet felt among the crystallographic community,[41] and was received within a political climate newly sensitized by an increasingly powerful women's movement. The story is brief, as was the life of this scientist who died at 37 of cancer. Born into a well-off North London Jewish family, Rosalind Franklin was sent to St Paul's, a fee-paying girls' school which prided itself on the educational performance of its pupils. She went to Cambridge to read science, did postgraduate work on the physical chemistry of coal, worked with the crystallographer Marcel Mathieu in Paris and then accepted a post-doctoral fellowship in the department of biophysics in King's College, London. The laboratory was one of a number interested in the structure of the giant molecule of DNA, which was already thought to be associated in some way with the genetic mechanisms of heredity, and both Franklin and another scientist, Maurice Wilkins, were engaged in making X-ray diffraction photographs of the rather intractable DNA crystals.

The relationship between the two was far from cordial; a matter not made easier by the anti-woman atmosphere at King's, which in the 1950s still excluded women from the common rooms as a matter of course; by the failure of John Randall as the head of department to clarify the lines of authority between the two researchers; and by the assumption of Maurice Wilkins that the woman scientist, who had more technical experience, was in some automatic sense his junior.[42] Lastly, Rosalind Franklin was regarded by a number of her contemporaries as a 'difficult' woman.

While feminism has commented with some sophistication on the construction of 'difficult women', not least in the context of independent and creative women such as Franklin, there has been little discussion in this otherwise much examined story concerning the extent of anti-semitism in educational institutions during the immediate postwar period, and what this meant to any Jewish person with a sense of cultural identity. We know that Rosalind Franklin and her family had such a sense. During the war her father worked with the Jewish Board of Deputies to help refugees, and she helped too during school holidays. At Cambridge she had become friends with the metallurgist and French Jewish refugee Adrienne Weill, who was responsible for Franklin working in Paris with Mathieu, who as a communist had egalitarian attitudes to women scientists and was a committed anti-fascist.

Coming to King's must have been something of a shock, not least after Mathieu's laboratory, for not only was King's very much a male bastion, it was also a bastion of the Church of England. The origins of King's were as a Church of England college established in direct opposition to University College, which had been founded by the Utilitarians to provide university education to unitarians, free thinkers and Jews. Androcentricity and Christian ethnocentrity were thus the twin hallmarks of institutions such as King's. But Christian ethnocentricity in the forties and fifties was not simply a matter of exclusionary or even hostile speech practices; there was also institutionalized anti-semitism, not least in education. A number of direct-grant schools, particularly those in areas where there was a considerable Jewish community, had a Jewish quota to prevent the stereotypically clever Jews flooding out the Christians. In the discourse of the time, Christians as the privileged group were unmarked; marking was reserved for the Jewish others. Nor was anti-semitism limited to negative speech and institutionalized exclusion; it also took violent forms, particularly in areas where the poorer sections of the Jewish community lived. Despite the death camps and the war, anti-semitism was still a virulent force on the streets and a taken-for-granted aspect of everyday British life.

Most of these cruder forms of anti-semitism faded as the objects of racist abuse were changed. The advent of the Caribbean and Asian migration into Britain resulted in Jews being replaced for some years as the scapegoats of racist fears. Because replacement rather than resistance weakened it, the phenomenon of anti-semitism within cultural life remains under-explored, but it was there for Jewish men and women who found a number of elite educational institutions difficult places to study and to work at. To be a woman scientist and Jewish during the immediate postwar period in any laboratory where there was no counter-ideology was to carry a double burden, none the less real for not yet being fully named. It is doubtful if it is even healthy not to be 'difficult' in such a situation.

In the context of the DNA project, success required the collaboration between theoreticians, or model builders, and experimentalists, who would take the X-ray diffraction photographs to provide the empirical evidence to sustain the models. The former, Francis Crick and James Watson, were based in Cambridge and the latter, Maurice Wilkins and Rosalind Franklin, in London. The crux of what was increasingly seen within crystallography as a shabby affair was that Franklin had made the key photographs which clearly indicated the helical form, but that these had been taken, without her permission, by Wilkins to show to the two Cambridge men with whom he was collaborating. In addition a Cambridge colleague, Max Perutz,[43] who was on the Medical Research Council committee which had received Franklin's research report, also showed this privately to Crick and Watson. Although the crucial papers published in *Nature* included one by Franklin and her colleagues, she did not know just how important her photograph had been to Crick and Watson.

For this and other reasons the situation at King's became intolerable and was resolved in the usual way; the woman, not the man, moved. Franklin went to work at Birkbeck with the crystallographer Desmond Bernal. Bernal's communism, like that of Mathieu, meant that his laboratory was a more congenial environment in which to work. She stayed there until her death, with Bernal writing her obituary memoir.

Thus Franklin was already dead when the Nobel Prize in Medicine and Physiology for the DNA work was awarded to Crick, Watson and Wilkins in 1962. Despite the centrality of her contribution, none of the Laureates made a reference in his Stockholm address to her published papers, and Wilkins only spoke of her in very general terms. In Jim Watson's best seller *The Double Helix*, written several years after both Franklin's death and the award of the prize, Rosalind Franklin appears as a bad fairy in the Watson fantasy of himself as artless young man stumbling on the double helix. Despite the enthusiasm shown by a number of men scientists for the 'Jack the Giantkiller' quality of Watson's book,[44] Crick considered suing him. Wilkins would have gone along with the action, but the matter was dropped. Similarly the London Science Museum's construction of the DNA story erased Franklin's contribution until her crystallographer friends and colleagues protested and ensured that her work was acknowledged. However, it was not until Sayre questioned Wilkins directly in 1970 as to the probity of taking the photographs to Cambridge that the masculinist appropriation of the work and the erasure of the woman scientist came into full view.

The interesting and unanswerable speculation must be what would have happened if Franklin had not died, given that the prize can by tradition only be shared between three, and that it was her photograph which provided the critical empirical support to the double helix model. For Franklin herself, gender, 'race' and cancer colluded to diminish her contribution, yet the combination of personal and scientific friends speaking out in the context of a rising women's moment, has meant that her name has become a warning beacon for any who contemplate the erasure of women scientists.

Maria Goeppert Mayer

One year after the DNA awards, Maria Goeppert Mayer, born in Kattowitz, was awarded the Nobel Prize in Physics. Although she was married to and had earlier collaborated with a physicist husband, the prize was awarded for her individual theoretical work on a model for the structure of the atomic nucleus – 'the shell model'. The prize was shared between Goeppert Mayer and Jensen, who had simultaneously but separately developed the model. While previous biographies are in the third person, this time the voice is personal – and the confidence of a privileged class background and a trained and recognized intellect mark every line. In a perverse rereading of Galton's view of hereditary genius, she describes herself solely on her father's side, as 'the seventh straight generation of university professors'.[45] Her father was a professor at Göttingen, where until her marriage she had spent most of her life and where, she says, in both the private and public schools she had very good teachers. Although it was accepted by both her parents and herself that she would go to university, she notes that it 'was not trivially easy for a woman to do so' and tells us how she attended a special school in Göttingen which prepared girls for the examination but which was forced to close because of the inflation of the time – although the teachers continued to instruct their pupils.

None the less Göttingen in the 1920s was a central institution of the golden age of German physics, a golden age in which many German Jews played a rich part and which was destroyed with the advent of the Nazis in 1933. Goeppert Mayer's autobiography is fully conscious of that special time and her acceptance within it; she both acknowledges her debt to Max Born for guiding her scientific education and recalls that along with Born on her doctoral committee there were also Franck and Windhaus, all three Nobel Prize-winners. It was Born, a family friend and, in these early years, among those men able to recognize great ability in a woman, who opened the door. During this period Maria met and married Joseph Mayer, a US physicist working for a year with James Franck, and returned with him to the States where he had a post at Johns Hopkins. With a certain matter-of-factness her biography reports, 'This was the time of the depression, and no university would think of employing the wife of a professor. But I kept working just for the fun of doing physics.'[46] She worked with her husband and with Karl Herzfeld and published a number of papers – which remain classics – in chemical physics.

Her husband, in a piece written for a glossy[47] put out by the Nobel Institution on the occasion of the 1963 prize, speaks of Herzfeld's kindness in arranging an assistantship for her. Mayer spends some energy on complaining about his own career in that in 1938 his Baltimore appointment was terminated. He continues with more than a little self-congratulation, 'Fortunately I soon received several offers and I had the gratification of being able to resign before the specified time limit to go to Columbia with a considerable salary increase.'[48] (While it may well cross the reader's mind that there is no physics for fun for Joseph, it is also interesting to speculate on what a similar article by a scientist wife of a Laureate – such as Mileva Einstein Maríc – might contain.) Maria Goeppert Mayer's job was at this time teaching at Sarah Lawrence College and researching with some difficulty at the atomic laboratory directed by Harold Urey. Despite the need for skilled physicists in atomic bomb research, Urey seems to have ignored her talents, directing her to work on side issues. Joseph reports this experience more positively, in much the same tone that he uses to describe Maria's equipment-building contribution to experimental physics, as being part of the ideology of romance when physics was done with 'love, string and sealing wax'.

There are two images of Goeppert Mayer published by the Nobel Institution; in the first she is answering a phone, looking like a busy, unfussed professional woman at work. All that is unusual is that there is a big bunch of roses in the foreground, as she has just received the news of the prize. The formal portrait is very different – here she smiles out, again in the glamourizing three-quarter profile, with meticulous lipstick and carefully waved hair, looking away from us. (See plate 8.) Femininity rather than scientific authority is portrayed.

For Goeppert Mayer herself, it was only when they moved to Chicago in 1946 that the magic of her early life in physics was recreated. 'This', she says with the clarity of the at last secure and recognized, 'was the first place where I was not considered a nuisance, but greeted with open arms. I was suddenly a Professor in the physics department and in the Institute for Nuclear Studies . . . with very little knowledge of Nuclear Physics!' What she does not record is that because of the nepotism rule this was an unpaid chair. It was only Argonne, as the defence establishment, which was willing to pay a half salary. She goes on to thank Edward Teller and especially Enrico Fermi for their help. L. M. Jones, a feminist historian of physics, sees this fulsome expression of gratitude to Fermi and informal style as enabling hostile critics to diminish her achievement.[49] Yet to argue this is not to recognize the extent to which the scientific elite values one another's judgements more than those of anyone else. Jensen, her co-theorist and co-prize-winner, explicitly recognized this, saying, 'I have convinced Heisenberg and Bohr, you have convinced Fermi. What do we care about the others?'[50]

Now aged 57, looking back in her Nobel lecture on the research which led to her prize, she says very little about the militarization of physics which facilitated her recruitment, but only that her model initiated a large field of research. It has served as the starting model for more refined calculations. There are enough nuclei to investigate so that the shell modelists will not soon be unemployed.'[51] What could be more different than the physicist Richard Feynman's account of his sexualized relationship to old theory?[52] To read Goeppert Mayer is to read the account of someone who knows both that she is amazingly talented and yet that she is in some sense lucky to be accepted by the gods of physics. Never for a moment does her precarious confidence convey that unambigous certainty of the elite men physicists. But then hers is a not unrealistic reading of her situation.

Dorothy Crowfoot Hodgkin

Although Goeppert Mayer mentions her mother – by contrast with many men Nobel Prize-winners, who seem either to have made the remarkable biological, to say nothing of social, achievement of being the sons only of men, or alternatively to have sprung fully formed from graduate schools as from the brow of Zeus – her origin story is one of a male lineage. It was left to Dorothy Crowfoot Hodgkin, awarded the Nobel Prize for her work in crystallography the following year, 1964, to claim her mother as a person in her own right,[53] noting that she had been involved in her husband's work as an archaeologist but that she 'became an authority in her own right on early weaving techniques.'[54]

At school in Beccles in Norfolk with her sisters, Dorothy Crowfoot was permitted to do chemistry with the boys at a neighbouring school. Her account moves graciously between the scientific and the personal, between her devotion to Margery Fry, principal of Somerville when she went to Oxford, her gratitude to an aunt for providing for her financially, and her appreciation for the academic guidance of Professors Robinson and Hinshelwood and the influence and friendship of Bernal and others. She talks of her historian husband and their talented offspring. Her Nobel lecture begins by describing the happy chance of meeting X-ray crystallography through reading, when she was 15, the book by the pioneer crystallographer W. H. Bragg: *Concerning the Nature of Things*, and goes on with a mixture of modesty and certainty to claim her own place while acknowledging her collaborators, 'without whose brains, hands and eyes very little would have been done'. While not unconscious of the honour of being awarded an unshared Nobel Prize, she recalls, when she first heard of the news of the prize when in Ghana with her husband, her sense of sadness that it was not shared with any of her close colleagues, particularly with Bernal who had been her mentor, friend and political and scientific ally.[55] Her extraordinary sense of confidence, as a scientist and as a woman, radiates from both her biography and her lecture. While originating in her biography of class and family, it is indisputably underpinned by the immense scientific recognition she had achieved at the young age of 37, when, in 1947, she was elected as one of the earliest women Fellows of the Royal Society for her work on the structure of penicillin (see the previous chapter). In the 1930s the Royal Society was not quite the bastion of political conservatism that it became in the 1970s and 1980s, and Hodgkin benefited from belonging to a particular generation of scientists, many of whose leading figures, especially those from Cambridge, were politically radical. For this generation, accepting women as having equal potential was a matter of socialist commitment; however, working to secure the advance of women was not.

When Hodgkin went to Cambridge as a postgraduate student in 1932, working for two immensely productive years under the supervision of Desmond Bernal, she found herself in a group which was a ferment of scientific and social ideas. When she returned to Oxford in 1934 she maintained her Cambridge connection, and as the social and political crises of the 1930s deepened she became a member, along with Dorothy Needham and Marjory Stephenson, of the Cambridge Scientists Anti War Group (CSAWG) in which Bernal was a leading figure. Anti-militarism has sometimes been misunderstood as a grouping synonymous with pacifism,[56] but although pacifism after the horrors of 1914–18 had unquestionably a strong presence, anti-militarism was intermingled with the class politics of opposition to nation-state war but support for 'class war'. CSAWG's efforts in showing how the government was making no attempt to defend the civilian population of Britain in the context of the coming air war was a project around which pacifists, socialists and communists could unite. When war came in 1939, the group members put aside their anti-militarism and became directly involved (among leading women scientists of this generation, only Kathleen Lonsdale remained a committed pacifist). They worked on the Manhattan project; on radar; on code cracking; on Mulberry Harbour (the floating docks which enabled the allied forces to land on the Normandy beaches); and on cures for wounds, of which penicillin, being developed in Oxford by Howard Florey, Edward Abraham and Ernst Chain, was the best candidate. Working on the structure of penicillin, as a biologically and socially significant molecule, represented a synthesis of Hodgkin's science and her political and social concerns. As a mother – and she lists children among her pleasures in *Who's Who* – she was not unaware of the capacity of penicillin to save child life as well as deal with infected wounds. Penicillin was followed by her work on the structure of vitamin B_{12} and insulin; her choice of molecules reflects a distinctive and consistent fusion of her social and scientific concerns. Yet with characteristic understatement Dorothy denies a commitment to feminism, while the genders have always been approximately evenly represented in her laboratory. She similarly refuses any suggestion that being a woman influenced her subject choice.

Yet Dorothy's spoken claims, or rather lack of claims, are very different from the messages from her practices, which created a laboratory that was an oasis of gender justice, and put her scientific skills to serve the people and swiftly to join her name to denunciations of the misuse of science in military aggression. Although she is not a religous person, her moral integrity constantly reminds me of the finest Quaker tradition among women, in which actions indeed speak louder than words. (The fact that everyone, at all levels, addresses her simply as Dorothy sustains this perception.) Talking about this lifetime of actions, so expressive of what my generation has claimed as feminist values in science, gave a sense of echoes between these interviews and those by Evelyn Keller with Barbara McClintock. While there was an obvious and important difference between US and British scientists of their generation, namely that the US scientists were not particularly radicalized by the thirties whereas the British were, there was also a similarity between these women. Their construction of their sciences offered hopes for the new, less violent science feminism longs for. In Hodgkin's case, this is a science serving human ends; in McClintock's, a holism which refuses the reductionism of DNA as the macho molecule.

As both a wife and a mother (unlike the single and sexually threatening Rosalind Franklin) and hence with both her scientific gifts and her sexuality safely defined, Hodgkin was admissible as a comrade and an equal into the scientific community of honorary men.[57] Is it by chance that, despite the photographic fashions, her portrait looks straight out at the viewer, her compromise with femininity, apart from the modestly waved hair, only the friendly slight smile? (See plate 9.) On the occasion of the prizegiving, she was 54, slightly younger than Goeppert Mayer, but with many more years of institutionally acknowledged success behind her.

In addition to being awarded these great scientific honours, Hodgkin was also the Chancellor of the University of Bristol, the first woman to be appointed to such a post on the basis of her academic distinction. (The only other women chancellors are members of the royal family, whose ascribed social status but lack of scholarship or any other achievement subtly worsens the politics of gender.) One prize was denied her. Fellows of the Royal Society report that there was at one time much discussion within the Society as to whether it should overcome its three centuries of androcentricity and elect her to the Presidency, with all that this post carries in terms of cultural influence. Needless to say, as I discussed in the last chapter, the gender conservatives won out and both the male lineage and the inner citadel of the new masculine knowledge were preserved. Yet despite this particular story Hodgkin is unique among women scientists this century, even among the women Laureates, in being accorded both peer recognition and great institutional recognition while she was at the height of her powers.

Rosalyn Yalow

It was thirteen years before a woman was again awarded the Nobel Prize. Rosalyn Yalow was the first woman Laureate who did not come from the world of inherited material or cultural wealth. Neither her mother, a childhood immigrant from Germany, nor her father, first-generation American, had even a high-school education, but like many Jews of their generation they were deeply committed to putting their children through college. Yalow's picture of her family world is one of an affectionate and mutually supportive environment. She makes her mother real to us by citing her long-standing joke that it was just as well her daughter's determination was fixed on an acceptable goal, as there was no way she could be deflected. Yalow describes a family world of few domestic cultural resources, but one which none the less encouraged her and her brother to find in New York's public school and library system the resources they needed.

She tells her story of her growing fascination with science and her educational development with something of gender reversal of the artless young man. 'By seventh grade I was committed to mathematics. A great chemistry teacher at Walton High School, Mr. Mondzack, excited my interest in chemistry, but when I went to Hunter [then the City of New York women's college], my interest was diverted to physics . . . especially by Professors Herbert N. Otis and Duane Roller.' In her praise of the New York public education system Yalow both echoes a number of her male counterparts who went on to win Nobel Prizes, and also conveys something of the excitement of science in the late thirties: 'It seemed as if every major experiment brought a Nobel Prize.' Eve Curie had just published the biography of her mother, which, Yalow says, 'should be a must on the reading list of every young aspiring female scientist'. She also describes the excitement of hearing Enrico Fermi speaking at Columbia on the newly discovered nuclear fission.

Yalow's entry into physics at graduate school was made the hard way, and it is the toughness of the social and economic environment as well as the special problems of women which her biography reports in unsparing detail. Her family thought that a safe job teaching in elementary school made the best sense; she had neither a husband to depend on, nor a wealthy father, and no monied aunt. The best idea that sympathetic academics could come up with was a proposal that she learn to type and become a part-time secretary in one of the university departments, and in this way secure backdoor entry into the crucial graduate courses. The secretarial course had clearly not been the only iron in the fire, as while she was on it she learnt that she had been offered a teaching assistantship at Illinois. Yalow was able to abandon typing and, again with the help of government-provided free tution in physics, prepare herself for her future.

At Champaign-Urbana in 1941 she found that she was the only woman among its 400 engineering staff. 'The Dean', she writes, 'congratulated me and told me that I was the first woman there since 1917. It is evident that the draft of young men into the armed forces, even prior to American entry into the war, had made possible my entrance to graduate school.' Unlike the account offered by Hodgkin, Yalow does not hide the problems that women scientists faced in the forties. She reports being supervised by Dr Maurice Goldhaber, later director of Brookhaven National Laboratories, and getting much encouragement from Dr Gertrude Goldhaber, to whom he was married. Although a distinguished nuclear physicist, Gertrude Goldhaber could not get a university position because of the nepotism rule.[58]

Rosalind Yalow's graduate-college story reflects both her pleasure in the intellectual demands of physics and an immense capacity for work, a capacity which then and later enabled her to use and/or create whatever chances there were. Her account of her marriage to Aaron Yalow, whom she met on her first day at the graduate school and married a couple of years later, hints at a number of subtle ways in which this relationship deviated from the dominant patterns of the time. She tells us for example that she returned to New York ahead of him, as her thesis was completed first; both his and her careers were taken seriously. By 1945, when they were both settled in New York, she notes that 'a fulltime teaching job . . . and a small house in the Bronx were hardly enough to occupy my time fully.' As her husband had moved into medical physics she was able, using his contacts, to meet a leading woman in this area, Edith Quimby, and through her good offices enter what was to become the medical radioisotope service. Yalow moved across to research on the applications of radioisotopes in blood-volume determination, the clinical diagnosis of thyroid disease and the kinetics of iodine metabolism. That she now had only two rather than three activities was maybe why she was able to write that 'during that period Aaron and I had two children'.

In her research work she began a collaboration with Solomon Berson which was to continue until his death in 1972. Yalow acknowledges Berson in a very distinctive way. He is more than a collaborator and friend; he is nothing less than the co-parent of radioimmunoassay (RIA). 'Together we gave birth, and nurtured [it] through its infancy . . . Would that he were here to share this moment.'[59] After Berson's death she arranged for the laboratory to be called the Solomon A. Berson Laboratory, so that her scientific papers would not go out without his name. Like the widow who keeps her husband's name in the telephone book, Yalow refuses to let Berson's name disappear from her life.[60]

She takes pleasure in the unconventionality of her and Berson's career, and says that neither followed a systematic post-doctoral training period, but taught one another. She suggests that their powerful technique of RIA came about 'not by directed design but more as a fall-out from our investigations into what might be considered an unrelated study'.[61] Following the hypothesis that maturity-onset diabetes might not be due to an insufficiency of insulin secretion but rather to abnormally rapid degradation of insulin by hepatic insulinase, and finding classical immunological techniques were inadequate, they pioneered the radioisotope technique to detect the presence of soluble antigen-antibody complexes. Using these techniques they were able to demonstrate the 'ubiquitous presence of insulin binding bodies in insulin-treated subjects'. The problem, as Yalow flatly reports it, was that 'This concept was not acceptable to the immunologists in the mid fifties.'[62] She takes some satisfaction in demonstrating this unacceptability and tells how the paper reporting her work was submitted to *Science* and rejected, then submitted to the *Journal of Clinical Investigation* and rejected initially there too. Yalow, not one to miss the turn of the screw, reproduces the rejection letter, dated 29 September 1955, from the editor. While such rejections are far from unknown (the paper reporting what was to become the Krebs cycle was rejected by *Nature* in the 1930s), it is rare for a woman scientist to take such public pleasure in being proved right. Yalow's definition of feminism is part of a liberal feminist project which means taking the men on at their own game and beating them. Her complaint is against the inequality of opportunity, not the game itself.

She has a strong sense of the importance of her work; authority, it seems, doth become a woman too. The final sentence of her Nobel lecture concludes: 'The first telescope opened the heavens; the first microscope the world of the microbes; radioisotopic methodology, as exemplified by RIA, has shown the potential for opening new vistas in science and medicine.'[63] The names against which she locates her achievement and experience are those of van Leeuwenhoek and Galileo. Yet in her Nobel portrait, formally dressed and carefully made up, Rosalyn Yalow smiles; the puzzle is whether we are to read the smile as in some way claiming femininity and propitiating the male gaze, or as a smile of triumph merely dressed in the trappings of the time. (See plate 10.)

Overdue recognition and its social and scientific implications

The next three women Laureates were awarded prizes for work which they had done between forty and thirty-five years earlier. Zuckerman's general point that women scientists are recognized later for their work is now made almost grotesque. Very few men, other than the ethologists Konrad Lorenz and Niko Tinbergen, have received prizes in their seventies; and their late recognition was intended to flag the new field of ethology, which was seen as of great scientific interest but which the rules had hitherto precluded. In fact little was said about the new field at their prize-giving ceremony. Tinbergen used the occasion to ramble on about the Alexander method while Lorenz chose to explain/explain away his erstwhile support for the Nazis. (Actually he was rather more active than he indicated in his speech, as he was a member of the Nazi Party, a detail which the revisionist history of science omitted in his obituaries.) By contrast the three women prize-winners neither wandered into therapeutic enthusiasms nor used the occasion to explain away unfortunate political associations. They are intensely professional, each speaking technically and elegantly about her science. Only Rita Levi-Montalcini directly confronted the time gap between her science and its recognition, but not even she, on the occasion of the prize or in her subsequent biography, chose to examine the social and scientific meaning of her late recognition.

Barbara McClintock

The first of the three, Barbara McClintock, presents her biography in a highly detached manner, only touching the events which were 'by far the most influential in my scientific life' in an enigmatic text. None of the social or economic sensitivities which scatter both Hodgkin's and Yalow's biographies appear in this intensely impersonal account. McClintock comes into the world as a student, attending the only course in genetics open to undergraduates at Cornell. For a more intimate account of her childhood and young womanhood we have to read Evelyn Fox Keller's widely read and highly sympathetic biography, which was published just before the prize was awarded.[64]

In both her autobiography and Keller's study we are given a picture of an unusually independent and intellectually purposeful young woman; thus by the time she graduates her research direction is set. Whatever problems there were for some in the economic climate of the 1920s, her self-account gives away nothing. Unlike the earlier generation of US Nobel Laureates, McClintock was American-born. Unlike Yalow and Elion she came from a privileged background and despite the harsh times was essentially naive socially and politically, so that her research fellowship to Germany in 1933, where she encountered Nazism and Aryan genetics, was traumatic, and she fled back to Cornell.

Scientifically the biography is a story of a coherent intellectual and academic trajectory, unusual among women and only achievable where women are either without children or have such resources that others take adequate care of them. She reports that she completed her PhD and began a collaborative study locating maize genes to the appropriate one of the ten maize chromosomes. It is as if it is at this point that her history as a scientist begins, and it is the only moment where the dry impersonal prose becomes suffused with the warmth of remembered friendship: 'a sequence of events occurred of great significance to me. It began with the appearance in the fall of 1927 of George W. Beadle (a Nobel laureate) . . . to start studies for his PhD degree with Professor Rollins A. Emerson.' She then goes on to describe the close-knit group which grew up and which drew in any interested graduate students. 'For each of us this was an extraordinary period . . . Over the years members of this group have retained the warm personal relationship that our early association generated. The communal experience profoundly affected each one of us.' We are, from very early on in the autobiography, flagged that this scientist is working as an accepted group member within an elite setting.

Despite a widespread reading of Keller's biographical study as implying that in some way McClintock was not adequately recognized in science, there is little solid evidence of this, except that she did not receive the accolade of a Nobel Prize until she was 81 (perhaps not insignificantly, shortly after the publication of the acclaimed Keller biography). Yet McClintock had long been an acknowledged member of the scientific elite, and she was, as Keller points out, early spoken of as a 'genius' – a compliment which is more rarely made by one scientist about another than by the media. She was the third woman to be admitted to the National Academy in 1944, when she was 42, for the work for which the Nobel committee honoured her almost forty years later in 1983. At the time of her election to the National Academy there were, despite the scale of the US scientific community, rather under 1,000 members; thus the distinction of recognition is considerable. An early recipient of the Association of American University Women's prestigious prize, she had no less than twelve honorary doctorates, from Rochester in 1947 to three in 1983, the year she won the Nobel Prize. Such a biography speaks of McClintock's extraordinarily self-sufficiency as part of the small ultra-elite within science. Such people are rare – perhaps particularly so among women, for whom having sufficient privacy in which to be creative is more commonly a problem.[65] McClintock's isolation was not entirely self-chosen, for her work was not easy to communicate and her ideas on the mobility of genes within each chromosome ('transposition') commanded little support. Despite her early recognition, for many years she was relatively isolated in her Cold Spring Harbor laboratory, but – and it is an important but – never without research resources.

Keller interprets this isolation as a problem of language, of the difficulty that McClintock experienced in trying to communicate what she 'saw'. Keller argues that 'seeing' is crucial to many intensely creative scientists; the problem is that appealing to the 'seen' when there is no pre-existing understanding about what is out there to be 'seen' cannot provide empirical support. In this situation, when the geneticist Joshua Lederberg observed that 'the woman is mad or a genius', he was only articulating publicly what many geneticists more privately thought.

But while Keller lets the reader share the scientist's self-doubt at her failure to communicate her theories to her satisfaction, the outside world had the strong suspicion that she was a genius, and scientific honours continued to be bestowed on her, from a non-residential chair at Cornell in 1965 to the Kimber Genetics Medal of the National Academy in 1967 and, in 1970, the National Medal of Science. By the mid-seventies her ideas about transposition, which potentially challenge the central dogma of the fixity of the genome, began to be understood more widely and became influential in shaping the directions of new work. By this time, conventional molecular biological wisdom had already begun to question the earlier seemingly inviolable concept of the stability of the genome, not in the sense that there was a challenge to the understanding of genetic reproduction, but that the genome itself can, under a number of conditions, undergo rearrangement. There was considerable excitement about such 'jumping genes' as the flexibility they gave was seen to endow their bearer, whether the salmonella bacteria or maize, with a distinct evolutionary advantage. By transposition McClintock wished to draw attention to the general occurrence of cellular mechanisms which restructure the genome, mechanisms which are called into action by external or internal stress. DNA, far from being the stable macho molecule of the 1962 Watson–Crick prize story, becomes a structure of complex dynamic equilibrium. Such a complex dynamic structure has echoes of Laura Balbo's quilt-making metaphor to describe women's work in maintaining everyday life.[66]

A number of critics have suggested that Keller's account excessively celebrates McClintock's mysticism as if this was some undeclared dimension of femininity, or essentialist feminism, yet such criticism diminishes the very real difficulties in talking about the creative process, of understanding how an alternative vision is developed, how it is possible to 'see' something not seen before in nature. The brave attempt by Koestler with his book *The Sleepwalkers*, and the autobiographical accounts of scientists from Einstein to Richard

Feynman, go some way towards discussing this process, but Keller attempts to make clearer what Dorothy Hodgkin spoke of when she thanked her colleagues for their 'eyes, brains and hands'. However, part of the charge of mysticism lies in McClintock's distinctive relationship with nature itself, for her conception constitutes a return to an earlier tradition when nature was seen as active, not passive. Vitalism, however discussed, in the context of the macho reductionist language of contemporary molecular biology (which I describe in more detail in chapter 8), with a nature drained of all subjectivity, would be all too likely to sound like mysticism.

The detached style of McClintock's Nobel lecture makes no genuflections to the occasion, expressing neither pleasure nor gratitude; she neither notes the delay between the date of her work nor its subsequent recognition – yet this could be read as a matter of forty years. The nearest she gets to Yalow-like celebration of the certainty of her vision is when she reports offering 'my suggestion to the geneticists at Berkeley who then sent me an amused reply. My suggestion,' she says rather mildly, 'however, was not without logical support.' The lecture, essentially an overview of her work in genetics, describes the crucial experiments, almost entirely during the 1940s, showing how 'a genome may react to conditions for which it is unprepared, but to which it responds in a totally unexpected manner.' Her view of future research is that 'attention will undoubtedly be centred on the genome, and with greater appreciation of its significance as a highly sensitive organ of the cell, monitoring genomic activities and correcting common errors, sensing the unusual and unexpected events, and responding to them, often by restructuring the genome. We know nothing, however,' she concludes, 'about how the cell senses danger and initiates responses to it that are often truly remarkable.'[67] This activist conception of the cell is kin to Lovelock's Gaia, but where he seeks a popular audience, she is primarily concerned with her invisible college.[68]

McClintock's photographic portrait is of a piece with her prose. Despite the grandeur of the occasion her portrait shows her wearing the uniform of East Coast women intellectuals, a shirt collar over a woollen jersey making no concessions. She looks away from the camera as if she is really looking at something else; her lined face has a slight, detached smile. (See plate 11.)

Rita Levi-Montalcini

Nothing could be more marked than the contrast with Rita Levi-Montalcini's portrait, which speaks of an agreement between photographer and subject – that this is to be an exceptional statement. (See plate 12.) And indeed she does cut an exceptional figure amongst women scientists. Dressed with silken elegance, she poses with her hand to her chin. On her wrist is a rich bracelet which acts, and has been chosen to act, as a foil to the eyes. Everything about her conveys a theatrical consciousness of her beauty and her presence. While the autobiography she writes does not begin, as did a rather earlier one, 'I was crawling out over the palace roof to rescue my kitten',[69] the world of high culture and wealth is evident in every aspect of her presentation of self and work.

She describes an intellectual dynasty of mutually admiring and affectionate people. Her father is described as a 'gifted mathematician,' her mother 'a talented painter and an exquisite human being'.[70] Her three siblings are all named and praised either for their achievements or, if these are not particularly evident, for their good taste. While she describes a domestic world governed by the father, not least in terms of secondary education, where he held strong views about the suitable subjects for girls, the larger context of the Italian university system had different and more liberal traditions of bourgeois women studying and researching from those of the Anglo-Saxon one. As a teenager Rita describes herself as isolated, directionless, uninterested in young men, and spending her time reading Selma Lagerlöf. From a very early age her construction of her own femininity excluded wifehood and motherhood: 'My experience in childhood and adolescence had convinced me that I was not cut out to be a wife. Babies did not attract me and I was altogether without the maternal sense so highly developed in small and adolescent girls.'

The death of a loved governess turned her towards medicine, and together with her cousin Eugenia she set about preparing herself for university admission. She gives a graphic account, not unlike a story from the eighteenth-century Edinburgh medical grave-snatchers Burke and Hare, of the means by which research students of the brain gained access to human material.[71] She describes travelling on a Rome bus with the corpse of a two-day old baby wrapped inadequately in newspaper. Her reflections as she sees that a small foot is sticking out are solely those of embarrassment from the construction that might be placed on the sight of a young woman carrying a dead baby. The lesson that she derives is not to carry such experimental material on public transport, but what is interesting is the confidence – not to say arrogance – that permits the retelling of this story, without any reflections on either the nature of the material, or how it had been secured.[72]

During the 1930s class privilege was only a partial protection from Italian fascism. Her increasingly tenuous place as a Jewish woman scientist in a developing fascistic context became non-viable once Mussolini's 1936 manifesto against Jewish scientists and professionals had been declared. Now the family was left with 'two alternatives . . . to emigrate to the States, or to pursue some activity which needed neither support nor connection with the outside Aryan world where we lived. My family chose this second alternative. I then decided to build a small research unit at home and installed it in my bedroom.' She then describes how the Jewish biologist Giuseppe Levi, who at university had taught both her (and also Salvador Luria and Renato Dulbecco, both Nobel Prize-winners and her lifelong friends), came to work with her as the universities gradually expelled the Jews.

As the situation became more stringent, even this existence, a scientific Garden of the Finzi Continis, could not be continued. After 1943, Italy was occupied by the German army and the family went underground; where Italian political culture did not take anti-semitism entirely seriously, the German did. In 1945 Rita Levi-Montalcini and her family returned to Turin where she was restored to her university post. By 1947 she was involved in collaborative work with the St Louis based Viktor Hamburger, a collaboration which lasted thirty years. During this period she held a professorship at the University of Washington from 1956 to her retirement in 1977. With the enthusiastic support of the Italian Science Research Council she established a research unit in Rome in 1962, and divided her time between the two continents. This engagement in the science of both countries may have cost her something at the US end, and certainly her biography, unlike McClintock's, lists few scientific honours, but it did ensure a strong Italian lobby for her Nobel award, and there was long and open discussion about how signficant this would be for the morale of Italian science.[73] Undoubtedly *Unita*, the Italian Communist Party newspaper, long anticipated her Laureateship, referring to her as 'our Nobelist'.

The title of her Nobel lecture, 'The Nerve Growth Factor: Thirty Five Years Later', makes her scientific claim and political point rally. She then provides a historical perspective so that the audience may share the frustrations experienced by experimental embryologists of the 1940s, despite an earlier period, during the 1920s and 1930s, which had seemed to promise the early resolution of the paradoxes of development. Her work with Hamburger built from her earlier work with Levi, although she continued to suffer technical problems in resolving these immensely complex neurogenetic systems. She ushers in the next phase of the work with the subhead: 'The unexpected break: a gift from malignant tissues'. But as we read on we learn that the gift came from the imaginative experimental work of one of Hamburger's students; thus it was the created luck of science rather than the accident of fortune, but this is a scientific voice that enjoys story telling.

She explains how the development of the research work was initially blocked because the group lacked the expertise with tissue culture. This was, however, being developed in Brazil by Hertha Levi, working at the Rio de Janeiro institute directed by Carlos Chagas. In a passage which fuses images of science and femininity, Levi-Montalcini explains that she was invited by Chagas and so 'boarded a plane for Rio de Janeiro, carrying in my handbag two mice bearing transplants of mouse sarcomas'.[74]

Despite her vivid reporting that it was in Rio de Janeiro that the nerve growth factor 'revealed itself . . . in a grand and theatrical way', Levi-Montalcini and her colleagues had difficulty in convincing others. She does not report a story of results refused publication in prestige scientific journals, but infers this failure to convince from the evidence that few followed her down what she saw as an exciting path. Her problem in this respect was similar to that of McClintock during the fifties and sixties. Yet where McClintock had a second immensely creative and communicating period in the 1970s, Levi-Montalcini's significant work was concentrated thirty-five years ago; the gap between achievement and recognition is in her case even harder to explain within the terms of the institution of the prize.

Gertrude Elion

The most recent woman to be honoured as a Nobel Laureate was Gertrude Elion[75] in 1988. Her parents were both first-generation US immigrants from Europe. Her father had qualified as a dentist, but was bankrupted through the stock-market crash in 1929. None the less her parents were able to help her financially within four years of the bankruptcy. She recalls a lost world of the Bronx as a good environment for childrearing, with good public schools and unrivalled opportunities for free tertiary education. She describes herself as 'a child with an insatiable thirst for knowledge and remember[s] enjoying all of [her] courses almost equally.'[76] She speculates that her affection for her grandfather, who died of cancer when she was 15, motivated her towards medical research, so that when she entered Hunter College she planned to major in science and especially chemistry. (Again, as with Yalow, the New York public educational system of the time showed its strength, demonstrating that it could be an effective substitute for the educational and cultural privilege of class.)

After college Elion had a bleak time searching for support to do graduate work or even merely to get a laboratory job. She describes a world of systematic and taken-for-granted discrimination in which progress was painfully slow. 'Jobs were scarce and the few positions that existed in laboratories were not available to women.' She describes one teaching job she had – biochemistry for nurses – which ran for three months out of the year. She then describes how a chemist offered to take her into his laboratory for no pay; she accepted for the experience. After eighteen months he was paying her 'the magnificent sum of $20 a week'. In 1933, some six years after entering undergraduate study, she was able, with the help of her parents, to enter graduate school at New York University. Having completed course work she trained as a teacher, then worked as a substitute teacher by day, researching at nights and weekends, completing her master's by 1941.

It was the outbreak of World War II with its demand for chemists by industrial laboratories which gave Elion, along with many of her generation and gender, the chance of a research job. From the inauspicious foothold of a job in a food industry laboratory she secured an assistantship with George Hitchens at the Burroughs Wellcome research laboratories. This was the first time, now almost ten years after entering Hunter, that Elion had a job where she could develop herself as a scientist. At the same time she also began a PhD at Brooklyn Polytechnic Institute, but the crunch came when she had to choose between going full time to complete the PhD, and abandoning it for her industrial research.

Elion tells us that after she had received three honorary doctorates from the Universities of George Washington, Brown and Michigan she felt that she had made the right choice. By carefully reciting this arduous story of getting into research, and of the fact that she has achieved so much without a PhD, she reminds us of just how exceptional her story is. For a man to get so far without a PhD would be surprising, for a woman it is little short of astonishing. It was the context of one of the most powerful US industrial laboratories, not academia with its passion for credentialism, which made this possible.

At Burroughs Wellcome she began that relationship to her work which enabled her to look back and characterize it as both 'my vocation and my avocation'. Although she began as an organic chemist she was never restricted to the single discipline. She became interested in microbiology and in the biological activities of the compounds she was synthesizing. Thus over the years she worked in biochemistry, pharmacology, immunology and eventually virology. In her Nobel address she sets about reporting forty years of work with no hint of complaint or criticism; instead she describes the research in which she and her colleagues have been engaged as a coherent set of scientific developments achieved over a period of time, which have consistently resulted in producing major thera-peutic agents. One of these, Acyclovir, was a pioneering anti-viral for herpes, and also paved the way for other anti-virals, not least Retrovir or AZT, also produced by the Wellcome laboratories. It is a matter of some note that she describes this highly innovatory work, often developing drugs for patients with then fatal diseases, without using the language beloved of today's clinical researchers, in which they 'aggressively treat'. Elion's prose concerning her research, and one feels Elion's laboratory, goes capably on, not minimizing painful matters or glossing over clinical testing, but not glamourizing it with violent metaphor either. Her official portrait echoes this capable good sense. (See plate 13.)

Such a voice, describing the complex task of basic science directed very closely towards clinical objectives and in active collaboration with the clinical treatment of patients, is rare in the Nobel proceedings. More typically, research, even when it has a considerable pay-off for medicine, is described within the science and not hand-in-hand with its applications. It is often only when the joke is made that had the research not been carried out on such a socially significant compound, then the honour and recognition currently being enjoyed by the researcher would not have been forthcoming, that we can see the boundary line between social and scientific esteem being gently moved around. Elion is refreshing in that she ignores these delicate boundary games between social and scientific prestige systems, and talks about the nature of contemporary medical research when it is done by good scientists committed to patient care in the best industrial laboratories.

Where are the future women prize-winners?

Perhaps it is not by chance that many women biomedical researchers say that it is easier to work and to be promoted in an industrial laboratory than in an academic setting. Maybe it is there that the Nobel committees should look for more potential women prize-winners, or among the observational sciences such as ethology or astronomy where numbers of women are currently eminent.[77] Can committees and procedures predominantly composed of men scientists under immense pressure to recognize other men scientists acknowledge the contribution of women unless they open their committee structures themselves to women, who are in an age of gender consciousness less likely to be gender blind? Otherwise it seems that the Nobel Prize system is unlikely to escape an even more age-linked construction of women of gold than operated in Plato's Republic. For today's Nobel committees it seems that women have to be at least 70, the age of the wise woman, the symbolic grandmother, to achieve recognition. Is there an unstated anxiety that, by recognizing women at the height of their creativity and with the social and political commitments of their generation, the committee might begin to disturb the networks of power? Perhaps women scientists in a period of feminist consciousness cannot be trusted to sustain the politics of prudential acquiescence which have become increasingly the hallmark of the scientific elite?

And in fifty years' time, when the descendants of today's feminists have access to the records of this past decade, how will the correspondence and debates of the 1980s compare with the unambiguous evidence of the double standard deployed (too late and unsuccessfully) against Marie Curie at the beginning of the century, or the manipulation of women's access to the Royal Society during the 1940s? Will they have been equally 'man'aged?

8

Feminism and the Genetic Turn: Challenging Reproductive Technoscience

[The state] must see to it that only the healthy beget children; but there is only one disgrace: despite one's own sicknesses and deficiencies, to bring children into the world, and one's highest duty is to renounce doing so . . . [The state] must put the most modern medical means in the service of this knowledge. It must declare unfit for propagation all who are in any way visibly sick or who have inherited a disease.
                                        Adolf Hitler, *Mein Kampf*

It is of course true that, in 1990, we have no Nazi conspiracy to fear. All we have to fear today is our own complacency that there are some 'right hands' in which to invest this responsibility – above all, the responsibility for arbitrating normality.
Evelyn Fox Keller, *Nature, Nurture and the Human Genome Project*

The demand of feminism for reproductive rights has run into increasing difficulties, not least in terms of how we engage in the politics of reproductive science and technology, or technoscience, for this is an area where science and technology are bound together as one. In the context of a world-wide shift to the right – a shift which in the West celebrates possessive individualism – feminism's successful mobilizing language of rights, and above all the slogan 'A woman's right to choose' has been increasingly recognized as no longer entirely effective in challenging an imperializing technoscience which seeks to invade women's bodies and women's lives ever more intimately.[1] We find our claims of rights and choice increasingly recuperated by a proliferation of charters and marketing stategies, and we experience the return of the repressed as 'our' language reveals its roots in 'their' liberal democratic theory. But perhaps its time was over anyway, for the theory and practices of possessive individualism are increasingly brought into tension with an ecologically responsible feminism which, while refusing eugenicism, acknowledges the need for some form of democratic self-management in issues of reproduction.[2] It is hard to think globally and act locally about having or not having children, using the language of individual rights.[3]

But what has also helped weaken the slogan of the right to choose has been the transformation of the technoscience landscape. While feminism has been grappling with the fast-changing reproductive technologies, trying to stop them or at least to slow them down, the new technosciences – computer science, electronics, biotechnology, mass air travel – have shared in this dramatic acceleration, in which successive technological generations follow each other at tremendous speed. These others are focused neither so specifically on the bodies of women, nor so directly on everyday life, and in consequence there has been rather less attention given them by feminists.[4] None the less, these too have a powerful influence in opening or limiting new social and cultural possibilities. For example, will the new human/machine relation offer the rather optimistic picture of Donna Haraway's cyborg figure, or instead will the feminisms choose from among the diverse possibilities of 'virtual reality', leaving the reality of everyday life to care for itself?[5]

In its concern with the in vitro fertilization (IVF) debate and the speculation over the implications for women of such seemingly remote prospects as ectogenesis and parthenogenesis, and in its battles with sociobiology, feminism has been rather slower in grasping the depth of the genetic turn within scientific research which was begun in the 1970s and which shows no sign of slackening. In areas as diverse as explanations of schooling failure – the notorious IQ debate – alcoholism, cancer, heart disease and psychiatric illness, there has been an immense turn away from the environment, particularly the social environment which had been addressed by governments in the sixties and by international agencies such as the World Health Organization in the seventies and early eighties, into a search within our genetic make-up to explain who will succeed and who will fail, who will get sick and who will die.[6] The Human Genome Project, a global research initiative whose intent is to map and sequence the 3 billion nucleotide pairs of the entire human genetic sequence, serves as the icon of this turn. 'Genes', it claims, 'R, us'.

The Genome Project signals a dramatic change in the politics of late twentieth-century genetics, from the biological politics of acquiescence to those of interventionism. The genetic turn which opened (reopened, I should write, for this is a refrain in a long scientific ballad) with the sociobiology of the seventies insisted that biology was destiny. That political demand for a quietist submission to the conservative laws of nature gave way in the eighties to the political interventionism of the new genetics. This new genetics, a product of the alliance between an aggressively entrepreneurial culture and the life sciences, fused the conservatism of biology as destiny with the modernist philosophy of genetic manipulation. The new endless frontier of science is within both green and human living nature.

This enhanced capacity of the life sciences offers not merely to determine and detect genetic disorder in human nature but also to manipulate and modify human genetic structure. Technological difficulty and cultural constraints hold back (though how far and how long?) such developments from being applied to humans, but already the vegetables available in most supermarkets bear witness to an increasing power to engineer nature. Those tomatoes and straw-berries with extraordinary resilience to being transported, the endless, identically sized flawless – albeit tasteless – apples, are the new vegetables and the first fruits of biotechnology. If green nature can be redesigned to fit better into the demands of the market, how do culture and society restrain and perhaps exclude such redesigning of human nature? Marx, writing over a hundred years ago, presciently observed: 'Animals and plants, which we are accustomed to think of as nature, are in their present form, not only the products of, say last year's labour, but the result of gradual transformation, continued through many generations.'[7] None the less, what is occurring today is a gigantic speed-up of those transformations of plants and animals – including human animals – which makes ecofeminists feel that we are in some sense defending 'nature' against culture. Nature, and its sister concept the environment, stand in as symbols of resistance. Given women's marked place within physiolo-gical reproduction, what does this dramatic genetic turn in both culture and material production mean for women, in all our historical specificities of class, race and sexuality?

Hybrid locations: hybrid forms

The new technologies do not advance unaided. They have been assiduously fostered by governments that still see growth, despite the criticism from the new social movements of the environment and of feminism, as a one-dimensional economic phenomenon. Scientists talk up the curative and commercial possibilities, for public support means public cash to sustain the laboratories. They actively contribute to that process of creating the need, searching out, making and creating the necessary marketing niches for innovation to take place. Economic growth and competition structure these changes, fostered by venture capital and supranational groupings; increasingly important are the Framework Programmes of the European Union, which seek to enable Europe to catch up technologically and economically with Japan and the US. That the European Union now has its own Genome Programme, along with those of the US, Japan and Russia, is part of this embrace of technoeconomism. Because of the centrality of technology in contemporary economy and society, or maybe even more because of the overwhelming belief in its centrality, this rate of technological change is likely to be speeded up through the conversion from military research and development which since 1989 is grudgingly following the collapse of the arms race between the superpowers. Every year since 1945, Britain for example has spent more than half its state science budget on military research; were this to be redirected into civilian research the rate of technological change in everyday life would accelerate.[8]

As happened for chemistry in the nineteenth century and physics in the mid-twentieth, it has now become the turn of significant sections of the life sciences to enter the process of industrialization. Changes in the production system of the life sciences which had been foreshadowed in the sixties with the advent of automatic analysers in biochemical pathology were extended in the eighties, particularly in genetics, as the techniques of molecular biology gathered strength and showed increasing potential application to medicine, agriculture, crime detection, the military and the food industry.[9]

Integral to the development of this new area of 'biotechnology' came a change in the organization of research. Where the industrialization of chemistry and physics had largely taken place outside the university, the former in industry and the latter initially in the huge military establishments associated with the production of the bomb, the new trio of biotechnology, computer science and electronics – the technosciences – began to appear within, or slightly adjacent to, the university research system. Hybrid developments such as science parks began to occupy a new geographical space near or even on university campuses, where a new breed of academic entrepreneurs could operate. Such new space offered a means for the new entrepreneurs to share both in the prestige system of university science and also in the immense financial rewards possible within industry. Such mutant developments, where commercial secrecy and open academic work make uncomfortable neighbours, are taken for granted in the higher education structures of the eighties and beyond. However, as is abundantly evident in the long-drawn-out fight in the Genome Project over patenting DNA sequences, with the resignation/firing of the Director of the National Institutes of Health programme, Jim Watson,[10] such new organizational forms have had their own contradictions, culminating in a series of cases of 'conflict of interest' which both have formidable implications for the management of scientific research through peer review, and also intensify the commodification of nature and knowledge.

Such double commodification has been employed with remarkable intensity in human reproduction. Birthing as a 'natural' process had in the West already come increasingly under medical men's dominance but was still relatively undercapitalized, and as such was an ideal location for the new complex of interests and power. This new phenomenon, the 'biomedical industrial reproduction complex', now invades women's reproductive lives at a historically unprecedented level, and without fierce resistance is only likely to intensify. Such tremendous changes around reproduction have disturbing echoes of the dramatic passage in *Capital* in which Marx describes how, as the machine enters industrial production in the nineteenth century, the old craft skills fall away; the worker is no longer in charge of the machine but has become its 'mere appendage'.[11] Instead of feeling satisfaction in the product, the worker feels alienation. The prenatal tests of chorionic villus screening, amniocentesis and ultrasonic foetal imaging during pregnancy, followed by the hospital birth with foetal monitors, and an increasing rate of caesarian sections, all too commonly give women a sense of loss of control, of being merely attached to some gigantic birth technology.[12] When medical interventions are particularly heroic, as with infertility treatments, foetal surgery and embryo genetic screening, women can find themselves having 'chosen', yet feeling that it was not 'this' that they wanted.

Framing the debate

While there has been an immense feminist literature studying, analysing, debating the new technologies, searching for a new politics, it has done so largely in a framework which rather rarely explicitly draws on theories of (science and) technology in society.[13] Here I want to pull out three theoretical approaches which have framed discussion in the mainstream debates. Although in the mainstream they are treated as alternative approaches, here I see them as resources which need to be woven together.

The approach most frequently drawn on by feminists is that of social constructionism. This focuses on the network of actors who, through their practices, construct scientific facts and technological artefacts. Despite the charges that can be laid against mainstream social constructionism, feminist constructionist approaches to reproductive technology are neither philosophically relativist nor politically evasive, not least because they have drawn on what I call the second and third approaches.

The second approach is that of 'externalism' which raises the issue of power and the production of knowledge: first in its classical form (discussed in chapter 1), in which the interests of the dominant class were seen as determining science and technology, then in a more general structural externalist determinism, in which theorists of feminism and imperialism have pointed to the inescapable dimensions of gender and race along with those of class in shaping the direction of science and technology.

The last is that of technological determinism, whether of the kind denounced by Jacques Ellul,[14] or in marxist theories of the relative autonomy of science and technology as forces of production. In the sense that the new reproductive technology feels as if it is pursuing its own unstoppable masculinist logic, strands within feminism share more than a little of this strong determinism. Yet feminism rarely collapses into that masculinist but would-be neutral technological determinism so well caught by David Noble:

> our culture objectifies technology and sets it apart and above human affairs. Here technology has come to be viewed as an autonomous process, having a life of its own which proceeds automatically, and almost naturally, along a singular path. Supposedly, self defining and independent of social power and purpose, technology appears to be an external force impinging on society, as it were, from the outside, determining events to which people must forever adjust.[15]

Feminist social constructionists have drawn on 'externalism' and its preoccupation with power. In particular they have paid attention both to the powerful actors recognized in the mainstream accounts, and also to the excluded, not infrequently silenced by the mainstream. In bringing these excluded actors into the production of its accounts, feminist social constructionism has a more inclusivist relationship with both technological determinist and structural determinist accounts, for what unites all three is their feminism; that is, their common recognition that any adequate theory of science is necessarily political.[16]

However, in so far as feminism's debate has often focused on the technology as if it were separable from the science, it reproduces that ideological binary split between science and technology – preserving the 'purity' of science while admitting the 'dirty' worldly nature of technology. In much of modern science, and particularly in the molecular biology of the new genetics, it is necessary to insist on their intimate connection. Nowhere in the sciences is it clearer than in the new genetics that there is no useful line to be drawn between science and technology. Modern – that is, Western, patriarchal, heterosexist and capitalist – science is above all a technoscience; it does not simply contemplate nature, but seeks to dominate and exploit it.

PLATE 1
Marie Curie, 1903

PLATE 2
Pierre Curie, 1903

PLATE 3 Marie Curie, 1911

PLATE 4 Frédéric Joliot, 1935

PLATE 5 Irène Curie, 1935

PLATE 6 Carl Cori, 1947

PLATE 7 Gerty Cori, 1947

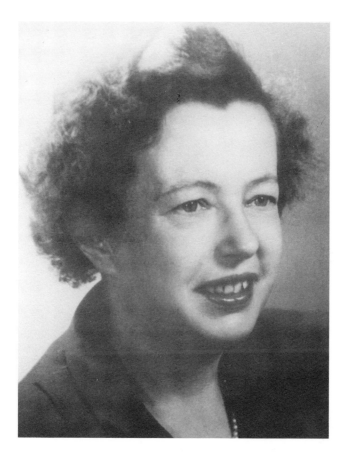

PLATE 8 Maria Goeppert Mayer, 1963

PLATE 9 Dorothy Crowfoot Hodgkin, 1964

PLATE 10 Rosalyn Yalow, 1977

PLATE 11 Barbara McClintock, 1983

PLATE 12 Rita Levi-Montalcini, 1986

PLATE 13 Gertrude Elion, 1988

Currently feminism confronts two very separate challenges stemming from the reproductive technosciences. First, there is a series of extremely detailed debates concerning the new reproductive technologies and human embryo research. These debates do not take place within some universal abstraction, but are located in very specific political and cultural contexts, so it is not by chance that feminism does not speak with one voice. (And where feminists have tried to offer one universal response it retreats into the technophobia of feminist fundamentalism.) But secondly, feminism is confronted by reproductive technoscience not only as dynamic interventionist genetics, but also as a dominant and global culture, and it is this deeper, more fundamental genetic turn that has been neglected in the preoccupation with the immediate and the challenge of reproductive technology. In this chapter I want to bring the immediate and the deeper cultural turn together by interrogating the reproductive technology debate and by setting this in the context of the Human Genome Project.

The Petri dish, the patriarchy and the private market

It was the birth of the first 'test-tube baby', Louise Brown, in 1978 which precipitated feminism's intense debate over the new reproductive technology. Despite the hopes some of us had held earlier that decade that feminism might see and effectively resist the manifestly impending transfer of IVF to humans,[17] it is now securely established among a burgeoning number of new reproductive technologies, and many thousands of assisted conceptions have resulted in babies. From this first successful assisted conception, the new technologies and genetic prenatal screening have together offered unprecedented powers to biomedicine about who is to mother and which foetus is to be permitted to survive.

Such a fusion of powers was symbolized in the agreement that Lesley Brown was required to sign as a condition of her being treated with the new experimental techniques, that she would have an abortion if the foetus was abnormal. When the child, Louise, was born, at the private Bourne End clinic, the clinician Patrick Steptoe and the biologist Robert Edwards were hailed by the media as the 'fathers' of the world's first 'test-tube baby'. Among fathers in Britain that year this metaphorical pair had, with the aid of venture capital, demonstrated unprecedented and eugenicist power over the possibility of motherhood for Lesley and life for the embryo which was to become her daughter. The year 1978 was an important one for the Petri dish, patriarchal power and the private market.

Thus while the British Medical Association gave Edwards and

Steptoe a standing ovation for their achievement, there was a hostile recognition by feminists both of the new power reproductive science and technology had given to clinicians over women,[18] and also of feminism's profound difficulty, for this was a real child 'chosen' by a real woman. This difficulty has remained. There was, too, a general cultural concern which feminists shared, in that the birth had given society the possibility of a third and historically new concept of motherhood.[19] To the culturally understood categories of the biological and the care mother was added that of the carry or surrogate mother, who was no longer necessarily either the biological mother or the care mother. The unease at these developments was very strong, providing the fast-developing specialism of bioethics with a new area for deliberation, giving the media the possibility of running a number of horror stories, and setting feminists into a passionate and increasingly international debate about the rights and wrongs of IVF and genetic screening. In these early phases, thinking about IVF was quite difficult, disturbing those taken-for-granted self-concepts of bodily integrity much in the way that blood transfusions and, rather later, organ transplants had disturbed that intimate and deeply cultural sense of 'This is my body, that is alien.' But there was a fundamental difference; where the technologies of blood transfusion and organ transplant had invaded both men's and women's sense of bodily integrity, this new heroic medicine takes place exclusively on the bodies of women.

Today IVF is an increasingly routine, albeit expensive and invasive technology, typically located in private medicine. But in the early 1970s it was far from clear that the technology was going to be realized. It is important to insist on this, for not every technology which is possible necessarily finds its expression as a product.[20] The progress of IVF technology was unaided by the state in either Britain or the US; instead, private venture capital was crucial. The debate about the social implications of embryological research, not least sex selection, was firmly on the public agenda in the US,[21] and public research funds for human embryological research were cut back, not from doubts about sex selection, but rather as part of anti-abortion politics.

However, the point is that the US government did – negatively – steer research. For rather different reasons a similar negative steering situation prevailed in the UK. The then main source of public funds for medical research, the Medical Research Council (MRC), which had supported basic embryology in non-humans, such as amphibia and mice, was unwilling to fund the Cambridge group of biologist Robert Edwards and clinician Patrick Steptoe for human IVF and embryo transfer. Within the scientific elite the animal work at Oxford was acclaimed as a scientific breakthrough, while the proposed work on humans at Cambridge was seen as merely a further replication. Scientifically, changing species was unoriginal.[22] But whatever the

MRC, despite its medical brief, thought of the social implications of the proposed IVF being applied to humans, it did not choose to share such deliberations with the public. The efforts of the radical science movement to urge that scientists speak out responsibly about the possible implications of their work moved the embryologists more than the MRC.[23] Within Britain such matters were to be deemed 'ethical' and the concern of a newly emergent occupational group of ethicists drawn largely but not entirely from philosophy.

Something of the scientific establishment's continuing hostility to the project was shown, not only by the refusal to fund the research, but also by the response of the Royal Society to the immense media acclaim given to Edwards and Steptoe. With some political ingenuity the Society elected Steptoe on the grounds of his pioneering laparoscopy work while leaving Edwards outside, as a mere replicator of the basic science carried out by more original others. But this small token of the scientific elite's disdain for the medical and media brouhaha was part of the old closed politics of science; the new approach to maintaining science's legitimacy through both 'ethics' and what was called the 'public understanding of science' was yet to come.[24]

Such closed establishment science politics was unable to postpone or resist the new technology, for Steptoe and Edwards turned with success to private funding sources, and established a private clinic at Bourne End. The fact that their work, and the profuse flowering of clinical activity in IVF and embryo transfer work in the US, Australia and elsewhere, could be sustained by venture capital points to the changing character of biological research during the seventies.

'Ethics' and the regulation of the new reproductive technologies

With venture capital entering the structure of research in biology, the old forms of steering science needed reworking.[25] For the biomedical sciences the solution which became increasingly clear over the course of the eighties was the deeper institutionalization of 'ethics' as a means of guiding public regulation. Research ethics committees, initially a response to social criticism of biomedical research without patients' informed consent, developed in most major teaching and researching hospitals and have been increasingly extended. A number of philosophers had become interested in biomedical ethics; in the US, research institutions, notably the Hastings Center, and in the UK the Nuffield Foundation, supported a number of developments including the *Bulletin of Medical Ethics*.[26] In many ways this institutionalization of 'ethics', as well as incidentally providing a new occupation for philosophers, can be read as the response of democratic liberalism to a growing social criticism of science as having an ethic of no ethics. The introduction of a new specialism of bioethics also gracefully sidestepped the increasing criticism from the social sciences of professional dominance, and from feminism, of masculinist dominance – but power is always a difficult concept for the discourse of liberalism.[27] The introduction of philosophy produced a certain clarity in the discussion, but because it did so largely from within the philosophical canon it was unable to gain much critical purchase, and indeed all too frequently proved a close ally of the social conservatism of the biomedical profession.[28] It is unclear whether the ethicists have modified, or simply given a new gloss to, professional power.

For the reproductive technology debate in Britain the report of the Committee of Inquiry into Human Fertilisation and Emybryology, chaired by the philosopher Mary Warnock, was to become a foundation text framing subsequent debate and policy making. For this reason it remains well worth reinterrogation. First, the composition of such public committees is organized so as to secure a particular kind of consensus; in consequence 'key' interest groups have to be represented. As the issue related to 'the family', this required that the various churches (of white not black Britain) would be strongly represented; as it related to scientific research, then the Royal Society, as the pre-eminent scientific institution, would be asked to suggest a scientist with suitable expertise. What was path-breaking in the committee's composition was the number of women (six in a committee of fifteen). Such political recognition that issues of infertility and human embryology particularly affected women was indisputably assisted by the presence of the feminist movement.[29] However, social innovation was not to be permitted to get out of hand; none was publicly identified as a feminist or even a member of women's organizations such as the Townswomen's Guild or the Women's Institute, who might be expected to promote a gender, if not a feminist, perspective.

But if the composition of the committee was innovatory, the assumptions were not. The strong presence of the obstetricians and, in Anne McLaren, the presence of a distinguished developmental biologist well used to collaboration with the clinicians[30] meant that it was easy to accept clinical medicine's rhetoric of helping the infertile (a newly acknowledged group in society). From being something of a pariah group, associated with messy and undignified procedures, infertile women and men were now the potential recipients of glamorous, highly scientized medicine.[31] Such rhetoric, elaborated particularly by Edwards and amplified by the media, drew strongly on that ideology of motherhood in which a woman without a child is incomplete. At the same time there was silence concerning the price of this assistance to infertile women for 'choosing' a massively invasive technology. Women's longing for a child was seen as such an unstoppable force that their capacity to endure distress and pain to realize their biological destiny was seen as natural and therefore negligible. Such a celebration of natural desires for motherhood fitted into that generalized celebration of 'the family' which was fostered in the opening years of the Thatcher government, when the prime minister claimed that only the 'individual' and 'the family' existed and that 'there was no such thing as society.'

Despite the apparent sensitivity of the committee report, which broke new ground in its carefully gendered pronoun usage, the abstractness of the ethical discourse it employed served to render invisible both the distinctly messy corporeality of existing infertility treatments and the invasiveness of the new. In the name of equality Warnock rendered 'his' and 'her' gamete donation the same, rhetorically masking with abstraction the more than slight difference between masturbating into a glass and having surgery.[32] That the ideologies of motherhood and parenthood also insisted that the woman had to be in a heterosexual and stable partnership was evidenced first by the practices of the clinics (and was flagged from the beginning by the Brown family story) and was brought explictly into view by Warnock and then by the 1990 legislation. Motherhood by assisted conception was not to be permitted without fatherhood.

Nor was the new reproductive technology to foster or even be neutral to any of the new social models of motherhood. Thus, within the UK a long struggle had slowly secured the right of single women to parent. No longer did the welfare agencies, the state and a woman's own family bear down to make such women give up their babies for adoption.[33] No longer was she told that out of her love for the child she must give it up so that it might be reared by a married couple. However, as the Lord Chancellor made clear on the third reading of the embryology bill, unmarried heterosexual partners were to be surrounded by conditionality:

> if it is to remain possible for unmarried couples to receive the benefit of treatment to bring a child into being, both should have imposed upon them the responsibility for the child. I am most concerned that this proposal should not be seen as encouraging unmarried people to use infertility treatments, thus leading to children having unsuitable social fathers because of the difficulty of distinguishing partners of stable relationships from more transitory ones.[34]

Those heterosexual relationships deemed by the medical profession to be 'unstable,' all lesbian relationships and all single women were thus to be denied technological benefit.

Ideological enthusiasm to restrict assisted conception to stable heterosexual couples, exemplified in Warnock, has led a number of countries not merely to limit access to IVF and similar sophisticated infertility treatments but to legislate to control 'artificial insemination' (AI), a technology which had been practised extensively in animal husbandry (sic) and rather modestly by infertility clinics since the

1920s.[35] This willingness to bring the law into contempt by passing manifestly unenforceable legislation – for AI only needs a man willing to provide sperm and a woman ready to inseminate herself with no more elaborate technology than a turkey baster or a wine glass – speaks more of a punitive refusal to give medical benefit to those deemed unfit to mother than any real ability to stop AI. The interests of the children who would be conceived were less important than policing maternity and prioritizing nuclear family forms.[36]

Policing assisted maternity was none the less to be given into the charge of the doctors.[37] Thus the Warnock Report, while appearing to affirm universal rights, withdrew them in the same sentence:

> However notwithstanding our view that every patient is entitled to advice and investigation of his or her fertility, we can see occasions where the consultant may, after discussion with his or her colleagues, consider that there are valid reasons why infertility treatment would not be in the best interests of the patient, the child that may be born following treatment, or the patient's immediate family.[38]

In this return to professional control and refusal to enter the political economy of infertility, both the report and the subsequent legislation in 1990 disregard the cash-limited control of health care services which was so central to health policy making in the eighties. Nowhere was there going to be an attempt to try to estimate the incidence of the problem, to establish, however tentatively, how many people were childless or childfree, or to consider the most appropriate models of services. What was offered was an 'ethical' debate around treatment which assumed and constructed a universal longing for children, and which assumed and constructed a universal provision of infertility treatment without even the most rudimentary financial calculations. In the context of public sector cuts it was inevitable that the expense of assisted conception, an extremely modest success rate (just over 10 per cent according to the Interim Licensing Authority's 1989 Report), and the possibility of profitability ensured that the majority of this high-tech medical care takes place outside the National Health Service. Even here, as in a number of other private medical treatments, the costs to the patient are subsidized in that the costs of drugs are typically provided through the NHS. But subsidizing the better off and the socially aspirant acquired new respectability in the eighties.

Embryo research: admitting personness and resisting personhood

The focus of much of Warnock's attention was directed to the regulation of human embryo research, and concluded (with some dissent) that the limit of the age to which human embryos should be kept alive in vitro was fourteen days. A predictable furore followed this recommendation, though relatively little comment was been made on others of the technologies – such as hybrids – which also seemed close at hand. Edwards and some other researchers and medical doctors urged that the fourteen-day limit be lifted, whilst in Parliament supporters of the rights of the 'unborn child' demanded a total ban, encapsulated in the eventually defeated Powell bill. What the furore masked was the fact that even almost a decade later the fourteen day limit is still not feasible technically.

In this notion of the embryo having a hint of sentient life at fourteen days with the development of the embryological feature identifiable as a precursor of the nervous system, the 'primitive streak', the Warnock Committee and subsequent public debate and legislation began to construct the embryo as having more personness at fifteen days than at thirteen. In defence of experimental research, the category of the 'primitive' or 'neural' streak which appears at the fourteenth day was recruited to draw the lines between the acceptable and the non-acceptable. While the supporters of this legislation were also supporters of abortion, and thus opposed what the US courts spoke of as personhood for the foetus until independent survival was possible, this new boundary admitted a softer notion of personness much earlier within development.[39] Inconsistent in that it admitted that what was little more than a cluster of cells had some claims as having a special status – what I have here called 'personness', – it made an acknowledgement of the widespread anxieties that human embryo research arouses. Whilst there was an agreement that the production of in vitro or in vivo embryos purely for experimental purposes was unacceptable, the construction of the 'pre-embryo' and primitive streak offered a way for British legislators to admit limited research. Developmental biology had written a moral economy of the embryo and the foetus.[40]

This moral economy/fourteen-day marker has been extensively debated by British feminists, and some, like Sarah Franklin, are both critical and self-critical in that feminism is failing to influence the public debate.[41] By contrast I read the developments more optimistically as I think that feminists have influenced the debate (although less than I would like) not in some general abstract and universal sense but in the specifics of British sexual biopolitics. Because I think and feel that terminating a pregnancy is not quite like removing diseased tissue and requires a political commitment to a woman's reproductive freedom,[42] like many feminists[43] I still want in some measure to privilege both the human foetus and gametes as potential research material over and above other tissue, even while resisting claims of personhood with enforceable rights.

Granting personhood to the foetus must lead to opposition to abortion, and indeed some Australian feminists opposed to IVF, with its entailed embryo research, have lined up with anti-abortion groups. In some other countries, for example Norway, it has been so far politically possible to keep the issue of abortion rights separate from embryological research. In the UK, however, anti-abortion forces had connected the abortion and embryo research; in consequence the nuanced account of foetal development offered a guarded but effective defence to both. In the British context a public affirmation of feminism is rarely made by women scientists (and perhaps is not easily made); there is, though, a need to be sensitive to women-friendly biological constructions. They rarely occur without friendly agents.[44]

Kin, inheritance and the persistence of feudalism

The very peculiar structure of the British political system, with an unelected second house composed of hereditary and life peers, bishops and senior judges, cannot be ignored, as these predominantly elderly white men represent a fundamentally feudal structure left over from a pre-modern era. This still has immense influence in British political life, giving British feminism a very difficult task, as it has to confront simultaneously both pre-capitalist and capitalist forms of patriarchy in which different constructions of the 'family' still compete. Thus the debate of the House of Lords' third reading of the bill, saw it as a threat to overthrow the 'blood line' and the laws of inheritance. Without a blush as to the peculiarly privileged world of wealth that they inhabit, the Lords waxed passionate about the evident collapse of the morality of family and property were castles and estates to be inherited by children who were not biological kin.[45] Yet what is fascinating is how much the concerns of this world were anticipated in the discussions of Warnock. The more modern voices, represented by the Lord Chancellor, were less concerned with kin and inheritance and more with the need of a patriarchal state to ensure that the new technology did not produce family-less persons.

Though I suspect it was far from their lordships', and in particular the Lord Chancellor's, minds, this modern anxiety about producing family-less persons was precisely why Shulamith Firestone had in 1970 seized on the potential of test-tube babies to free women and destroy the family form.[46] Whatever the complexity of their personal and sexual lives, no one in this feudal institution had the courage publicly to defend new family/household forms.[47]

Feminist resistance

Feminism's trajectory of opposition has had a rather different character. Beginning with Firestone's early and unambiguous welcome as a means to resolve the problem of women's subordination, feminists have come full turn. The strong disquiet with the new reproductive technologies only appeared with the birth of the first IVF child in 1978. 'We' spoke with many voices: the voices of infertile women, disabled women, women who wanted to resist compulsory motherhood, women willing to carry a child for another, women who wanted to create a National Health Service rather than extend the National Sickness Service, were rightly unable to make a single set of demands.

However, one feminist grouping which entered the political arena in 1985 did have the single-minded agenda of denouncing and opposing the new reproductive technologies. Beginning with a meeting in Sweden, Finnrage (Feminist International Network of Resistance to Reproductive and Genetic Engineering) rapidly made itself felt within European politics, by stressing the eugenicism inherent in the new technologies and making alliances with the German Social Democratic Party, the Greens and a number of other European socialist and communist parties. Negatively, Finnrage campaigns so as to project infertile women and surrogate mothers as 'victims' of 'techno-patriarchy'. It uses horror stories (often fanned by the media, as in the case of the surrogate mothering of Baby Cotton) to denounce surrogacy, and even motherhood, as forms of 'prostitution'.[48] Foetal tissue to be used in the treatment of Parkinson's disease,[49] sex selection for 'femicide',[50] the new 'brothel model' of reproduction,[51] images of the Holocaust,[52] infertile women portrayed as intellectually lacking, who are 'too easily blinded by science',[53] the use of brain-dead women as surrogates,[54] all keep certain nightmare visions sharp and unambiguous.

Positively, these Finnrage tactics have served to put the issue of reproductive technology and genetic engineering onto the political agenda, but at the price of imposing a universal politics on an immense diversity of location. Similarly because the 'only' course of political action is total resistance, feminists who seek to regulate the new technologies are seen as 'handmaidens of mechanical science.[55] In no small measure this stems from the biological determinism of radical feminism, the most powerful theoretical current within Finnrage. Thus the origin story of the new reproductive technologies is told by Janice Raymond,[56] drawing on Mary O'Brien's analysis,[57] as a story of men's relentless desire to overcome the alienation of man from his seed in the act of procreation, appropriating reproduction from women.

Some deepening of the Finnrage agenda is evident since its 1989 meeting in Bangladesh, at which the environmental implications of genetic engineering for reducing biodiversity were also included.[58] The *Declaration of Comilla* which came from this meeting, while enriched by this ecofeminism linking the concerns of North and South feminists, none the less places its first political demand as halting the new genetic and reproductive engineering. In Maria Mies's words, the central struggle for a united North and South feminism is against reproductive technology for this as 'anti-natalist technology is mainly targeted at poor women of the South whereas pro-natalist technology is mainly targeted at white middle class women'.[59]

My difficulty with radical feminism, particularly that associated with Finnrage, is not that it admits the body, for I welcome this,[60] but that in a peculiar mirror image of the patriarchal ideology it opposes, it frequently reduces women (and men) to nothing but biology, in which, in this particular struggle, reproduction is the central function. In other struggles, for example around pornography and violence, sexuality becomes central, and similar but not identical biologically determinist arguments are set in train. In this binary world the challenge is which sex is to control women's reproduction. In this insistence that the new reproductive technologies are in themselves always patriarchal, there is no possibility, through changing the power relations of the context in which they are produced and used, of modifying either the technologies themselves or their gendered outcomes.

Having criticized Finnrage for biological reductionism and for reducing the many and complex struggles of feminism to this one issue, there is no case for being sanguine over either the new reproductive technologies or the genetic turn, and Finnrage's strong attack on both has publicized the dangers of eugenicism. IVF and its related technologies are unquestionably part of a general technologization of birth, which disempowers women. It is also a highly profitable enterprise, which, in those countries where there has been little public regulation, has fostered a runaway technology particularly hurtful to women. There are well-documented accounts of commercial fertility clinics in the US where no evidence of success is provided and where women anxious to conceive enter a programme of highly invasive, ineffectual and costly treatment. There is evidence of commercial surrogacy in Mexico. In India, there has been a consistent feminist outcry against amniocentesis as a sex-selection test extensively used to abort female foetuses.[61] Preconception sex-selection tests are a thriving business in Japan.[62] In Britain, despite an apparently high level of public regulation, even the most basic matters of informed consent were initially disgracefully heterosexist in that both the woman and her male partner were required to sign the model 'agreement' provided by the Interim Licensing Authority.[63] As Naomi Pfeffer observes, the purpose of 'his' (not legally required) signature was so that the ILA could reassure itself and the 'public' that only stable heterosexual couples are being treated. British legislation and practice, while certainly resistant to the radical critique made by Finnrage, can be seen as responding to the detailed criticisms stemming from feminist research. Willingness to take such criticisms on board is facilitated by the presence of a more fundamental critique, despite the political gap between the two approaches.

Technoscience futures and fantasies

Given that women have been simultaneously shut out and socialized to shut themselves out of science, it is politically vital that feminism develops a realistic appraisal of the new technosciences of the body. Which are an immediate threat, and which are relatively remote or almost certainly not feasible? Certainly beyond IVF, screening and therapeutic interventions other than abortion lies a scientific and technological horizon along which is ranged a variety of possible reproductive interventions. Would it be possible to rear a foetus from fertilization to independent 'birth' entirely in vitro (ectogenesis)? To clone identical copies of individuals from single cells or 'gene libraries'? To rear a human embryo in the uterus of a non-human creature or even make human–non-human hybrids? To provide a technique which would enable women to give birth without the need for sperm to fertilize their eggs (parthenogenesis, a form of cloning)? Could men bear babies? These prospects and others form the stuff of science-fantasy dreams, of serious dystopic concern and utopian hopes amongst feminists.

The Warnock Report, guided by the advice from developmental biologists, dismissed some of these possibilities as beyond the range of presently envisageable science and technology (for instance, cloning); others it saw as feasible in principle but to be controlled by criminalizing or licensing (trans-species fertilization; ectogenesis). So far feminism feels itself able to rely on a widespread revulsion from trans-species fertilization, yet there is a nineteenth-century history of racist speculation that black women are 'naturally receptive' to breeding with apes, and a hideous reality of a racist project proposed in the Soviet Union in the 1920s, supported by influential scientific and Party groups, to breed hybrids from black women and apes in French colonial Africa.[64] That the French colonialists refused to co-operate, from whatever motivation, gives an irony to the relief that the project did not take place. That it was seriously proposed when the Soviet Union was still young, so there are no possibilities of hiding behind the argument of a later Stalinist degeneracy, speaks of the power of patriarchal racism within an ideology which claimed human liberation. Such an appalling story serves to remind us that in the name of science and 'progress' unthinkable proposals are thinkable. Thus, even while cloning is forbidden, work on the embryo stem line continues apace, and it is this basic science which is a biological precondition for cloning techniques.[65] Without yielding to moral panic, feminists have to look at technoscience with pessimism of the intellect and optimism of the will,[66] for believing in the possibility of other and better futures is a social precondition for bringing them into existence. It is to the technoscience of the new genetics that I now turn.

The new genetics

From the moment of conception, of the fusion of egg and sperm, each individual carries two sets of genes, one inherited from each parent. As the embryo grows, its cells divide over and over again and, at each division, new copies of both sets of genes are made, so that each body cell of the foetus, and later of the child and adult, contains copies of the original genes which the fused egg and sperm contributed – about a hundred thousand different pairs of genes in all. The genes are present in long molecules of the 'genetic material' DNA, assembled into a number of individual strands, called chromosomes. Ordinarily humans have forty-six chromosomes in all, arranged in pairs. These are similar for males and females with one exception – females have two copies of the X-chromosome, whereas males have only one X-, and one Y-chromosome. This difference, as well as influencing the sex of the future child, also explains why some genetic disorders (for example colour blindness and haemophilia) can be carried by females but are predominantly expressed in males.

Genes form the chemical code for the production of proteins, which are the molecules of which cells and ultimately organisms are mainly constructed and which also carry out most of the day-to-day biological functions of being alive. Just how a foetus and later a child develop depends on the interaction, during its development, of the unique set of genes which it has inherited and the specific environment in which it develops. The properties of a gene depend on its chemical composition – its DNA. If the DNA composition changes for any reason (such changes are called mutations and can be produced, for instance, by radiation), then either the gene cannot be used to produce the protein it codes for, or the protein is changed. If the normal protein is essential, its absence, or a change in its structure due to changed coding, can lead to the loss of a pregnancy. (This accounts for many spontaneous abortions.) But if the protein is not absolutely necessary to life, or can function in its altered form, then the foetus will develop, but may do so with characteristic variants or 'abnormalities'. Whether the variants/abnormalities make any obvious difference depends in part on the fact that each foetus has two sets of each gene, a set from each parent (except in males for the genes on the sex chromosomes). If both parents transmit the same variant set, then the foetus is bound to inherit the genetic 'variation'. But if only one parent transmits the variant gene then each conception carries only a statistically calculable chance of inheriting the problem.

Some variant genes are dominant – that is, even one copy of the gene is enough to affect the development of the foetus. Examples include such conditions as hypercholesterolaemia, which in the heterozygous form (inherited from one parent) is associated with heart disease and premature death in mid-life, but in the homozygous form (inherited from both parents) is associated with severe heart disease and death in adolescence or very early adulthood. Other genes, however, are recessive; which means that in the presence of a normal copy of the same gene, the foetus will develop more or less normally. The blood disorder sickle-cell anaemia, common amongst black people of African origin (affecting about one in 500 African-Americans), thalassaemia amongst Mediterranean people, and cystic fibrosis as one of the most common serious genetic disorders amongst white Europeans are examples. However, the person with a single copy of a recessive gene remains a carrier for that gene and has a 50 per cent chance of transmitting it in turn to each child.

It is important to be clear that despite this alarming word 'abnormal', whether or not a gene is actually deleterious depends in part on the environment in which the foetus, child and finally adult grow and live. For instance, whilst the gene for sickle-cell anaemia can be deleterious in the US or northern Europe, where people with two copies of the gene *can* suffer from problems resulting from not being able to get enough oxygen into their body tissues, causing considerable pain from blood clots in the capillaries, a single copy of the gene seems to convey some immunity to malaria and is, therefore, an advantage in regions where malaria is common. But the *can* is not sufficiently well documented to introduce screening programmes which may leave one in twelve black people feeling that they have a potential 'disease'. Such moves in the Nixon era left many black Americans feeling genetically stigmatized.[67]

Similarly, there is evidence that much short-sightedness is genetically inherited. This must have been a disadvantage in, say, a hunter/gatherer society, but in societies where spectacles or contact lenses are readily available it no longer matters very much. More dramatically, phenylketonuria (PKU), a genetic abnormality which affects one in every 10,000 children born in Britain, and means that they are unable to metabolize the common amino acid phenylalanine, used to lead to irreversible mental retardation, until effective dietary management was introduced which eliminated all the phenylalanine in the diet (in meat proteins for example). Newborns are now routinely screened with a blood test, and dietary management is introduced where the phenylalanine metabolism is abnormal. The example of PKU, as a genetically transmitted abnormality whose outcome can be transformed (and partially, though not entirely, corrected) through modifying the environment without changing the genes, makes the point that there is no inevitable and direct line between the presence or absence of a particular gene and a particular outcome; genes respond to changes in their environment with a norm of reaction which makes simple deductions about 'genetic causes' fallacious. PKU is also a precious and rather rare example of a screening and intervention process with no losers, only winners. Familial hyper-cholesterolaemia, at least in the heterozygous form, is perhaps another.[68] The point is that there is not a simple direct line between gene and phenotype; how any gene is expressed depends on its environment.[69]

Such genetic information seemed, only a few years ago, to belong in the realm of scientific knowledge without much immediate technological application. It did, however, make genetic counselling and therapeutic abortion possible, as in the case of thalassaemia – which has steadily been reduced in 'at risk' population groups, in both Sardinia and Cyprus as well as among the Greek Cypriot community in Britain, where it was once common.[70] A similar development has occurred within the Ashkenazi Jewish community for Tay-Sachs disease, which produces blindness, mental deficiency and then death in children. But what has changed dramatically since the early 1970s is the explosive growth of the new science of molecular biology and its application as biotechnology. Molecular biology opened up not mere abstract genetic knowledge and increased certainty in diagnosis, but the possibility of manipulating genes. It became possible, for example, to develop methods for identifying genes – lengths of DNA – by means of specially synthesized molecular probes, to snip out the individual genes, from a human or other organism's DNA, and to make multiple copies – or clones – of the genes. Even more remarkably, it became possible to take a gene isolated and copied from humans, say, and insert it into bacteria, so that the bacteria would now begin to make the human protein – a technique now used to manufacture human insulin, the protein hormone which is lacking in some forms of diabetes. (That a number of people with diabetes find the new insulin less manageable suggests that even this technical gain has unanticipated costs for its claimed beneficiaries.)

Biotechnology entrepreneurs took these molecular biological meth-ods, coupled them with the techniques of chemical engineering originally developed by the brewing industry, and a multimillion-pound venture was born. In the early seventies, almost overawed by the potential of the new techniques, leading molecular biologists (such as the signatories to the Asilomar letter initiated by the American molecular biologist Paul Berg) themselves pointed to the dangers of 'accidentally' making a potentially harmful organism which might escape control, and called for tight state regulation of the new labs. In a rather short time the selfsame group seemed less perturbed at the potentially vast new profits which seemed to lie before them as almost overnight academic molecular biologists became entrepreneurs, forming new science-fiction sounding companies (Genentech, Celltech, Cetus and so on), and, bankrolled by big investors (not least the drug firms), rapidly became paper millionaires as the biotechnology bandwagon began to roll. By the 1980s, some of the more wildly optimistic prophecies of what the new biotechnology might produce, not least a new green revolution, had begun to wear distinctly thin, and the same biologists who cautioned against the hazards of the new techniques began to argue that the restrictions be lifted. But in the meantime biotechnology has opened a whole new phase of research within which the discussion about the new reproductive technologies must be set.

Predictive medicine or eugenicism by another name?

Predictive medicine for genetically transmitted or congenital disorders or psychological (mis)behaviours[71] raises problems of social meaning which go well beyond the boundaries of existing science and medicine. As Ruth Hubbard and Mary Sue Henifin argue, increased screening in the early stage of pregnancy intensifies the pressure on women to have abortions because of 'abnormalities', while at the same time there is an increased technical capability to preserve premature babies.[72] Biomedicine as both discourse and technology thus increasingly influences which foetus and which newborn shall survive.

Because preventative medicine was historically portrayed as the socially progressive alternative to curative medicine, advocates of the new genetic screening have tended to argue that the possible gains to be made are unproblematic. Yet in the names of prevention and prediction, coercive practices are increasingly being introduced. Already available are methods for screening pregnancies by amniocentesis and chorionic villus screening (CVS can be done as early as the eighth week) for a range of developmental and genetic conditions associated with disability or disease. Weighing up the 1 per cent or 2 per cent chance of losing a pregnancy simply through the test itself as against trying to decide whether to parent a disabled child is no small emotional task.[73] At present, the choices which can be offered to a pregnant woman as a result of these screening procedures revolve around the information that her foetus carries, or does not carry, any particular chromosomal or genetic condition. She can then, in principle, choose whether to have the pregnancy terminated.

In this sense the new techniques increase the choices for women by offering the possibility of at least attempting a pregnancy with the option of not giving birth to a potentially impaired child. But the 'choice' also throws into question the rights of disabled people to have been born and women to choose to bear a disabled child. There is now a 'good practice' consensus that the decision to have a test should be separated from the decision about an abortion. However, the situation reported by Wendy Farrant,[74] where testing and termination were linked if an 'abnormality' was found, has far from disappeared.

Genetic knowledge and screening technology now play a significant part in the decision whether to conceive. To take the most common form of congenital mental impairment, Down's syndrome, which accounts for 35–40 per cent of all children born with mental impairment, 'at-risk' women themselves take steps to reduce the chances of giving birth to impaired children. As it is widely recognized that one of the key factors in producing a child with Down's syndrome is the age of the mother (and to a less clearly researched extent that of her partner), women over 40 have, over the past ten years, increasingly eschewed conception. As a result, of the 1,000 babies born each year with Down's only a quarter are born to older women.

Those in the group most at risk who, by chance or choice, do conceive are more likely than younger women to be offered cytogenetic (screening) services. In the United States access to amniocentesis is not always covered by Medicaid and is controlled by income; in Britain the existence of an NHS does not guarantee adequate provision. A study of the Northeast Thames area revealed that rather under half the women in the at-risk group for Down's received amniocentesis.[75] (Some among those not screened refused, for reasons varying from moral concern to unwillingness to accept the 1 per cent risk of miscarriage. For some the lack of information and language difficulties were a problem.) While the impact of genetic knowledge is evident, the effect of the availability/acceptability of amniocentesis in terms of elective abortions is quite modest, as it appears that only 50 to 100 Down's conceptions are actually terminated over the course of a year. It is this modest actual figure which, despite the heroic claims for screening, has given rise to more radical proposals for non-invasive screening procedures for the entire pregnant population. Technical feasibility and the acceptability of the unit cost – rather than the acceptability to pregnant women[76] – seem to be the two chief considerations in the discussion of these proposals,[77] with the issue of accuracy breaking through at intervals.[78]

But even where such screening programmes seek to protect women's rights to choose,[79] the problem for the intending parents is that while prenatal screening can provide reasonably accurate information about the presence or absence of Down's syndrome, what the tests cannot provide is any guidance to the mildness or severity of the effects of the condition on the child. Would-be parents have, in the context of poor public services for disabled people, to assume the worst outcome and consider whether they can cope. As counselling services supporting screening programmes which might help parents in this difficult decision making are underfinanced, the would-be parents are forcibly reminded that they will be left to cope alone. In these circumstances the action imperative built into any screening test which can only generate either/or answers works strongly towards the abortion decision. For this reason moves to non-invasive population-based screening for 'defects', even though such moves appear under the apparently benign banner of predictive medicine, are unacceptably coercive.[80] The state and professional regulation of reproduction enter as a 'friendly' eugenicism, not codified by law as in Nazi Germany, but none the less increasingly codified in the practice of an expanding clinical genetics.[81]

Whilst Down's is a chromosomal abnormality and has been recognizable by simple microscopic examination of cells since the 1950s, the molecular biology and genetics of the past two decades have made possible the recognition of at least 3,000 distinct 'conditions' which are transmitted from parent to child along straightforward genetic lines of inheritance.[82] These include serious and relatively widespread disorders which result, as yet inexorably, in early death, such as Tay-Sachs disease and the very rare conditions of Lesch-Nyhan syndrome. (Children born with this latter disorder, as they grow, develop a compulsive tendency to bite off their own fingers, lips and tongue.) Tay-Sachs and Lesch-Nyhan are conditions for which treatments are unknown, and may be beyond medical reach, but many more of the long list of known, common or rare genetic disorders are relatively mild or not easily predictable in their effect on a person born with them. As genetic knowledge increases, and more and more human genes are studied, more and more potentially 'deleterious' genes are likely to be recognized, and it is already clear that most people carry a proportion of such genes, without necessarily being in any way disabled.

Whilst the new genetics has theoretically identified the genes, the new molecular biology has opened the way to their detection by gene probes in tissue removed by amniocentesis or CVS. On the cards, then, is the possibility of detecting an increasing number of conditions during pregnancy which may or may not have disabling effects on the child or later adult. But the potential goes far beyond this.

Predictive medicine and biomedicine's perfect body

In many ways the arguments for mass screening for genetic defects constitute a remarkable re-run of the history of the biochemical screening debates for medical patients in the sixties. Then all sorts of promises were made concerning the advantages of mass screening for the early identification and management of disease; the interests of the biochemical pathologists in extending their domain became as one with the sales talk of the firms selling huge auto-analysers. Medical insurance programmes in the US pioneered the biochemical screening approach, and the NHS trailed after. Problems thrown up then within health-care circles are echoed today by the much more public debates around genetic screening. Just what is the right course of action if a biochemical assay used in a screening programme unequivocally points to a number of people as 'diabetic' even though some or all of them do not feel any 'diabetic' symptoms? What has happened to the concept of a 'normal' measurement, and how are these so-called 'abnormal' findings to be interpreted? Just what is the right action if would-be parents are told that they are carriers for a disorder but do not know whether the condition in any specific offspring is likely to be mild or serious in form? Meanwhile clinical genetics, in the name of extending choice, takes an increasing slice of the NHS pie.

But the social implications are much wider, for employers may wish only to hire those with a 'good' genetic profile, while those lacking such a profile may be relegated to the unemployable. The pressure in hazardous employment may be to search for the particularly robust worker rather than to clean up the hazard. Given the way the reasoning and the reproductive technoscience move from animals to people, and the fact that, for example, fish molecular biologists are currently employed to engineer salmon which can cope with the polluted environment of fish farming rather than ecologists being employed to make the salmon's environment more habitable, future pressure to modify human beings cannot be entirely discounted.

Basically, the political ploy of the would-be mass genetic screeners is to talk up the most cruel medical cases, such as Tay-Sachs, Lesch-Nyhan, cystic fibrosis, Huntington's or the acute form of thalassaemia, where the issues are so painful that relatively few women would choose to carry the foetus to full term. These predictive doctors – though eugenicists would be the truer concept – seek to transfer this apparent cruel clarity onto other areas of genetic impairment, without acknowledging that the issues, not least when they involve behavioural manifestations, as with mental illness, are much more complex.

However, it has been the highly publicized story of the search for the gene implicated in one of these conditions – Huntington's disease, where irreversible dementia develops, usually in mid-life, followed by death within ten or fifteen years – which has played a dramatic part in educating professionals and public alike about the cost of genetic knowledge where there is no therapy. What is special in this story of the search for the Huntington's gene is that a key story teller is a potential patient and therefore silenced in the usual accounting practices of biomedicine. For the response of psychologist Nancy Wexler to the information that her mother had Huntington's was, with the support of her sister and the financial resources of her father, to throw herself into the pursuit of the gene, on the grounds that only the identification of the gene would yield certainty.

The tenacity of Wexler and clinical geneticist James Gusella resulted in the location of a marker associated with the disease in the early 1980s, and, together with a much larger team, the gene itself in 1993. But the new tests confronted Wexler with a new predicament – uncertainty and hope may be better than certainty without hope. Even now, finding the gene removes the first problem but not the second. Finding a gene associated with a condition does not necessarily – or even readily – provide any clue to how to treat the condition, other, of course, than abortion.

There is as yet no treatment in sight for Huntington's and Wexler herself has steadfastly refused to say whether she has been tested,[83] instead campaigning for public education to help people become aware of the meaning of screening.[84] Many people who share Wexler's possible genetic status have elected not to be tested, and of those that do elect, more are unaffected than would be statistically predicted. Clinicians working closely with such families suggest that those electing to be tested may be aware of subtle differences between themselves and affected relatives which increase their likelihood of being 'cleared'.

Wexler's participation in the construction of the Huntington account does three things. First, it underlines the complexities which confront those who are at risk of inheriting or transmitting such serious disorders. Second, it points to how, in establishing a clear diagnosis for one individual, the status of other family members becomes evident, so that people who have no desire to know risk learning their genetic risk status. The genetic link none the less means that the concept of the autonomous individual, so central to liberal theory, is seriously weakened. In good clinical practice the pregnant woman is recognized as the individual who makes the choice about keeping or not keeping a pregnancy, but these new diagnostic procedures throw such simple individualism into question.[85] To find out the risk may mean more knowledge than she or others biologically close to her want to have. Third, and very important for a radical social construction of science which wants to include historically excluded social actors in the process of creating the facts of science, Wexler actively constructs knowledge and is no longer a passive victim of genetic reductionism. She is currently an important figure in the politics of predictive medicine and has been recruited to chair the Ethical Legal and Social Implications (ELSI) of the Genome programme – with a budget of 5 per cent of the total cost of the project. Positively, her compelling account of the social meaning of testing has been able to challenge the previously technicist discourse. The arrogance and eugenicism of predictive medicine, which thought that screening could be constructed as a simple matter of accuracy and cost, has been moved towards a more humane discourse by a combination of her persistent campaigning and the receptiveness of the research community to her arguments.

While finding the gene for Huntington's is likely to change the predicament around testing for that particular disorder, the story underlines the questionable value of knowledge without a therapeutic response. That issue may be more difficult for the perspectives within which ELSI operates and points to difficulties ahead, even for Wexler, as ELSI seems directed towards getting the public to accept the Genome programme.

In less painful cases, where the diagnostics may be also less dangerous, the emphasis on identifying the transmission of genetic variation or identifying genetically impaired foetuses has the negative effect of devaluing people with disabilities. Should they have been born? Were their parents – above all was their mother – failing in her duty in giving birth to a disabled child? Listening to the discourse of predictive medicine, the underlying message is 'screen, abort or fix'. For this discourse, both the rights of disabled people and the complexity of decision are undervalued. The old and powerful mechanism of devaluing the victims and blaming mothers has been activated, and can run and run.

Gene therapy: the imperfect made perfect?

Many molecular biologists now believe – and their belief is being backed by funding from drug companies, venture capital and the state, as part of the biotechnology boom – that it is not sufficient merely to detect a potential genetic variation whilst the foetus is in the womb and to offer the choice of abortion. Now that it is possible to clone and alter genes in the laboratory, the way would seem, in principle, open to offer to 'correct' a faulty gene in an unborn child. The idea would be to diagnose the variation as soon after conception as possible and then to introduce the new genetic material in a form that could replace or over-ride the faulty gene in a number of cells in the tissue, whose functioning would ensure normal development. This is somatic cell manipulation. In the early eighties this was seen as more or less imminent, but, as the decade went on, both the technical difficulties and the personal complexities became more apparent. Molecular reductionism was not enough.

Progress, despite the intensity of research, has been relatively slow; for example the gene implicated in cystic fibrosis, one of the most intensively researched, has been located, but its mechanism of action, and hence any potential therapeutic intervention, remain elusive. Furthermore it turns out that a great many different alterations of the DNA in this gene are implicated in mediating the symptoms of CF in different individuals. The magic-bullet thinking behind 'gene therapy' was too simple-minded for the complexity of the disorder. However, by January 1992 somatic gene manipulation was sufficiently developed for the US Food and Drug Administration, on the grounds that risk levels now met treatment guidelines, to agree to a small number of clinical trials. Related tests are now under way in Britain as well.

The second, and even more ambitious, objective of the new genetics seeks nothing less than modifying the gene pool itself (germ-line manipulation). This is not argued for publicly for the human genome, but the potential inference and the technological transferability from green nature to human nature lies beneath the surface of the discourse. First, the new genetics sees itself as able to locate 'faulty' genes. By replacing these, in the early embryo, in the fertilized egg and in the sex cells, the undesirable trait would no longer be propagated in the offspring. In this eugenicist's dream the deficient gene is thus gradually eliminated from the population (though even in theory this elimination would only be feasible for dominant genes). Because of the large and frightening measure of uncertainty, most regulatory systems have moved to criminalize germ-line manipulation in humans.

But these techniques have been used, with increasing technical success, in non-human animals and in plants, and are now relatively straightforward. Mice, for instance, have had genes inserted for the production of growth hormone derived from humans or rats; the result has been the production of extra-large super-mice. New forms of plants have been produced. Biodiversity has become, as the result of the work of the bioengineers, simultaneously more predictable and diminished; the refusal of Bush, as president of the world's leading biotechnology power, to sign the Rio Biodiversity agreement in 1992 sent a clear signal that the US state is deeply committed to technoeconomism. Clinton's subsequent signature with commercial caveats is a less than convincing harbinger of change.

Molecular biology's holy grail: mapping and sequencing the human genome

The development of the Human Genome Project is a product of the political economy of the eighties and unquestionably marked the self-inserted entry of molecular biologists into runaway industrialized science. The idea of 'mapping' the human genome had been articulated within the scientific community since the late 1970s. At that stage the project was seen as the gradual making of a map of all the genes on the human chromosomes, so that it would be possible to read from the map that genetic disorder *A* was located at a particular location on chromosome 19, much as Leeds is located in England or Bombay in India. The map was to be built up through the coordinated activities of many small research groups; this co-operation would extend across national frontiers and build on agreements made at the regular international genetics conference, where such matters as the numbering of the chromosomes was agreed. Such international co-ordination, setting standards and agreeing categories, has been intrinsic to the modern global production of academic scientific knowledge and, as piecemeal co-operation, would have gradually built up a detailed map of the important genetic sites, but might well have left large tracts rather sketchily explored and certainly unse-quenced. Such co-operation is possible when science is relatively autonomous or 'far from market', and for that matter 'far from defence', as the market and military concerns for secrecy and commercial or national advantage work as much against this as against any other form of international co-operation.

However, the proposal, made at a meeting of molecular biologists at Santa Cruz in 1985, that the entire human sequence should be both *mapped* and *sequenced* represented a very deliberate move on the part of a number of elite-based molecular biologists into 'big science', with all that that entails in terms of a very visible relationship with industry and capital[86], and a much less visible relationship with the military.[87] For sequencing entails determining the order of the nuc-leotides of the long double helix molecules of DNA located within the nucleus of every cell of the human (and non-human) organism. Because the double-stranded DNA molecules are copies of the molecules inherited from each parent organism, they carry the genetic code; that is, according to the molecular biologists, nothing less than the information and instructions from the past to the future by which genes are turned on and off, determining the form, functioning and behaviour of the organism. Molecular biologists do more than draw attention to the centrality of the DNA molecule, and make no small bid for authority, when they speak of this process as 'the central dogma'.[88]

Nature as text

Sequencing is both breathtakingly simple and complex, for each double strand of DNA is made up of only four nucleotides arranged in pairs of the bases adenine (A), cytosine (C), guanine (G) and thymidine (T). A sequence simply indicates the order in which these four nucleotides are arranged along the DNA. But ACGT can in principle be almost endlessly arranged and rearranged within a strand of DNA; even one gene lying along a section of the DNA can be made up of a sequence of 10,000 nucleotides stuttering its AACTGCCTATTG along its length. Sequencing the entire genome, which includes many genes, of any species is a formidable task. Sequencing the human genome, with its 3 billion nucleotide base pairs, including at least 100,000 distinct genes embedded amongst them, thus requires a mass-production technique to identify the order of the separate nucleotides and an immense information-handling capacity to manage the great chains of letters. The new genetics mirrors the postmodernist discourse and reduces the complexity of nature to text.

As in many areas of science, the initial sequencing of DNA molecules was a product of immense craft skill, marked by the award of the Nobel Prize in 1975 to the three molecular biologists who pioneered such techniques: Walter Gilbert, his fellow American Allan Maxam, and the Briton Fred Sanger – for whom it was a second Nobel (his first was for the first sequencing of a protein, insulin). What the 1980s offered was the possibility of automating sequencing, a process which, because of the collaboration between science and the scientific equipment industry producing new laboratory-scale technology, was already transforming the analytic tasks. Thus one small, computer-controlled machine, producing sequences automatically, replaces many hours of painstaking manual laboratory work. By the nineties this replacement of labour was to take full form in the automatic laboratories of the French company Généthon. This highly industrial-ized facility, located in the countryside outside Paris, has robots and technicians rather than scientists. Not by chance is the new sequencing factory described as 'blue-collar' and 'hands-off' science.[89]

The original push for sequencing DNA and thence for industrial production came most strongly from within the US (currently spending some $154 million) with the Europeans by and large being more committed to mapping but prepared to acknowledge that sequencing around specific locations within the map was valuable. These pressures for sequencing were self-consciously developed by a relatively small number of men, whose scientific pre-eminence as leading molecular biologists endowed them with significant cultural capital. It is the possession of this intellectual capital which enables such men to make further alliances with power, so acquiring the financial and political resources which enable them to influence both the direction and the content of science and culture.

Reading the accounts of the pressures and counter-pressures for sequencing and mapping, there is a strong sense of a scientific and political debate which has two levels of instrumentality. First, the sequencers wanted to push for massive new resources and to break away from the piecemeal mapping tradition; second, while mappers thought that theirs was the more scientifically justifiable approach, and better served the interests of families confronted by major genetic disorders, they were far from unhappy at the expansion of resources into their fields while science generally was protesting about underfunding (not least in Britain, with the erosion of its research base under the Conservatives and the increasing outcry from the Save British Science group).

Something of this account of almost stage-managed divisions and accommodations is revealed in the part played by the committee of the international Human Genome Organization (HUGO) set up to co-ordinate the international effort. HUGO, as a self-selected merito-cracy of the great and the good molecular biologists, was the brainchild of Sydney Brenner, the director of the Cambridge (England) Laboratory of Molecular Biology. (Brenner himself, while an early supporter of the Human Genome Project, argued for sequencing simpler genomes, such as those of yeast or the worm *Caenorhabditis elegans*, on which his own team works, first.) HUGO, through its first two presidents, was to play a crucial part in creating ideological space for both sequencing and mapping. Thus Victor McKusick, the first HUGO president, in his influential report to the powerful US National Academy of Science, argued that both mapping *and* sequencing were necessary and appropriate. The London-based Walter Bodmer, as the second president of HUGO, continued to use a rhetoric which skilfully wove together the language of international co-operation of the 'small-science' mappers with that of the mass-production approach of the 'big-science' sequencers – a rhetoric which made him highly acceptable on both sides of the Atlantic. But before the HUGO presidents could create their rhetoric of reconciliation between the mappers and the sequencers, the latter had to intensify the political interest in sequencing as the key to the cornucopia to be brought by industrial-ization. Again, while HUGO claimed to be interested in the social, ethical and legal implications of Genome, by late 1992 there were reports of only one meeting, no clear plan as to how these implications might be explored, and no indication that the molecular biologists were prepared to let non-biologists explore such matters.

Within the US the Genome Project came rather curiously to have two sponsoring homes: one, the National Institutes of Health (NIH), comes as no surprise, but the other, the Department of Energy (DOE), appears more unlikely, and is only to be understood within the specifics of US science politics. To British eyes the DOE appears an odd location for a move towards big biology, yet in terms of its own institutional history the move is all too explicable, as the DOE is also the sponsoring agency for the nuclear programme. When the Nixon administration confronted the oil crisis in the seventies, the old Atomic Energy Commission, which had presided over both atoms for war and atoms for peace, was placed inside the new department, whose brief was to solve the energy problem. A decade and more later, with oil almost a glut on the world market and a very different politics around energy consumption, the DOE was an organization with a taste for 'big science', and a need for a new agenda acceptable to the exceedingly ungreen administration of Bush. Further, it had a second crucial resource for managing the data mountain thrown up by sequencing – the computing facilities of the nuclear programme at Lawrence Livermore. As a third resource, the laboratory also held the data concerning the genetic mutations from the testing and use of nuclear weaponry. The Human Genome Project – whatever its scientific merits or demerits – was tailor-made for the DOE.

While each potential Genome sponsor gathered its own scientific luminaries in support, the entire project developed its rhetoric of justification using a metaphor employed by Walter Gilbert, as a prime mover within the entire research field, as a call to action. Thus, although the discussion of mapping and sequencing – especially the latter – began at a meeting in 1985 in Santa Cruz, convened by the molecular biologist Robert Sinsheimer and attended by Walter Gilbert, David Botstein and Leroy Hood, it was at a DOE meeting in 1986, called by biophysicist Charles Delisis, that Gilbert drew on the religious metaphor of the genetic code as the 'grail', which inspired the meeting, and which has continued to surround and sustain the project. That Gilbert invoked the pursuit of the mystery of medieval Christendom to explicate the twentieth-century search for genetic explanation might be written off as a personal taste for medieval metaphor, but when a professional community where, as Richard Lewontin[90] ironically notes, there is an unusually high concentration of Jews and atheists, continues to echo that metaphor, then something more than personal idiosyncrasy is being displayed. Even Jim Watson's ostensibly less religous metaphor of Genome as 'the Book of Life' carries an undertone of religiosity for the Peoples of the Book.

For this massive investment in molecular biology to take place it was crucial to capture the interest of 'the public', a concept better understood as several publics, notably the politicians, industry and finance capital, as resources. For capital, what was on offer was the potential profits to the pharmaceutical industry as gene sequencing identified the potential sites of what were portrayed as disease-threatening genes. Diagnostic kits to detect the new genes in routine screening were to be followed by therapeutic genetic interventions. The criminal justice system was offered nothing less than absolute truth through the so-called genetic fingerprinting.[91] The commercial thrust of the programme was made transparent when, in 1991, an NIH researcher, Craig Ventner, isolated a group of gene fragments (of unknown significance) from DNA from human brain. The Director of the NIH, Bernadine Healy, moved swiftly to attempt to patent the sequences, to the consternation of many of HUGO's luminaries (including James Watson), who argued that to do so would destroy the fragilely maintained international nature of the collaborative effort. Watson lost his fight with Healy, and resigned/was fired as director of the US end of the project.

Predictably, in Britain the MRC responded by taking protective action to hold the copyright on its own sequences. As late as April 1993 the US Patent Office was opposing NIH's patent claims, but the final outcome is uncertain. Meantime, the desire of the molecular biology elite for an alliance with industry, to both bring rich rewards and take away the routinizable aspect of sequencing, is demonstrated in the enthusiastic welcome given to the news that Ventner's sequencing group (at the centre of the patenting conflict) was leaving the NIH to form a new, highly automated and robotically controlled, non-profit facility. That this non-profit facility was to be financed by venture capital was indicative of the social mutant created by the elite researchers and the commercial world.[92]

Meanwhile, the metaphor of the grail fosters a sense of mystery about to be revealed to the true knights of sequencing; sustaining these knights-errant as they pursue the grail becomes a privilege, for the knowledge of the mystery will reveal the ultimate truth about ourselves.[93] It works to re-romanticize scientific research, not least for the scientists themselves, for in an age when the romance of science has been rubbed off by historical experience, the metaphor speaks of science as a heroic, moral and deeply masculine activity. As such the metaphor reinvigorates the ideology of science, for it returns to the spirit of the foundational text of molecular biology, the story of the discovery of *The Double Helix*, in which three knights, but above all the youngest and most nobly innocent, take part in a successful quest. That James Watson's text was received with tremendous enthusiasm by the scientific community as telling about science 'how it is' speaks of how men scientists see themselves, both as willingly bound by the chivalrous code of scientific method and also as the embodiments of daring masculinity.

In a very straightforward sense the knights of the human genetics grail are as historically gendered as the knights of the medieval grail. It is not by chance that all the most public players in the Genome Project are men, and their shared metaphor reflects and constitutes their view of the world and themselves in it. But perhaps what is less evident is just how deeply the metaphor links gender, estate and sexual violence, for the rules of chivalry, among other matters of knightly conduct, detail the conditions under which it is entirely appropriate for a 'parfit gentil knight' to rape a woman and those when it is appropriate to wear a woman's favour. The crucial distinction is one of birth, being born unfree or being nobly born. The historian Mark Adams has reminded us that 'eugenics' is literally the 'well-born science'[94] and his colleague Daniel Kevles[95] has drawn attention to the slippage between genetics and eugenics. The metaphor of the grail offers to smooth the path.

Whilst putting the mystery back into science was crucial to mobilize scientists, particularly as they were poised to enter with rather less publicity the serious business of making money, the other great legitimator of the Genome project was, for politicians and the public alike, illness and the risk of dying – above all from cancer. Cancer as a major killer within contemporary Western society is also a powerful generator of research funds. Cancer researchers were not to be left out of joining this well-resourced move within the scientific culture which looked inwards, rather – as with tobacco or other pollutants – than outwards to the environment. Thus the influential Renato Dulbecco of the Salk Institute declared strong support for sequencing and mapping the human genome, arguing that the cancer research community must concentrate on the genetics of cancer. Dulbecco sets up the choices as between, 'either to try to discover the genes important in a malignancy by a piecemeal approach or to sequence the whole genome of a selected animal species'.[96]

In endorsing the taken-for-granted genetic explanation of cancer, in which the only choice is either piecemeal or wholesale sequencing, the possibility that cancer-producing environments and life styles have some part in the explanation is put outside consideration. With one rhetorical move the entire tobacco industry, with its responsibility for lung cancer and other smoking-related heart and chest diseases,[97] disappears. As always the hint of a cure for cancer is a power move for science; it reaches people's fears, not least of men in high places, and reminds industry of immense potential markets. Unquestionably Dulbecco's intervention in the US and the early association of Walter Bodmer (head of the Imperial Cancer Research Fund) in the UK gave a medical legitimacy to the Genome Project.

Opposition to the Genome Project

The initial, and indeed continuing, response to the programme from within the biological community was divided, and nowhere has the debate been expressed more sharply than within the US; the scientific claims of the genome project have been questioned, its consumption of an excessively large slice of the biology budget criticized, and philosophically its biological determinism challenged. Many have been or are disturbed by the project's eugenicist implications. In *Science* in 1989, MIT biologist and Nobel Laureate Salvador Luria argued that the Genome Project 'has been promoted by a small coterie of power-seeking enthusiasts'. He went on to denounce the eugenic implications of Genome, implications previously hinted at by *Science*'s editor Daniel Koshland[98] but presented as benign, as potentially helpful to the aged, infirm and homeless. Luria asked, 'Will the Nazi programme to eradicate Jewish or otherwise "inferior" genes by mass murder be translated here into a kinder gentler programme to "perfect" human individuals by "correcting" their genes in conformity to an ideal "white Judeo-Christian, economically successful" genotype?'[99]

Feminists, including Anne Fausto Sterling, Haraway, Keller and Hubbard, have been strongly represented among the US science critics ranged against the Human Genome Project. Hubbard suggests why it is that US scientists and the US public feel that they have a special responsibility, in that it is within the the US that most of the research is being carried out. However, while these critics and also the US-based Council for Responsible Genetics[100] (which in many ways is an inheritor of the radical science movement) have played a significant part in criticizing the Genome Project and helping to contribute to an informed and reasoned public understanding of science, they are having to take on both science journalism and the public relations accounts of Genome's leading protagonists. The former has produced a number of well-informed accounts, which, however, largely reproduce the views of the project's leading figures. Thus Joel Davis invites the reader to believe that 'Mapping the human genome will be the greatest scientific and technical achievement of this century, greater even than the invention of the atomic bomb.'[101]

Such journalistic forays make no attempt to penetrate the sociomedical rationale bred by the biologists committed to the project; by contrast, they emphasize that Genome costs *only* $3 billion – cheap compared with much of physics – and assert that the Human Genome Project is a cost-effective way of improving health. In doing this the journalists ignore the interrogation by historians and epidemiologists of medicine's contribution to public health, in which clinical intervention is seen as contributing very much less than the social and natural environment to health outcomes.[102] Where the new public health looks *outside* to the context of everyday life to explain why human beings get sick or stay well, live shorter or longer lives, the new genetics looks *within* to the determining code.

Thus the Human Genome, as the highly visible face of the new genetics, serves to devalue any public health attempt to bring about improvements in the contexts in which human beings realize or fail to realize their health potential (though, in the usual contradictory way, individual scientists may be associated with both). The overlarge claims of Genome stand as a negative icon against the global objectives of the Health for All programme of the World Health Organization, with its generous discourse of 'more years to life and more life to years'. Striking a radically different note from either the genetic turn or for that matter the individualistic healthism which has obsessed the eighties, WHO indicts war and want[103] as fundamental impediments to the realization of health, and seeks to foster healthier life styles through healthy public policy in transport, food, agriculture, energy and not least economics.[104] The project of securing greater public health requires political, economic and environmental policies,[105] with rather little by way of medical intervention.

To take just one increasingly visible target, tackling the tobacco industry in a coherent way (which the UK government failed to do yet again in its 1992 Health of the Nation report, instead relying on exhortation as its key weapon), from the time and place at which tobacco is grown to the point at which people get sick and die from its use, would require a major political rather than a primarily medical or biological effort. A sustainable policy would require alternative crops for tobacco farmers, cigarette factories diversified, advertising stopped, taxes maximized and places to smoke limited – all still without declaring tobacco, unlike narcotics, illegal – but these actions would mean taking on the economically powerful. In that third-world women are disadvantaged by cash crops like tobacco and that it is young women who are increasingly taking up smoking, feminism has particular stakes in this struggle. By contrast, Dulbecco's reductionist claim for molecular biology, that the complete sequencing of CAGT offers *the* way to eliminate cancer, borders on the Cruel And Grindingly Trivial.

But the public health arguments are lost in a journalistic celebration of technological geewhizzery, in which scientists heroically compete to map the mutant genes that lead to cystic fibrosis, Huntington's, cancer etc., mixed in with mawkishly sentimental individual case studies. (This mix, in different proportions, seems to be the central formula for science journalism's many book-length accounts of Genome.) In these accounts, brave mothers of profoundly disabled children, and noble child and courageous adult candidates for screening, provide the moral motive force for the heroic pursuit of science. Yet this gendered sentimentality can mask the complexity and the costs. One such case study concerns Betty Le Blanc, with no fewer than three children with Friedrich's ataxia. Her appeal to genetic clinicians for help led them to recruit this far from well-off woman to fundraising to sustain the hunt for the gene. But in telling this as a success story, the message that Nancy Wexler had painfully learned is ignored, for finding a marker or even the gene is not the same as finding either cure or control. That can be many years away. Indeed the limited technical achievement of prenatal diagnosis can only lead to the offer of therapeutic abortion, a particularly ambiguous 'help' to a woman who, like Betty Le Blanc, happens to be Catholic.[106] But the fusion of technomania and sentimental tales is yet another powerful rhetorical device for biomedical science, and what it masks is the danger of producing diagnostic tools, with no other remedies than abortion, which serve to intensify an ideology of eugenicism.

As Evelyn Fox Keller demonstrates in her examination of the rhetoric of the leading ideologues, they have with frightening success proposed a model of the world in which the Master Molecule of DNA controls not only the body physical but also the body politic. Nothing, not even the drug and alchohol dependency, mental illness and widespread homelessness which deforms the body politic escapes DNA's explanatory power; thus even the homeless have a stake in the success of the Genome Project.[107]

Science in its place

Away from the inflated discourse of the Genome project's elite ideologues and their media acolytes, down among the working geneticists and molecular biologists, the picture is at once less glamourized and more oriented towards understanding how biological processes work and what therapeutic measures might be devised to help individuals and families facing intractable biologically transmitted challenges. Perhaps unsurprisingly, in these less heroic circles there are substantial numbers of women researchers, who, in the context of a still influential women's movement, often have a clear sense of themselves as 'women' and as such as being more open to relational issues.[108] Some, echoing the fractured identities of postmodernism, speak of feeling inconsistent, of feeling like different people, both enchanted by the technical complexity and also concerned at the relational and moral issues. But here among the non-elite scientists, whether women or men, the term 'gene therapy' is rather rarely used, for this is more redolent of the earlier proselytizing phase of innovation, when a talking up of the therapeutic and knowledge promises by the masculinist ideologues was integral to attracting resources. Here, more modestly and probably more honestly, gene therapy is seen as reserved for the exceptional situation, as many, even most, genetic disorders lose the magical simplicity of the sales pitch and recede into tremendous biological complexity. The science of genetics here takes its proper and modest place amongst a range of knowledges and techniques (its emphasis on the inevitability of difference resharpening the case for the necessity for a national health service), which together might contribute towards the goal of Health for All.[109]

Gene therapy, that dramatic promise embedded in the meta-discourse of a handful of extremely powerful men biologists as they sought yet more power, including economic power, is none the less receding from their public discourse as the Genome Programme develops and issues of genetic complexity become inescapable. The rhetoric of the elite ideologues, particularly in Europe (the US rhetoric is almost unmodified), begins to speak more of uncertainty, less of genetic silver bullets and more of cautious gains and the possibility of more effective pharmacological intervention.[110] Molecular biology as a technoscience has enthusiastically embraced the market, but it is learning, like other entrepreneurs, to be cautious about the resistance of the 'consumers' of its products, whose opposition has been mobilized by the powerful movements and discourses of feminism and environmentalism.[111] One of those critical discourses has been provided by the burgeoning of feminist science fiction, which has taken the issues raised by science in our daily lives to a much wider readership than those ever reached directly by the feminist critics of science. It is to this rich genre that I now turn.

9

Dreaming the Future: Other Wor(l)ds

Prose can push the frontiers of knowledge about ourselves
forward. It can keep awake in us the memory of the future that
we must not abandon on pain of destruction.
                        Christa Wolf, *The Reader and the Writer*

While literary criticism has only relatively recently taken science
fiction seriously,[1] both scientists and critics of science have long had a
close and much less discussed relationship with SF.[2] Literary critics
often see the thirties and forties as the golden age of science fiction,
but they are slow to recognize that this was also a golden age of a
science criticism which took writing futurist accounts of science, if not
actually SF, as a seriously pleasurable task. Desmond Bernal wrote
both *The World, the Flesh and the Devil* and also the influential *Social
Function of Science*.[3] Both the journalist and feminist Charlotte
Haldane and her husband J. B. S. wrote SF scenarios concerning
genetic engineering.[4] She concentrated on the negative implications
of sex selection for the future of women while he celebrated the
possibilities of cloning clever men and (uniquely among male
scientific futurists) clever women. For that matter Katherine Burdekin
used the form in *Swastika Night* to draw attention to the entirely
dystopic future involved in the cult of masculinity inherent in the
new national socialism, while Dora Russell's *Hypatia, or Woman and
Knowledge* offered a feminist utopia of a non-authoritarian and non-
technocratic society, sharply at odds with the prevailing masculinist
and technocratic enthusiasms.[5]

Some feminists viewing feminist SF from a literary perspective, such as Lucie Armitt,[6] while first paying respect to Mary Shelley as the founding foremother of SF, see its flowering over the last two decades – as part of a general flowering of women's writing. Others, above all Joanna Russ, have proposed SF as a genre peculiarly appropriate for feminist writing as it provides a vehicle for exploring our pressing anxieties and experiences concerning science and technology, in a way which is not possible within more traditional genres such as the novel, where what is expected limits the feminist imagination.[7] That Russ argued this in the early days of 1974 was prescient; she went on herself to become one of the leading feminist SF writers. Since then feminist science fiction at it best has been a means of raising social and political issues around science. Located at the margins of literary culture, it has provided an ideal framework for the perspective of marginal people whose experiences and concerns have no place in the dominant androcentric culture and its forms. Paradoxically, through the creation of extravagantly fictitious worlds in which everyday reality becomes strange, there emerges the possibility of dreaming of (or having nightmares about) different and other futures, of writing new myths which will enable us to take a part in shaping our futures.

For feminism the fact that, in Britain and to a lesser extent in the US, SF is both innovatory and from the wrong side of the literary tracks has meant that, like other new and low-status forms, whether in the arts or in the sciences, it has been a relatively accessible genre for women writers. SF in the seventies and eighties, like crystallo-graphy or biochemistry in the forties, has been relatively open to creative women. Its success in the seventies and through the eighties has both been reflective of and constitutive of the feminist critique of science. Arguably it has been feminism's golden age of SF. Yet the current recovery of SF by literary criticism and cultural studies, which is part of an important and welcome attempt to dissolve the divide between popular and high culture, has often underplayed the close relationship between science criticism and SF, not least within feminism. It is as if, while taking down that cultural divide, another between the arts and the sciences is allowed to reproduce itself uncriticized. It is this division, a sort of replay of Snow's two cultures, even though the categories themselves constantly shift, that I want to see removed.

But before discussing feminism's dreams of the future as reflected within SF, I want respectfully to dislodge the priority claims made for the writer Mary Shelley (for she is my heroine too) as the foremother of SF and replace them by those of the natural scientist and philosopher Margaret Cavendish. Not only was Cavendish an immensely prolific thinker and writer who published books on a number of scientific topics and played an influential role in the formation of mechanical philosophy in mid-seventeenth-century England, she was also a duchess, and, as a Cavendish, a member of a particularly powerful dynasty (see my choice of epigram for chapter 7). However, neither her scientific contribution nor her noble rank could offset her sex, and the Royal Society excluded her from membership. Cavendish responded by writing the first feminist science fiction. Her utopia, *Blazing-World*,[8] is an island which she reaches as the sole survivor of a shipwreck. Here she marries the emperor and, as Margaret the First, presides over the development of an educational and research system whose students and scholars are part beast, part man, and where she at last receives the love and knowledge denied her in the real world. *Blazing-World* is written not simply *against* the Royal Society and its founding project of the new masculine science which would conceptualize both women and nature as to be subdued, but *for* a new scientific community where love and the acquisition and production of knowledge are reconciled, and where the lines between nature and culture are softened.[9] Totally egocentric, it none the less passionately speaks of this woman's longing for education and knowledge.[10]

Although literary criticism offers diverse interpretations of Mary Shelley's Monster, ranging from the representation of the proletariat to that of the second sex, I want not to enter into the essentially literary discussion as to why all such subjective readings are possible and not competitive, but rather to point to the powerful ambiguity of the dreaming which triggers these rich interpretations and which she invites us to share.[11] As in a dream, one interpretation, one representation transmutes effortlessly and seamlessly into another; everything is possible, but what is consistent is the feeling of pain and suffering of the Monster and the revulsion it inspires in its creator. The Monster understands that the scientist has created it, but has done so without love, and remonstrates with Frankenstein, 'I ought to be thy Adam: but I am rather the fallen Angel, whom thou drivest from joy for no misdeed.' With surely a fitting historical irony, the Monster has become in everyday naming 'Frankenstein'; the patriarchal scientist and his monster have fused. Yet in that separation between love and knowledge which Shelley saw as at the heart of the Promethean myth she creates the dystopic counterpart to Cavendish's utopia.

Mary Shelley's fictious world is constructed through three concentric stories: the first told by a stranger, the second by the creator Frankenstein, and the third by the Monster himself. Despite the inexperience of the writer, her powerful intelligence and imagination compel the reader to share the horror and pathos of the scientist's Promethean creation. In a new introduction written thirteen years after *Frankenstein* was published, Mary, now widowed and a professional writer with a son to keep, described the genesis of the book. Mary and Percy Shelley were in Switzerland at the time as part of a house party including Polidori and Byron. At the latter's suggestion each agreed to write a ghost story. Mary, at the time only

19, described how she spent the evenings listening to the philosophical debates between the men, including the possibilities of life being
generated through electricity. She was also reading scientific books
including that of Humphry Davy. She describes how she suddenly
realized that this Promethean theme provided the material and the
inspiration to write her ghost story. In considerable measure she is
politically and philosophically her parents' child, and she fuses her
mother's feminism and her father's theory of the necessity of social
benevolence.[12] Her Gothic story today, one hundred and seventy five
years on, still creates an indelibly ambiguous and disturbing
elsewhere, like and unlike daily life. Her murderous Monster is the
product of a masculinist science which denies women and love their
parts in creation.

Within very early first-wave feminism Shelley makes that connection between science and gender that second-wave feminism had to
recover. In *A Room of One's Own* Virginia Woolf reminded us of the
historical importance of the change which came about at the end of
the eighteenth century, 'I should think of greater importance than the
Crusaders or the War of the Roses. The middle class woman began to
write.' As part of this momentous change Mary Shelley created in her
story of Frankenstein her icon of a dawning science/gender system.

Confessions of a compulsive reader

Science fiction has often been criticized as mere escapism, and if I look
back at my own changing relationship to the genre, I can see the
justice of that criticism, not least in terms of my own graduation from
being a compulsive escapist consumer, to being a refusenik, and
much later becoming an interested reader – and perhaps more – in
feminist science fiction. As a very young woman I was a compulsive
reader. I suspect that the class antagonism I experienced as a
'scholarship girl' attending a snobbish, upper-middle-class school in
the immediate post-war period meant that voyaging within my head
was less precarious than everyday life. In pursuit of a nameless but
insatiable hunger, I indiscriminately consumed (surely never 'read' in
the sense my English literary friends use) an incredible quantity of
books. Actually that is not quite true. I read about some things with a
single-minded determination to make sense of them. It was the
poetry and novel reading where the nameless hunger ruled supreme:
nineteenth-century realism, eighteenth-century precursors, fantasy
and SF, crime, romance, modernism – all demanded that I read them.

The external hierarchy between genres or writers made little impact
on these activities. Particular authors, sometimes just individual
books, mattered. *Orlando* I read and reread, whereas *To the Lighthouse*
or *Mrs. Dalloway* could only hold my attention once or so round.
Other novels written by less grand writers mattered to me in a way

that, at the time and for a long while after, I did not have the words to explain. It has been in the reissue lists of the feminist publishing houses, where I recognize like old friends those books from forty-plus years back, that I can glimpse the then inarticulate principles that were operating. In saying this I do not want to claim a celebration of some naturalistic 'coming home',[13] of some essential self, but rather convey that sense of pleasure and empowerment given by books reread with a new critical consciousness.[14]

Such subjective eclecticism led individual authors and indeed whole genres to crash. They produced only a blank screen, their power to insist that I read them gone. Thus detective stories – despite being so frequently written by women – went except when I had flu. (It was only later that I discovered the wonderful products of feminist crime writers). Romance, another refuge, became boring. Fantasy had a hard time, as with a certain puritanism (not to say priggishness) I thought the task was to change the world, not dream. But the blankest screen belonged to SF. The rejection went beyond merely not wanting to read it, to feeling that it was actively hostile and unpleasant. The pervading quality was a macho enthusiasm for the technology of domination, and its equation of technological advance with progress. Generally politically reactionary, it was frequently racist and almost invariably sexist. As Ursula Le Guin wrote in *Language of the Night*:

> The only social change presented by most SF has been towards authoritarianism, the domination of ignorant masses by a powerful elite – sometimes presented as a warning, but often quite complacently . . . In general American SF has assumed a permanent hierarchy of superiors and inferiors, with rich, ambitious, aggressive males at the top, then a great gap, and then at the bottom the poor, the undereducated, the faceless masses, and all the women.[15]

While Le Guin's own work was an important exception to the dominant tradition of SF her analysis spoke to my revulsion. To no little extent SF seemed to be part and parcel of the genocidal wars that the imperial powers, notably the US, were waging, in Latin America in the 1950s, Southeast Asia in the late sixties, or the Persian Gulf in the 1990s. In SF and in reality, Man saw himself as infinitely irresponsible, always able to move on, to find and conquer new worlds, brutally and carelessly vandalizing and laying waste to 'his' environment. There was always another planet, another third-world country, out there.

Yet simultaneously, against the technocratic dystopias offered by reality and by SF, there was the confidence of the political generation of the sixties which believed that it was in the process of making a new and beautiful society. Whether on the scale of the vast revolutionary movements of national liberation which swept the

world, the Cultural Revolution in China, or the new social movements rising in the old capitalist societies, the sixties were above all an
age of practical utopianism. Maybe in periods of immense social
confidence, when those who have been cast as having no part to play
in the making of history suddenly move on stage as historical
subjects, the need for fictional utopias is less acute. When we feel as
in 1789, 'Bliss was it in that dawn to be alive', or as the graffiti of 1968
urged, 'Let the imagination take power!', then our dreams and
desires and our everyday lives converge; each of us is a poet and our
imaginations are in charge of our futures.

Discovering feminist SF

It was only in the mid-seventies that I began to discover the
burgeoning wealth of feminist SF. At first I was reluctant to like it, for
its evident utopian strand was surely a distraction to stop us realizing
our dreams. I caved in over Marge Piercy's *Woman on the Edge of
Time*.[16] Moved by her meticulous account of Connie's all too painful
existence at home and in the psychiatric hospital, Connie's time
travelling in which she could even 'fly back into the past and make it
all come out right' read initially as a cop-out. But the book drew me
back and I began to read it differently, as a hope-giving construction
of an alternative reality which took off in a positive way from the
everyday life in which Connie was trapped. Whether it was the meta-
reality of the mirrors on the Greenham fence reflecting back the
powers of imperial America; or the result of being able to recognize –
not without a struggle – that Piercy had written of a still problematic
outcome, of dystopic and utopic alternatives; or whether I simply
gave up the uneven struggle against an inner need for dream is
unimportant.

Certainly that tension around fantasy writing – and most feminist
SF novels have strong fantasy elements – is not unique to my
experience. Precisely because the writer can make it safe for the
reader simultaneously to think about – and transcend – men's violent
relationship to nature as well as the structures of gender, class (and
more rarely racist) domination, the analysis and the change are likely
to be contained within our heads. The richness of the alternative
projects of Piercy's Mattapoisett, Russ's Whileaway and Gearhart's
Wanderground[17] may be so attractive that perhaps our everyday
selves feel condemned to live in a grubby reality, with neither the
courage – nor the strength to oppose it. Should we concentrate our
energies on more immediate matters? The only sensible reply that I
can make is that nourishing imagination, making it possible through
the creation of new myths both to see and see through the structures
and inevitabilities of this society, is rather practical and also
profoundly subversive of the dominant culture.

But there is a fundamental difference between the best of the new
SF fantasy writing and the old – and it is not only the dimensions of
gender and race. Whereas the old dystopia or utopia was complete,
fixed and final in its gloomy inexorability or its boring perfection, the
new accepts that struggle is continuous and interesting. As Luciente,
a future person in Piercy's Mattapoisett puts it, the conflict in the
future is changed, not ended. Per (Piercy's post-gendered pronoun)
says:

> It's that race between technology in the service of those who control,
> and insurgency – those who want to change the society in our
> direction. In your time the physical sciences had delivered the weapons
> technology. But the crux we think, is in the biological sciences. Control
> of genetics. Technology of brain control. Birth to death surveillance.
> Chemical control through psychoactive drugs and neurotransmitters.[18]

Feminist SF writers explore and raise in the imagination issues of
overwhelming importance to women in culture and society and, from
within diverse political and theoretical frames, propose solutions.
Because a continuing preoccupation is with the construction of
gender and the connections between masculinity and the technosciences of exploitation and control, regardless of political perspective
or epistemological stance, whether the writer is realist, postmodernist
or propagandist,[19] the novels return again and again to Luciente's
crux. The significance of the SF feminist writing as an intervention in
popular culture needs underlining, not least when we think about
what was happening – and initially not happening – within left and
feminist politics in terms of their capacity to ignore science and
technology in the many ways I have discussed in previous chapters.

But feminist SF does this and much more than this, for it offers us
dreams and nightmares of different and other futures in which we
can see and feel ourselves in all our diversity. The ambiguity of Mary
Shelley's Monster and those of the feminist SF projects of the almost
two centuries following her extraordinary innovation continue to
nourish our capacity to dream. The very fact that SF is popular means
that these successful novels sell and are read by hundreds and
thousands of women and men. Some of the problems of SF flow from
the publishing industry's sales techniques, of niche marketing and
the like, so that a book is read as SF if it is sold as SF, even though it is
manifestly possible to read it as romance, history or even just a
novel.[20] None the less, the marginality of SF's claims to be part of the
discussion of *the* novel makes it a freer arena for innovative writing –
the way that mainstream literary critics went rather quiet when Doris
Lessing[21] produced the *Canopus in Argos Archives* series suggests that
marginality has gains.

Rather than being only a handicap, being a low-status/no-status
genre may well also be functional for the prolific and richly visionary

outpouring of feminist SF writers. They can speculate with technological change as an integral and necessary element so as to propose other futures – even multiple and simultaneous other futures, for the linearity of time[22] is itself uncertain – in which dreams and desires, those needs silenced by the orders of domination, find partial expression. Because it can bring together a multiplicity of critiques and alternatives – which we have few, or no, way of putting together in other forms – the feminist SF novel makes a peculiarly well-fitted vehicle for conveying the complex stuff of contemporary social thought. Even where the novelists' style is realist, as with Piercy, the categories of sex and gender, and what constitutes the natural, implode into new understandings within her utopia. It is only within her dystopia that the concepts and relations are the same as – only worse than – today. For Piercy's dystopia is recognizably debating with masculinist SF. Socially frozen, technologically dominated, more or less pornographic, mainstream patriarchal SF touches us as in no way a literature of desire, only one of technological and sexual voyeurism. Piercy's dystopia stands with the text as a prophetic warning, the exterminist opposite of Mattapoisett.

Men's utopias: women's dystopias?

The dialogue between science and science-fiction has produced two very different traditions in main/male-stream SF. One is technicist, a socially unreflecting mediation; the other is preoccupied with the social implications of science and technology and takes the form of utopian essays or/and dystopic warnings. Technicist science fiction can manage perfectly well without a utopian thought in its head, acting instead simply as a magnifying mirror to existing society, its boyish enthusiasm for toys all too evident.

Obsessively concerned with technology, with men's relationships with things, this kind of SF leaves untouched the social order within which that technology is embedded. If men could travel at the speed of light, if the firepower of their weapons could multiply the destruction of Hiroshima a billion-fold, if they could communicate by telepathy with robots who would obey their every wish, the wishes, it would appear, are not to be greatly different from those of the *Dallas* soap: riches, power, domination over territory, the crushing of enemies – and available, beautiful and sexually compliant women.

The overwhelming character of these imagined worlds is not merely that they are made by men but that they are almost exclusively populated by men. Such relationships as exist other than between men and machines are between men and men as comrades or locked in mortal conflict. Women are either invisible or reduced to the passive beauty of the fairy-tale princess.[23] Even when, as in the film *Star Wars*, at a pinch the beauteous princess (incidentally the single

identified woman in the film except for the aunt who spends all her
time in the kitchen) can fire a ray gun, she too is soon confined to her
state role of seeing off and welcoming back the heroes.

The locker-room world of dominant SF comes into dramatic
visibility in the reactions to Joanna Russ winning the Nebula Award
for *When it Changed:*

> Yet it [the novel] was also severely criticized in some science fiction
> publications. It is a bit odd that readers should feel threatened by a
> story in which well-characterized likeable women can get along without
> men, when there is such an abundance of science fiction in which well-
> characterized likeable men get along without women.[24]

But if mainstream masculinist SF has little utopian element for either
men or women, what does the overtly utopian tradition offer? In that
utopian or dystopian speculations identify dreams and nightmares to
encourage us to struggle for or against particular futures, just who is
the 'us'? The immediate trouble, as any swift but gender-conscious
survey of the utopian literature indicates, is that by and large men's
utopias are women's dystopias. *The Republic, Utopia* itself (whether
Thomas More's or H. G. Wells's novel of the same name), *The New
Atlantis, Brave New World, 1984,* all present a dispiriting prospect to
women.

Nor, despite the ostensible egalitarian conception of women as
equals and comrades, did the left men scientists in the 1930s' social
relations of science movement escape (with the exception of J. B. S
Haldane) this taken-for-granted androcentricity. Their enthusiastic
interest in the possibility of scientific intervention in human reproduction led them to propose a glorious future of the endless cloning
of Lenin, Einstein and other suitably heroic figures of left masculinist
politics and science.[25] Alternatively, like Bernal, they fetishized
abstract intelligence itself, reducing humanity into a vast disembodied brain.[26] Bernal, who was himself to experience a stroke and
become an intelligence trapped within a body which could no longer
respond, wrote as a young man with enthusiasm of a future
disembodied intelligence:

> The new life would be more plastic, more directly controllable, and at
> the same time more variable and more permanent than that produced
> by the triumphant opportunism of nature. Bit by bit the heritage in the
> direct line of mankind – the heritage of the original life emerging on the
> face of the world – would dwindle, and in the end disappear
> effectively, being preserved perhaps as some curious relic, while the
> new life which conserves none of the substance and all of the spirit of
> the old would take its place and continue its development . . .
> consciousness itself may end or vanish in a humanity that has become
> completely etherialised, losing the close-knit organism, becoming

masses of atoms in space communicating by radiation and ultimately perhaps resolving itself entirely into light.[27]

Such masculinist enthusiasm for eugenics and reproductive technologies reads bizarrely to a post-Nazi and post-'test-tube baby' eye, and Bernal's dream of the abstract, disembodied intelligence stands as the antithesis of feminism's dreams of an embodied situated knowledge.

Bolshevik SF: anticipations of ecofeminism?

But if marxist scientists *outside* the socialist revolution were both sex-blind in their writings and Baconian in their concern to dominate and exploit nature, there is encouraging evidence from the science fiction of Alexander Bogdanov that a revolutionary analysis of science did not have to reduce all social struggles to those of class. Bogdanov's twin novels *Red Star* and *Engineer Menni*, originally published before the revolution and republished in 1918 (with *Red Star* performed as a play by Proletcult in 1922), remind us of the richness and diversity of cultural politics in the early period of the Bolshevik revolution.[28] Although eventually the entire debate was to be crushed by the monolithic structures of Stalinism, Bogdanov remains a kindred spirit to contemporary feminist and ecological writers in his appreciation of biodiversity – whether in nature or people – and his willingness to tolerate political and theoretical uncertainty.

This stems in good part from Bogdanov's sympathy with, and concern to interpret within a materialistic framework, a subjectivist philosophy of science.[29] His lack of closure is expressed in his treatment of post-revolutionary Mars, the *Red Star* of his title. At one level the story is of revolutionary inspiration, for the new society of Mars no longer experiences the class and social oppressions of Earth; at another there are still grave problems between nature and culture. Industrial success has fostered ecological crisis; all too successful medicine has so extended age that for a Martian to be able to die requires voluntary suicide.

Bogdanov makes a Martian woman engineer, Netti (and after the masculinity of the previous texts that alone is more than a relief), speak for a libertarian marxism, which is sensitive to nature, and recognizes the multiplicity of social divisions between people – including both gender and race – indeed welcomes diversity, seeing in this strength, not weakness. Even though Earth is barbaric compared with Mars, Netti argues for its free development, which the Martians should help. She has the responsibility of opposing Stern at the critical meeting at the Central Institute of Statistics where Mars's policy towards Earth is to be determined. Stern is also an engineer and is given the objectivist part to play. He argues that the gathering ecological crisis of Mars can only be resolved by colonizing Earth, a process which objectively requires the extermination of human beings, a matter to be regretted in so far that this will also require the extermination of the rather small handful of conscious socialists. Against this, Netti says;

> He would drain forever this stormy but beautiful ocean of life! . . . We must answer him firmly and decisively NEVER! We must lay the foundations for our future alliance with the people of Earth. We cannot significantly accelerate their transition to a free order but we must do the little we can to facilitate that development.[30]

She goes on to talk of the other way of managing Mars's crisis:

> We must increase our efforts to find synthetic proteins . . . If we fail to solve these problems in the little time we have left, we must temporarily check the birth-rate. What intelligent midwife would not sacrifice the life of an unborn child in order to save the mother? If necessary, we must sacrifice a part of our life that has not yet come into being for the sake of the lives of others who already exist and are developing. The union of our worlds will repay us endlessly for this sacrifice. The unity of Life is our highest goal, and love is the highest expression of intelligence![31]

It is our contemporary sensitivity to ecological problems, as well as our interest in utopias which take women's dreams of freedom as seriously as those of men, which makes the recovery of Bogdanov's work particularly valuable. It serves as a reminder that the complexity of cultural and political traditions is often more diverse than the monolithic history which has not infrequently been laid over them.

More vivid feminist utopias?

But while looking for pro-woman utopias in the marxist SF library may be regarded by some, to mix my metaphors, as a search for a needle in a haystack, life in the feminist culture itself has become incredibly rich. As a very young woman, I used to hug *Coming of Age in Samoa* to myself as not only reporting but offering a different culture, where children and parents would no longer be trapped in what I felt was excessive intimacy. Today feminism has made it possible to appreciate that this was not a wonderful aberration, but part of the systematic construction of an alternative world-view. When Margaret Mead[32] argued for 'more vivid utopias', she was restating the fact that there has been a long-standing feminist commitment to the making of alternative myths.[33]

As the result of the efforts of many feminists over the last decade or so, it has no longer been necessary for women to feel that the utopian project is inherently masculinist and that it is only exceptional good fortune which brings women's fictitious worlds into published and accessible form. Charlotte Perkins Gilman's hilarious text *Herland* reports a separatist utopia visited by three men.[34] Each adventurer is a distinctive social type, enabling Gilman to poke fun at masculinist scientific rationality, at romantic chivalry, and at macho-man. The super-rationalist and sociologist Vandyck notes that Herland's society is civilized and that, therefore, it logically must include men. But while Vandyck is slowly educated, not least emotionally, by Herland,[35] Terry, the super macho-man, attempts to rape Alima, the woman he loves, on the usual argument that women really want to be mastered. Terry's eventual punishment is banishment. The third man, the gentle and romantic Jeff, after a relatively easy resocialization settles down to live in Herland. Gilman's text, not least in the account of Alima successfully first beating off her would-be rapist, then together with another woman trussing him up so that he could cause no further harm until he could be dealt with by due process, radiates the self-reliant, confident good sense of this community of women. *Herland* stands as the antithesis to *The Yellow Wallpaper*, Gilman's autobiographical account of a woman confined in body and spirit to the point of destruction.[36] Herland, by contrast, is nurturant and just, and its inhabitants are strong, capable and gentle. These values come out of mothering, and in Herland the Mothers are a non-hierarchical counterpart to the usual city fathers. Motherhood as moral precept and as political theory avoids the compulsory intimacy of the mother–child dyad – indeed Gilman was, from personal experience, all too conscious that not all women are cut out to be mothers. In Herland a 'different voice' prevails.

Gilman's arcadian vision not only solves the oppression of women, but sees this as also solving class, town planning, clothes design, disease, dogs fouling streets, militarism, reproduction without males, violence, agriculture and relationships with animals. While the unquestionably most important thing to do with Herland is to have fun reading this wonderful 'vivid utopia', there is a problem with this heaven: it is static. There is no sense in which it will ever be necessary or desirable to go beyond Herland. Yet there is unfinished business, for Gilman's matter-of-fact ethnocentricity grates, not least because all else is incandescent with her powerful sociological imagination. The closure and finality of her vision as against the openness and uncertainty of the visions of present-day feminist utopian novels marks a central difference. With the exception of Bogdanov, this static quality haunts both the utopic and dystopic texts of the early and mid-century. In a way we have to go back to Mary Shelley to find a story where we learn what is missing – love – although without being offered any fixed solutions.

More direct struggles around science and technology and their dystopic or utopian possibilities initially take place between books rather than within them. Huxley's novel *Brave New World* was deliberately written to refute Haldane's optimistic view of the new reproductive biology.[37] While the left biologists had their sunny hopes of cloning lots of little Lenins, Huxley saw extra-uterine birth as promising nothing less than the death of the family – and with that civilization. But at best the debate was between a male left and a male liberal construction of the future, both in the interests of men. Feminism had to wait until Shulamith Firestone's brilliant – and profoundly flawed – intervention decisively shifted the ground rules of the debate towards women.[38] Her achievement was to take Huxley and turn him on his patriarchal head, seizing from his dystopia the contradiction which enabled her to reveal its feminist utopian possibilities. So where Huxley's 'test-tube baby' signalled the death of the family and the end of a male-defined civilization, Firestone saw the event as indeed the death of the family, and, precisely because of that, offering the possibility of a truly civilized society in which women could gain their liberation.

It is true that Firestone failed to understand that science and technology were dependent on the social formation within which they are produced and therefore could not be mobilized as the neutral instrument of women's liberation, but the point I want to emphasize here is the gain from Firestone's dramatic intervention into the politics of science. Her influential writing encouraged women to think imaginatively about the science and technology of reproduction, to think about the science question within feminism. Firestone was part of a process of breaking the theoretical and political log jam which resulted from the indifference and/or hostility to science and technology within the women's movement. On the face of things, Firestone's totalizing polemic against motherhood and Gilman's sanctification diametrically oppose each other, yet underneath this opposition lies a shared political commitment to putting women's needs at the centre of their gynocentric utopias. Each identifies what she sees as the critical link which must be broken in the chain binding women's sexuality, reproduction and oppression, but each does so in different historical and technological circumstances.

Gilman's solution is to downplay sexuality and desire throughout her writing, nowhere more clearly than within *Herland*. Unless my late-twentieth-century reading is too graphic and I am missing some understanding self-evident to Gilman's generation, conception occurs through the collective wish of women. Desire (which is seen as only heterosexual) is to be sublimated, it seems, for the male partners of the Herlanders have to be educated towards understanding and practising a higher love. Yet separating the masculinist equation of love and sex made space for a feminist definition of love and of sexuality. Such a separation was also helpful in a period when abstinence from sex offered the most practical form of contraception for women. (Again and again in the voices of the past, women have described a good husband as one who is 'considerate', meaning that he minimized his sexual demands, for restraint was often the only effective fertility control available.) By contrast, Firestone is writing after the contraceptive revolution which (even though the technologies still leave a good deal to be desired) made it possible for heterosexual women to realize their sexuality independently from their choices about having children. Firestone is therefore able to contemplate breaking the link between women's childbearing and women's oppression and still meet the societal need for the next generation.

Le Guin and humanist SF

These themes of sexuality and reproduction are strongly present in the concerns of the tidal wave of feminist SF which began with the generation writing in the 1960s, above all Ursula Le Guin, whose almost single voice was first raised against the dominant androcentric current of the time. Le Guin has a profound distaste for the contemporary capitalist and hierarchical US, which she criticizes from a humanist rather than a feminist perspective. In *The Left Hand of Darkness* she provides on Winter a gentle, anarchistic and androgynous alternative.[39] Gethenians are androgynous, during kemmer, the sexually active period, choosing to be either mother or father. Yet she has been criticized[40] both for there being so few women in her novels, and because those there are are stereotypically portrayed. Yet given the time at which she was writing, this seems unjust as she systematically works to break down gendered dichotomies: 'On Winter there is no division of humanity into strong and weak halves, protectors/protected, dominant/submissive, owner/chattel, active/passive. In fact the whole tendency to dualism that pervades human thinking may be found to be lessened or changed'.[41] In Annares, the utopian society of *The Dispossessed*, equality between the sexes prevails; names do not indicate the sex of their bearer and language has been emptied of proprietorial relations, so terms like 'wife' and 'husband' are not used and even the possessive pronoun relating to 'my' as against 'the' partner is unknown.[42] Indeed there is no great pressure for everyone to live in partnerships. In Shevek and Takver's relationship, love, regard and sexual desire are equally shared. Desire is not limited to the young and beautiful. After a lengthy separation the partners remeet. 'Shavek saw clearly that Takver had lost her young grace and looked a plain tired woman near the middle of her life . . . The acuteness of his sexual desire grew abruptly so that for a moment he was dizzy.'[43]

Similarly, she sees him as looking exhausted and equally desirable. From Annares we learn of the possibility of heterosexual utopia, and a myth which was important to Le Guin's generation, although unsatisfactory to a subsequent, offered what seemed to be a different and better way of living.

Piercy, Russ and Gearhart: second-wave feminist SF

Marge Piercy, Joanna Russ and Sally Gearhart take a more radical and less forgiving stance to the task of overcoming patriarchy.[44] Russ and Gearhart are unwilling to accept men on any terms; even Gearhart's Gentles, themselves an echo of Gilman's Jeff, are unacceptable. Russ's utopia, Whileaway, and Gearhart's Wandergound are separatist. Piercy does accept men within Mattapoisett, but her new men are not merely socially reconstructed like Gilman's – they are also biologically modified. Fantasy and mysticism around procreation cede to a robust biological interventionism.

Piercy's heroine Connie is within white feminism what Netti is within men's socialism. For Connie is a woman of colour, poor, forcibly detained within a psychiatric hospital, a child abuser who as part of her struggle to free herself plans to poison the medical staff who care for her. And we identify with her.

As a socialist feminist Piercy is sensitive to issues of class; her involvement with the US radical science movement, engaged in the struggle against scientific racism, added anti-racism to her political agenda long before white feminism began to give up its taken-for-granted Eurocentricity, and gave a sophistication to her analysis of science and technology lacking in Firestone. In Mattapoisett the link between biological motherhood and the societal need for new people is broken by turning, just as Firestone does, to extra-uterine conception. The difference is that where Firestone saw technology as outside society and as an autonomous and neutral agent of change, Piercy locates the technology as an integral part of the new classless and raceless society organized around a post-gendered system. In Mattapoisett she plays with a genetic engineering that outside utopia most of us would find hard to take, remembering the long, brutal and far from complete history of eugenics which even today confronts us in new ways. She accepts, for example, that human beings take pleasure and comfort in seeing the features of someone they love reproduced in a new being, and has no qualms about proposing that the replacement for the killed Jackrabbit bears some of the beloved's genes.

Neither sex nor gender is immutable in Mattapoisett. Connie meets Luciente as a man, then when she discovers that he has breasts is profoundly shocked. Piercy's realism enables the reader to share this feeling through Connie, and, more positively, to move beyond shock through her growing friendship with Luciente. Piercy deploys a similar device when we meet a man tenderly breastfeeding a tiny baby from the brooder. In Mattapoisett both men and women can lactate so that any one of a child's three co-parents can nurture like women. Some feminists have suggested that women pay the hidden price of this change, in that this implies that women must give up exclusive rights to an area of pleasure. Speaking personally, I am less sure; while experiencing pleasure and closeness in breastfeeding, I had problems with the twenty-four-hourness and learnt with some envy of the breastfeeding collectives developed by a subsequent generation of women.

Piercy understands that not only is nature modified continuously by culture – and that includes our own nature – but that our conceptions of what is natural and what is cultural themselves undergo subtle changes. Her SF novel anticipates the point made and remade by the feminist critique of science over the last two decades. Thus her utopia is ecologically sensitive and people live in harmony with their environment. They also develop new relations with animals (or rather some animals, in that conversation between cats and people is possible, whereas a dinner of roast goose remains in prevegetarian carnivorous glory). Sexuality is polymorphously permissive, where desire includes love, liking, pleasure and comfort.

Yet Mattapoisett is fundamentally different from earlier utopias. Written from within what Jeff Nuttall calls *Bomb Culture* [45] there is a recognition that uncertainty is the only thing we can be certain about. Even utopia has to contest its space for survival and renewal. Parallel with Mattapoisett is a society where what is gross within contemporary society is amplified a thousand-fold. Connie's time travelling to this other place, where women exist as subhuman sex objects, makes it clear why people from Mattapoisett will even die to defend their society. By the side of Piercy's dream of the future is the unequivocal declaration that unless the future defends itself, the nightmare alternative always exists, waiting to overwhelm utopia.

Joanna Russ is less optimistic about the chance of a new society which either includes new men or has a reconstructed sex–gender – let alone post-gendered – system. Her utopia is called Whileaway, because the men are 'away'. It is necessarily and pleasurably separatist. 'Men', writes Russ, 'hog the good things of this world.' But where Piercy writes as a realist, Russ is a postmodernist, not simply acknowledging fractured identity but deliberately splitting herself into four – whose names all begin with J – to play out four simultaneous worlds, yet with the Js meeting and travelling together through the hostile terrain of Manland. By contrast with Piercy who works through realism and identification with a heroine who bears the multilayered oppressions of America, Russ, except for brief moments (like the lovemaking between Janet and Laura which is privileged both politically and in literary style), precludes identification. Instead the reader is constantly distanced in an explosion of witty and savage writing.

Her title *The Female Man*, evokes all those women writers of SF from C. L. Moore to James Tiptree who wrote as female men. Two of the Js are drawn from the present or near-present – Joanna (the writer) and Jeannine, whose total existence revolves round 'the Man'; Janet comes from Whileaway where the men have died from a sex-linked disorder, and Jael is a warrior. In Manland, where there are no actual women but only modified men, the confrontation between Boss-man and Jael sends up an all too familiar story. Boss-man, overexcited by a real woman, talks first equality, then fucking, then proposes to rape her. As he demonstrably has failed to take no for an answer, Jael kills him. When one of the other Js asks if there was no alternative, Jael replies that she enjoyed killing him. With one move Russ decisively breaks the link between the male sex and violence. To a feminist the scene is both comic and emancipatory; a humanist has a harder time.[46]

Russ eschews that restricted variant of feminist science-fiction where the utopia merely ensures role reversal;[47] instead she burlesques it by creating a scene between Jael and her pretty boy live-in lover, only to reveal that Dave is in fact a robot. Janet's comment is left to sum up: 'Good Lord', she says 'Is *that* all?'[48] Compulsory heterosexuality is shown as integral to Jeannine's oppression, and as a mere robotic substitute within Jael's role-reversed world. By contrast, Le Guin and even Piercy read like stereotypical gender roles merely softened by a shift towards androgyny. Within Russ's utopia, the polymorphous permissiveness of Piercy is insufficient to deal with the problem of men. Only lesbianism offers the possibility of fusing mutual caring and passion. Men are by definition not includable within Whileaway. Russ justifies her separatist politics by showing the men in the lives of all four Js as repellent; she has no Jeff, no Gentles and no new men.

Gearhart's and Russ's silences in dealing with the phenomenon of feminist men suggests that the dualism they wish to maintain between heterosexual dystopias and separatist utopias contains all of dualism's usual problems for women. By contrast, Piercy's polymorphous permissiveness seems to offer a more open utopia where passion and nurturance are not inexorably tied to any privileged gender or sexuality.

None the less Gearhart's utopia spoke strongly to a movement which sought to avoid 'stars', for she has neither identifiable heroines, nor even recognizable individuals. The Hillwomen – like Greenham women – have a collective identity. Sensitive to the current distaste for high-tech reproduction, the Hillwomen reproduce by gene-merging, a process mingling mysticism and science. Gearhart's ecofeminism goes far beyond the relatively modest project of Piercy (in which conversation with cats is arguably a quiet personal squiggle on her canvas). Where Piercy wants green trees and clean rivers, the Hillwomen communicate with the sky and trees. Indeed the sense of where one species ends and another begins is blurred, a blurring that has developed politically with the rise of the animal liberation movement. Gearhart is concerned to establish not simply a respectful relationship to the environment, which is a common enough aspect of utopian writing, but nothing less than a common means of communication between nature and culture which will overcome the division between them. Her ecofeminism can only be sustained by the new natural/cultural system of the Wanderground. Her hope for an ecofeminist communication system which would overcome the divisions between human beings and nature echoes that longing that has often haunted feminist utopias, from Blazing-World onwards.[49]

With the single and exceptional voice of Marge Piercy, these white feminists do not go beyond a non-racist stance. Even Gearhart, a profoundly political writer, who has been described as a propagandist who writes SF cannot take anti-racism on board. She praises Piercy's work and goes on to reflect on her own: 'For me moving myself out of a non-racist stance and into an anti-racist one is like trying to push an idle steam roller. I can't get moving, it seems hopeless, and it's easier to do something that I have more passion about, more success in doing.'[50] Indeed, that the two women of colour who appear in the Wanderground are rescued by white Hillwomen is not in itself an overly encouraging signal to black women. This may be the price of a concept of sisterhood that claims a solidarity which, beginning by de-emphasizing individuality, ends by meeting racism with a replay of the white woman's burden. A utopia created from within a Eurocentric feminist perspective, in which the distinctive experiences of women of colour and black women are erased, can only be dystopic. A white feminist utopia becomes a black feminist dystopia.[51]

The distinctive voice of Octavia Butler

While there is a strong strand of utopian writing within contemporary black feminist fiction and poetry,[52] only when Octavia Butler's SF/fantasy novels began to be read in feminist circles did a strongly African–American perspective enter the feminist SF world. Beginning in *Wild Seed* with themes of slavery, time travelling and taking over bodies, Butler's trilogy *Xenogenesis* tells of survival in a postnuclear and postcolonial world where the boundaries between people, animals, machines and beings from outer space are permeable.[53]

In *Dawn,* the central figure is Lilith (not by chance named for the traditional conflation of the Hebrew demon who destroys children with Adam's first wife, who was created separately from him and refused to have sex missionary style), rescued from the aftermath of nuclear holocaust by extraterrestrial beings, the Oankali. This young African-American woman,[54] already a widow through an accident in which her son was also killed, becomes the non-consenting reproductive partner of the Oankali. While the extraterrestrials try to repair Earth, the human beings are stored in a state of hibernation on the Oankali spaceship – itself a living structure. They are awakened in small groups to learn how to deal with their new lives on the ship. These survivors, including Lilith, find themselves faced with the choice of either privileging their humanity and becoming Resisters who refuse to collaborate with the Oankali – effectively a policy of suicide – or entering into some sort of connection in which human autonomy is threatened to the point of extinction.

Careful to make a number of the Resisters sympathetic figures, though showing the deadly nature of their refusal as their genetic inheritance commits them to hierarchical behaviour (in this showing herself an unequivocal biological determinist), Butler pulls the reader's sympathies towards the far from free exchange that Lilith enters into with the Oankali. The exchange reaches new and disturbing levels of intimacy, for the Oankali are gene-traders. They constantly sense, modify and exchange genes. Initally horrified by humans as inherently drawn to lethal violence, the co-operative Oankali become fascinated by humans' genetic complexity and danger. Where cancer is a symbol of terror for the humans, to the Oankali cancer-bearing cells such as those of Lilith are a resource for technologies of metamorphosis and regeneration.

Lilith is chosen to take care of the awakening of human beings from their stored condition, but because of her pairing – a sort of molecular mentorship – with a young Oankali, Nikanj, she is seen as a traitor by the Resisters. By the end of *Dawn,* Lilith has been made pregnant by Nikanj without her consent, by a mixture of genes derived from several Oankali and the sperm of Joseph, her newly married Chinese husband, a fellow survivor, who was killed early on by humans. In the sequel, *Adulthood Rites,* we meet Lilith's hybrid children, and follow the wanderings of Akin as the first human-born, Oankali 'construct' male, whose self-given task is to support his human kin.

To make Akin and his siblings more acceptable to their mother they initially look like human children, but metamorphosis, as the physiological transition to adulthood, transforms them to Medusa-like figures with sensitive tentacles replacing hair. Butler lets the reader share both Lilith's revulsion at being pregnant with a non-human child and her learning an unpossessive affection for her strange, multiply hybrid offspring. Nor have the Oankali retained their original caterpillar form; genetic interchange with the humans has modified their form too, so that they can share the human communication system of words as well as their own non-verbal means. But, as Akin gradually realizes, their changes are chosen; those of the humans are unfree. As he puts it, 'we', and by that 'we' he means that part of him which is Oankali, 'consume' human beings, 'they' do not consume 'us'. Yet Akin, as part human, also understands the desire for autonomy of the Resisters and the anger they feel against the Oankali for denying them.

Where Mary Shelley's Monster was a creation from a feminist reading of early nineteeth-century science in which electricity generated life, Butler's Lilith and her son Akin are located in a feminist and postcolonial reading of the embryology and genetics of the late twentieth century, set in a post-holocaust world. Yet where Shelley's patriarchal science excluded love, thus creating a fallen angel who became a devil – an ultimately dystopic vision – Butler's alternative Oankali science creates hybrid beings who do have some sort of feeling relationship with others – and begin to pick their way. While it would be too strong to say Butler's spaceship is a utopia, it is not without hope.

We are led to feel that 'constructs' matter as individuals, even though their capacity for connectedness and distaste for violence are non-human qualities, and to understand that survival in a destroyed Earth probably requires that the old lines between nature and culture, between species, and between structures and living things have to change. Social practices – racism, as much as sexuality and reproduction – which were part of the past become irrelevant. Butler opens a kind of Pandora's box where she shows us that the worst things, global nuclear holocaust and its terrible consequences, have indeed flown out, but so also has hope. Lilith, and even more her construct son Akin, have to negotiate both.

But in *Adulthood Rites* Butler resists telling us that the future can only be a postmodernist world. Unlike Donna Haraway's celebration of cyborgs, which unhesitatingly embraces the taking down of the boundaries between people, animals and machines,[55] Butler has hesitations and leaves open the possibility that autonomous human beings can create a future. Thus, although because of his special gifts as a hybrid Akin is able to see and respect the longing of the Resisters, he has himself to become more Oankali-like in order to help. Butler's Akin has more than a little of the compassion of Netti for the Earth people, but with the new dimension that he is akin to both. While some readers, notably Haraway, see Butler as sharing her enthusiasm for pregnancy with another species,[56] I read her more ambiguously as suggesting that hybrids may indeed be the future but that she also understands the longing for species identity and the ongoing dream of a human future. Butler, quintessentially a voice of the late twentieth century, leaves Akin to negotiate this still undecided future.

The feminist laboratory of dreams

By contrast with the masculinist utopias and science fiction, these feminist visions of transformed natural and social worlds emphasize again and again that technoscience is not inevitable; as part of culture it is socially constructed and can be reconstructed. Thus, however powerful the present genetic/eugenic turn, it is not the only possible technoscience. Feminist myth-making through marginal genres offers a means of re-visioning the present so as to make other futures possible. At a time when high technology threatens us with its power and destructiveness, it is this fusion of creativity and courage in facing the unthinkable which goes some way to explaining the significance and lure of this writing. For feminist science fiction has created a privileged space – a sort of dream laboratory – where feminisms may try out different wonderful and/or terrifying social projects. In these vivid u/dystopias the reader is invited to play safely and seriously with social possibilities that are otherwise excluded by the immediacy of daily life, by the conventions of the dominant culture and by fear.

It has been above all questions of reproduction, both human and global, that are addressed by these contemporary feminist fictions.[57] This emphasis is shared with the newly dangerous social context in which the books themselves are written, which cries out for all our imagination to grasp implications we can only begin to anticipate. There is a long historical experience of violent transformations in the techniques, mode, means and location of production; in consequence, without diminishing the pain of this convulsive process as production is currently relocated from the North to the South, and experienced by the North as deindustrialization, it is at the same time but the latest convulsion in an old story whose basic plot is known. What is historically new is the possibility of global ecological disaster and the experience of being the bodily site of transformations of human reproduction. Thinking about such technological and social futures, which are at once profoundly intimate and global, is not only difficult but frightening.

But the difficulty about thinking about nasty things is not limited to ecological disaster. Extra-uterine conception was on the technological agenda well before the early seventies, and it was clear then that the births of Louise Brown and the other IVF babies were simply a matter of time. Most thinking about its implications was carried out by men, including some conspicuous mysogynists, with the two exceptional femininist voices of Firestone and Piercy raising the issues within feminism. But feminism was loath to take the implications of feminist futurist speculation and SF seriously, and it was not until Louise was born that the movement began to consider what this did to the meaning of motherhood, embedded as it was and is within a construction of the natural. Until then it was difficult to admit how intimate, how painful are these issues of change, where new boundaries between the natural and the social are being negotiated. There are important lessons to be learned concerning our need to read the visionary writing of feminist SF as deeply political writing, as speaking through the text about living both inside and outside the text.

When the rape of nature threatens in a rather concrete way to turn into the death of nature, and as bodies of women increasingly become new sites of super-profits, feminism is surrounded by the unthinkable. More thinkable and sustainable futures are nurtured by these dreams and myths of other wor(l)ds; and feminists, whether working inside or outside the laboratories, have need of the laboratory of dreams.

Epilogue:
Women's Work is
Never Done

In turning to this, the last chapter of my book, I have a double sense both of the achievements of feminist science criticism and feminist science theory and of the immense tasks that remain. I have wanted to celebrate that criticism and theorizing which has been created by feminists working within and without the sciences over the past twenty years. But in celebrating the achievements I do not want to gloss over the strains, not least between those feminists who do science and those who engage in the social studies of science. The work of the social studies of science feminists can easily be portrayed as weakening to women scientists; the insistence that the specific historical location of the producers of knowledge is important in critically understanding specific representations of nature can be cast as somehow gnawing at the truth-claims of all and every scientific fact. In Britain, powerful journals such as *Nature* pitch themselves on the side of a correspondence theory of truth, and because they do not take the feminist social studies of science seriously, they simply place it in the same witty but nihilist location where much (not all) of the mainstream social studies of science and technology lies.[1]

Against this I want to insist that the feminist studies of science and technology are different in that, while they include almost every conceivable methodological and disciplinary approach,[2] they are profoundly committed to the possibility of making better and more reliable representations of nature informed by new values. As part of a long haul to change knowledge, they have a place (or rather several places, for cultural struggles are never tidy and linear) within Peggy MacIntosh's five-phase model, which she describes modestly as 'curriculum integration' but which requires nothing less than the entire transformation of knowledge.[3] Beginning with a pedagogic situation in which women and their concerns are totally absent from the mainstream curriculum, she sees the second phase as one where significant women are added to the curriculum. The third questions the absence of the realities of women's lives. The fourth moves beyond questioning to the generation of feminist knowledge; while the fifth, the last, seeks to produce a scholarship which is responsive to all of humanity. Of course while it is entirely possible, indeed common, for every phase to happen simultaneously, MacIntosh's model gives a framework for reviewing the progress of feminist approaches to science and technology.

Feminist critics and science theorists unquestionably have very different and important things to say, and they do so with increasing confidence. They share a common wish to challenge the ethic of no ethic, the culture of no culture which lies at the universalizing core of modern science (that first phase which MacIntosh invites us to leave behind) and to rebuild the sciences as respectful and responsible. For historical and political reasons a great deal of feminist effort in the sciences has been directed towards reconstructing the life sciences, particularly as they relate to human and to collective reproduction, not least by making new conceptual relationships with the social, but feminist projects for the sciences and technologies by no means stop there.[4] As the Biology and Gender Study Group wrote in 1988: 'A theory about life affects life. We become what biology tells us is the truth about life. Therefore feminist critique of biology is not only good for biology but for our society as well. Biology needs it both for itself and for fullfilling its social responsibilities.'[5]

Part of our difficulty in working for change lies in recognizing how long and how subtle are the fingers of past cultural values. The century moves to its end in the context of an immense, confused and inconsistent struggle against the deadly weight of that Judaeo-Christian inheritance in which Man was made in the image of God and was therefore given dominion over the Beasts and the garden of Eden, and as Man dominion over Woman. Even the heretical Darwinian story of evolution draws strength from and sustains this inheritance, though its materialism appears to challenge religiosity. Darwin and his circle certainly worried about the threat to religious faith.[6] None the less, I want to suggest that at a more profound level Darwinism (and the more I read of Darwin the less convincing is the distinction between Social Darwinism and Darwinism) shored up that inheritance of the domination of Nature and of Woman which has long shaped the accounts of men and their relationships to both. In this reading, Darwin's burial in Westminster Abbey was not some supreme expression of Victorian hypocrisy, but rather the recognition by church and state that at core the Darwinian theory of evolution sustained the Victorian rendering of the Judaeo-Christian belief in Western Man's right to treat nature, women and other others as his things. To recognize the genius of Darwin was to recognize the spirit of the age. Darwin's intense anxiety lest his science should disturb his social acceptability is, within a single biography, a metonym for the story of science as an institution both continuously seeking to represent nature and anxious about negotiating its place among the powers.[7]

Within this Judaeo-Christian and Western scientific tradition, nature, including women's nature, is there to be dominated, albeit these days there is a sort of defensive surface civility towards both, as the powerful movements of feminism and environmentalism insist on a greater respect. Within theology the notion of 'stewardship' has been proposed as a replacement for dominion, with God understood as both mother and father. Such images seek to foster new practices and new relationships. Yet the extent to which dominion is expressed every day in the laboratory is there in the language and practices which increasingly treat animals as if they were interchangeable with chemical reagents. As one example, the Harvard oncomouse was bred (though more and more the verbs typically and significantly used to describe the production of this and similar animals are 'constructed' or 'made') to facilitate cancer research, so that research can be speeded up. To talk and think in this way is to treat animals – living organisms – as if they were merely chemical reagents. (There are some minor satisfactions in that the oncomouse has so far failed to live up to its US patenters' hopes, and that Europe refused on ethical grounds to patent the oncomouse, as it is a living organism.)

For that matter it is quite complicated living with a biologist who uses (that is, kills) chicks in his research on learning and memory. Over the years I have moved from a relatively straightforward ethical support with some aesthetic revulsion, to feeling uncomfortable, as creating knowledge by killing animals seems increasingly suspect. Always an advocate of research ethics, not least in animal research, I find my longing for a knowledge not based in routine violence is intensified. It is not that I am opposed to killing animals in some absolute sense, for I see no great difference between my cat and myself eating meat, except that as far as I can see it only gives me ethical problems. In consequence the distortion of energy use in the production of meat horrifies me socially and politically, while the factory rearing of animals disgusts me both aesthetically and as contemptuous to another species. So the cat eats meat while for the most part I do not.

Like many feminist science critics I want a science where all the actors take part in the construction of nature, so while I find somewhat intractable the problem of how a day-old chick might freely participate in constructing the facts of learning and memory, I recognize that moving practically towards a deep respect for nature offers at present only inconsistent and provisional way stations.[8] For me one of the more important aspects of the challenge from the animal rights movement has been the questioning set in train among experimental life scientists and their students.

As I tried to show in chapter 4 in my discussion of feminist science theory, I think we can begin to see the possibility of new feminist philosophies of nature which do not abandon the task of making representations, of making scientific facts, but instead offer different approaches to the practice of science. Such critical work, although at times an uncomfortable ally, is not to be equated with anti-science feminism – a current which, while I may disagree with it, unquestionably exists within contemporary feminism. Instead, feminist critics of science are likely to be particularly respectful of the work of feminists within the sciences and their patient and technically demanding task of constructing new representations of nature sympathetic to feminist values. Despite the necessary differences between the feminist social studies of science theorists (necessary because we are not all looking at the sciences from the same place), my reading is that there is a commitment to creating facts about the natural world which build defensible, objective accounts of the real, accepting that the real is always understood through historically, geographically, politically located and embodied subjects. Thus who produces science is a powerful clue as to what science is produced.

Pragmatically, I also think that possible tensions between those of us who construct facts about nature and those of us who point to their social shaping are relatively easily overcome, not least because of issues of education, gender and generation. In our still highly inequitable academic labour market those women scientists and engineers who do manage to survive are likely to be extremely bright, and, as part of a generation still coming into its full creativity, are likely also to have been influenced by the feminist movement which has surrounded them for a great part of their lives. It is not surprising that there is common and increasingly unembarrassed talk among feminist scientists and engineers from different subjects and societies about how they feel they do science differently.[9] These are part of a generation with much greater self-confidence in being women; thinking dangerously and reflexively is wider spread. There were such exceptions in previous generations, but women – let alone feminists – were unlikely to secure more than the smallest toehold within the institutions of science.

What I see as a less discussed strain is that between the highly educated women of academic feminism (from whatever discipline) and primarily working-class women. There is an under-examined issue of class, in which 'race', because of the way it is entwined with class in women's lives, is also present. Thus while I entirely support Anne Fausto Sterling's criticisms of women's studies scholars and students who, while fluent in the higher reaches of feminist cultural, literary and pschoanalytic theory, are content to claim that biology is 'too difficult', and I share her longing for a two-way street, what worries me even more is the gap of privilege, intensified by the recession and the world-wide lurch to the right. The intensification of poverty in women's lives both in the old industrial societies and in the third world is brutally evident.

Take two current examples, derived from third and first world, environment and health, both struggles within the politics of reproduction, whose intensity and scale has been immense. Thus third-world women move to the defence of the environment when its wellbeing is the guarantor of everyday survival. We have learnt from Vandana Shiva's writing of the struggle of the Chipko women to save the trees, a practical struggle aided by the intellectual attempt to redefine the scientific problem from the perspective of those who are affected. More recently there has been the immense struggle waged by the Indian and international environmentalist movements to stop the building of the Sardar Sarovar dams which affect no less than three states, ultimately displacing 250,000 farming families. Environmentalist and feminist pressures eventually shamed the World Bank and the Japanese government into withdrawing funding, but the Indian government remained intransigent. A few months later, in the summer of 1993, India's best-known environmentalist, Medha Patkar, together with 500 of her fellow environmentalist activists, threatened mass suicide by drowning in the Narmada river, the site of the dam, to compel the government to come to its senses and stop the project. The government's response was to say that the dam would irrigate drought-ridden areas and provide electricity, and on the basis of this scientific claim it ordered the arrest and imprisonment of the protestors. However, it also climbed down, at least as far as ordering a review. In Kenya Wangari Maathai, founder and spokeswoman of the Green Belt Movement, has mobilized many thousands of women to plant literally millions of trees and build new lives, reversing human-made desertification.

Ironically, although many commentators, including myself, see these struggles as part of a profound ecofeminism and as a social movement which seeks to empower the dispossessed,[10] the women who play such leading roles within these struggles for the continued possibility of entire communities to reproduce themselves rarely claim themselves as feminist. Such a reflection troubles me: is it that feminism has been seen as too far away from the struggle for women's survival, too engaged, for example in the possibility of a feminism without women, to be an evident ally of actual dispossessed women? Or is it, as Janet Biehl, a founding ecofeminist who now rejects the name, sees it, a problem of the growth of irrationality within the ecofeminist movement, where goddess worship, a biological essentialized account of women's caring, and a monolithic rejection of science and technology as male now dominate, throwing out the baby of feminist rationality along with the bathwater of masculine rationality?[11]

The women's health movement has been another bridge in the construction of new realities, and it is not by chance that feminists located in those countries where the Genome Programme is best funded and where the new genetics and reproductive technologies make their loudest claims spend so much energy criticizing them. In the summer of 1993 the political achievement of securing from the British government an official endorsement of home confinements as 'safe', against the opposition of the Royal College of Obstetricians and Gynaecologists, but with the support of the midwives, speaks of a battle about the construction of reality. On the one side the institutional power of the Royal College insisted that home births were risky and that women should sacrifice their desire for a supportive environment for the sake of the baby, and on the other the home birth movement argued for the link between women's happiness and safety. It was a battle about whose knowledge, whose construction of reality, a conflict which necessarily extended the boundaries of what was to count as reality, not least because it was a battle over whether the discourses of mind and body could be integrated, or whether the Cartesian split would be perpetuated.[12] What was also precious in this struggle was the refusal of key women's groups such as the Maternity Alliance to abandon the claims of low income women, pointing out that if the Royal College cared so deeply for the safety of babies it would campaign for better welfare provision, as poverty was the greatest threat to their wellbeing.

In such struggles, Sandra Harding's question *Whose Science? Whose Knowledge?* becomes vivid. Where the stakes are survival itself, women's knowledge of what they need to survive comes into dramatic conflict with the constructions of a patriarchal science prejudicial to both the environment and the health of women and their families alike. Arguably, struggles on such a scale, requiring co-operation between many different groups, have something of the character of Peggy MacIntosh's fifth stage. Certainly, to reconstruct natural and social facts in this way, which is both 'true' and friendlier to the diversity and complexity of the lives of women, is to work with a significantly changed philosophy of nature and a significantly changed conception of who is to construct the new knowledge. Feminist representations of nature, as we saw in the work of the feminist primatologists or the environmentalists, seek to replace hostile – even deadly – representations without giving up on claims of objectivity. What is under siege from both feminists and deep green environmentalists is the power/knowledge couple, offered with zeal by Francis Bacon in the seventeenth century and critically exposed by

Michel Foucault in the late twentieth. Integral to that power/ knowledge couple is the cultural commitment of the West to the domination of nature, with technoscience serving as both culture and the technological means of domination. Against this a socialist ecofeminism can offer a respect for nature, a respect that is located in the embodied practices and values of caring – without slipping into celebrating those naturally nurturant women beloved by patriarchy and for that matter by mystical ecofeminism.

Love, power and knowledge in everyday life

The price of living in a scientific and technological culture is that most of the time we are enabled to do everyday extraordinary things without any understanding of how they work. I sit typing this text into a computer, believing that when I stop the cursor on the window for 'print' sheets of crisp white paper printed with beautiful black letters will, after sounds of ticking and clucking, fill the tray beside me. The manual typewriter, that artifact of marvellous transparent Victorian engineering, always seemed to me self-evident, the electric typewriter mysterious, and the electronic computer more or less magical. Virginia Woolf recognized this extraordinary development long ago in the height of the electrical revolution when she wrote about the mysterious lifts in Selfridges. The new revolutions in information handling, in mass transport, in biotechnology, are also magical to most users, so that the cultural relationship to techno-science increasingly becomes 'black-box' or 'press-button'.

In everyday life, people in a highly scientific and technological culture cope by what is called the press-button approach to the increasing proliferation of black goods – videos, cameras, televisions, camcorders and music centres – in our domestic lives. Few understand what goes on within the black boxes; the more modest ambition, differentially aged and gendered, is to know which buttons and in what order should be pushed. We want to make our black boxes work; few of us care to know what goes on within.

But even at the button-pushing level, while there is a general recognition of the extraordinary skills of most 10-year-olds as against those of many adults in terms of picking up how to use the new video, research among adults shows that women's more frequent lack of skill than that of their male counterparts is often a matter of being smart socially rather than dumb technologically. Women, it seems, resist learning how to programme the video because they have a shrewd idea about who will be expected to do the family's programming if they learn. Such resistance is not without costs, but the point is that knowledge, ignorance and resistance to knowledge are all both socially structured and achieved through intentional action.

Something rather similar operates around the avalanche of new conceptualizations of the body produced by the life sciences. In a post-AIDS world, while few people understand the immune system, many acknowledge its importance and genuflect to it in everyday speech. 'My immune system is down', 'These vitamins really give your immune system a boost', replace the older language of 'lowered resistance' to infection.

Despite this there is a tendency – which is why at intervals within this book I become angered by the thesis of the two cultures – to suggest that the biggest cultural division is between the sciences and technologies on the one hand and the humanities and the arts on the other, unalloyed by the intersections of class, gender and race. There is a linked discourse called 'the public understanding of science' produced by elite science, in which society is portrayed as composed of two groups: the scientists and the public. The former arbitrate as to what is scientific knowledge and the latter are to be assessed in their competence within it. Such a division is only very superficially and at best momentarily true. It ignores the changing constructions of what counts as science and as non-science (think of Watson's arrogant assertion that there is only one science – physics – and that all else is 'social work', or the contempt of 'real' scientists for what is done in industrial labs), and it sets aside, for example, the very considerable technological expertise of many creative artists. But even more seriously, it ignores the problem of the huge proliferation of knowledges and their interconnections. The polymath, the Leonardo da Vinci, the Hypatia, who could operate at the cutting edge of all the knowledges was historically possible; today we cope by focusing on what seems important to us, letting the rest blur into faith in the scientific and technological culture. This process of selection, black-boxing the rest, in order to achieve any sphere of cultural competence is integral to life in the global village. Expertise, even eminence, within one area of the sciences does not – despite the tendency of male Nobel Laureates to offer very public opinions on many matters – guarantee more than superficial knowledge of others.

Harbingers of hope

Scattered within this book are what I see as harbingers of hope. They are not the outriders of some unidirectional cultural movement, a revolution which knows pretty much where it is going, as revolution-aries in the past have claimed; instead the harbingers are myriads of small and large, apparently disconnected struggles which are expressive of an immense and fragmented movement within and against the dominant construction of science. What all these fragments have in common, from the search for a holistic science amongst some physicists and biologists, through the ethnosciences with their respect for non-Western thought, to feminism's preocccupation with new ways of knowing, is the deeply subversive understanding that science is socially shaped. This understanding, fashioned over the last quarter-century, is out there, set loose in everyday life, making it possible for 'other others' to challenge the powers of technoscience. Few, except those convinced by anti-science, run a totalizing critique; rather, within a knowledge-based society in which we black-box much of the technoculture, some bits are inescapable. Specific sciences and technologies enter our every-day lives and dealing with them requires that we have greater knowledge of them and understand their meaning. Resisting specific parts of the technoculture, or seeking to renegotiate their relevance, requires that the lines between black-boxing and understanding are redrawn by the resisters. The examples within this and other chapters – and they can be multiplied many times – of third-world women's environmental struggles, and first-world women's campaigns over Genome and reproduction, show the capacity to go beyond black-boxing, to enter the terrain of science and to construct new definitions of reality, infused with a feminist understanding of caring. Within these examples we see feminism bringing love to knowledge and power. It is love, as caring respect for both people and nature, that offers an ethic to reshape knowledge, and with it society.

Notes

Chapter 1 Introduction: is a feminist science possible?

1 Feminist critics of science constantly explore this question. For a survey of approaches, see chapter 12, S. Harding, *Whose Science? Whose Knowledge?*

2 Pornography makes the same reductionist move, focusing on one body part and denying the totality of the woman.

3 Vandana Shiva powerfully draws on nature as a political resource in the struggle of the women of the forest. *Staying Alive: Women, Ecology and Development.*

4 Recent sociological research carried out within the Economic and Social Research Council Public Understanding of Science programme, through a series of case studies of different publics, indicates that neither women nor men readily use the word 'science' when they are discussing problems of securing reliable knowledge about natural phenomena on issues of direct concern to them. See Irving and Wynne (eds), *Misunderstanding Science.*

5 'The symbolic struggle . . . over the power to produce and to impose the legitimate vision of the world [is central because] . . . classification . . . the words, the names, which construct social reality as much as they express it . . . [are] the stake *par excellence* of political struggle.' Bourdieu, 'Social Space and Symbolic Power', *Soc. Theor*, 7 (1 1989). bell hooks extends this argument by suggesting that self-naming can aid emancipatory struggle: *Talking Back*, p. 166. Numbers of feminist activists have renamed themselves as part of their struggle, from the nineteenth-century physicist Phoebe Marks, who chose 'Hertha' to express her solidarity with the struggle of Swedish women for political and civil rights, to second-wave feminist Betsy Warrior, who created her surname as integral to her activism within the battered women's refuge movement.

6 Lloyd, *The Man of Reason.*

7   Daly, *Gyn/Ecology*; Gilligan, *In a Different Voice; Balbo, 'Crazy Quilts', and Kari Waerness 'The Rationality of Caring'*, in Sassoon (ed.) *Women and the State*; Petchesky, *Abortion and Women's Choice*; Rothman, *Recreating Motherhood*; Card (ed.), *Feminist Ethics*; Fraser et al., *Ethics*; Cowan, 'Genetic Technology and Reproductive Choice', in Kevles and Hood (eds), *The Code of Codes*.

8   The difference between the objectives of feminists from the third and the first world have been sharply spelt out in confrontations over the course of the UN Decade for Women. Bonnie Mass's pioneering account of different relationships to birth control and abortion made the same point. *Population Target*.

9   Such a totalizing hostility to science was the thesis of two influential books by Theodore Roszak: *The Making of a Counter Culture* and *Where the Wasteland Ends*.

10   Elizabeth Fee suggests that the new science can only develop with the new society: 'Critiques of Modern Science', in Bleier (ed), *Feminist Approaches to Science*.

11   Brian Easlea makes some telling criticisms of 'marxist viriculture', which was certainly part of the problem: *Science and Sexual Oppression*.

12   Many of the activists and theorists within the international radical science movement contributed to two early collections grouped around these themes. H. Rose and Rose (eds), *Political Economy of Science* and *The Radicalisation of Science*. The journals *Radical Science Journal*, later *Science as Culture*. also explored these and other themes.

13   The initial response to the race/IQ issue raised by Jensen in the US and Eysenck in the UK was a campaign around the theme of Race, IQ and the Class Society – also the title of an early pamphlet produced in the US by Progressive Labor. In Britain, a pamphlet was prepared by Ken Richardson and Steven Rose in conjunction with the National Union of Teachers, 'Race, Class and Education' and copies were sent by the union to all NUT branches in the country; followed by *Race, Culture and Intelligence* edited by Ken Richardson and David Spears. These took a 'bad science' position, whilst a strong social constructionist position was embraced by Les Levidow, 'A Marxist Critique of the IQ Debate', *Rad. Sci. J.*, 6/7, 1978. Both positions were anti-canonical and therefore the authors were potential allies. Looking back, I think that criticizably (and self-criticizably, for I also took part in that debate) too much energy was spent in picking holes in the arguments of allies instead of constructing a politics of solidarity.

14   H. Rose, 'Hyper-reflexivity', in Nowotny and Rose (eds), *Countermovements in the Sciences*.

15   Even the Waldegrave 1993 White Paper on Science and Technology, which seeks to overcome the gap between the research system and industry (re-embracing Bernalist ideas once more – astonishingly for a free market government), fails to tackle the ludicrous overspend on military research.

16   During the forties and fifties, relativity and quantum theory were denounced as bourgeois idealism within the Soviet Union rather as they had been denounced, during the thirties and forties, as non-Aryan science by Nazi Germany.

17   The Lysenko affair was neglected for almost thirty years by marxists. In the mid-seventies the American biologists Richard Lewontin and Richard Levins and the French philosopher Dominic Lecourt broke the long silence. Lewontin and Levins, 'The Problem of Lysenkoism', in H. Rose and Rose (eds), *Radicalisation of Science*; Lecourt, *Proletarian Science?*

18   Needham, 'History and Human Values', in H. Rose and Rose (eds), *Radicalisation of Science*.

19 Cultural studies in the UK were particularly attracted to Althusserian structuralism, with a strong interest in dissolving the boundary between 'high' and 'popular' culture. However, science, in a way which embarrassingly reproduced the thesis of C.P. Snow, was left to the scientists. Only in the eighties did this change and cultural studies begin to be interested in science, but from a largely poststructuralist perspective.

20 There were a number of ways the radical scientists sought to restore agency, starting with the clumsily but accurately titled British Society for Social Responsibility in Science founded in 1969. In the early 1970s Bill Williams led a Science in Social Context (SISCON) science education project, which laid siege to traditional and discovery models of science education. As part of this Kate Hinton produced a pioneering reader, *Women in Science*.

21 The terms internalism and externalism distinguished bourgeois (Western academic) from marxist accounts explaining the growth of science. Here I use the term externalism in a broader sense to flag any theory of scientific growth which looks outside science to explain growth.

22 The former project has been particularly associated with the Edinburgh school, notably Barry Barnes and David Bloor, and the 'strong' programme in the sociology of knowledge. Michael Mulkay has been associated with the latter.

23 Price, *Little Science: Big Science*.

24 Ravetz, *Science and its Social Problems*.

25 Popper, *Conjectures and Refutations*.

26 In this rescuing a hint made much earlier by Zilsell, 'The Sociological Roots of Science', *Am. J. Soc.*, 47 (4: 1942).

27 H. Rose and Rose, 'The Myth of the Neutrality of Science', *Impact*, (21 1977).

28 Cicotti et al. 'The Production of Science in Advanced Capitalist Countries', in H. Rose and Rose (eds) *Political Economy of Science*.

29 Noble, *America by Design*.

30 Goonatilake, *Aborted Discovery*, especially pp. 91–119.

31 See chapter 5.

32 Although feminist studies are predominantly craft production they are still subject to the overall political economy of knowledge, which goes some way to explaining the powerful influence of US work.

33 The initial theoretical strength of the German Greens was supported by this tradition, particularly by theorists associated with 'finalization theory'. Bohme et al., 'Finalization in Science', *Soc. Sci. Inf.*, (15 1976).

34 Sohn Rethel, *Intellectual and Manual Labour*.

35 Goonatilake suggests that new technologies rather than new sciences resulted from the experience. *Aborted Discovery*, p. 150.

36 Anderson, 'Science, Technology and Black Liberation', in H. Rose and Rose (eds), *Radicalisation of Science*.

37 Jordan, *Moving Towards Home*, p. 63.

38 During the seventies the new Black British found both employers and the trade union movement joined in hostility to their participation at other than the lower levels. It was not until 1991, with the election of Bill Morris as Transport and General Workers Union General Secretary, that a black person held a leading trade union position.

39 The capacity of capital to commodify both critical theory and socially useful technologies was once more evident – a version of the road–rail bus became a capitalist venture of the eighties.

40 Cooley, *Architect or Bee?* Prior to the abolition of the radical Greater London Council by Thatcher Cooley was appointed to develop socially useful products through technology networks. For an account of the difficulties of

changing stereotypical behaviour within the networks, see Lin, 'Gender Stereotypes, Technology Stereotypes', in McNeil (ed.) *Gender and Expertise*.

41 The dominant discourse continues to tell a story of science as cradled in classical Greek society and flowering in seventeenth-century Europe, even while there have a number of historians, such as Sarton, *A History of Science*, and J. D. Bernal, *Science in History*, who have pointed to the existence of scientific and technological expertise in Oriental and African society. However, even when credit is given to African and Oriental production, it is seen as technological – even Goonatilake does this. Ideologically 'pure science' is privileged and seen as a distinctively Greek accomplishment, mere technology as the inferior achievement of the inferior other and thus dismissable. While some historians, such as Struik, *A Concise History of Mathematics*, have always insisted on the connections between a slave-holding society and the mathematical pleasures of an idle class (and I would add gender), the connections between slavery and the question of who is to do science have been pressed with particular insistence by Rodney, *How Europe Underdeveloped Africa*. and by the contributors to Van Sertima (ed.) *Blacks in Science*. Martin Bernal's *Black Athena* centrally confronts the eurocentric origin story of science.

42 Something of this difficulty is shown in the much-cited, Van den Daele, 'The Social Construction of Knowledge', in Everett Mendelsohn et al. (eds), *The Social Production of Scientific Knowledge*. Even Sandra Harding's otherwise nuanced critique treats Kuhn in an internal perspective: 'Most historians of science in the early 1960s thought that the legitimacy of intellectual histories was clearly in the ascendant and thus that the debate about the competing programmes had come to an end, until the publication of Kuhn's *Structure of Scientific Revolutions* in 1962 re-opened the discussion.' S. Harding, *The Science Question in Feminism*, p. 210.

43 Denise Riley's *Am I that Name?* seems to me an example of reading theories/texts in this context-free way, in that she paradoxically erases the feminist historians who have made it possible to read those texts, and more negatively returns to mere intellectual history. Reading Catherine Hall's review in *Gender and History* 3 (2 1991), I found a remarkable congruence with my own responses.

44 Elizabeth Fee's work is an important exception. See 'Critiques of Modern Science'.

45 Harding, *Switched Off*; Rosser, *Female Friendly Science*.

46 For some time the focus was on the connections between technology, labour and gender; now feminists such as Judy Wajcman write across the science and technology divide. Wajcman's *Feminism Confronts Technology* both makes the theoretical connections and provides a superb overview of feminist work in this linked area.

47 Susan Griffin, *Made from this Earth: Selections from Her Writings*, was an early influential link between feminism and environmentalism. In addition to reclaiming Rachel Carson, *Silent Spring*, as a significant precursor, feminist environmentalism is now strong and multivoiced.

48 Rossi, 'Women in Science: Why So Few?'

49 Rather than enter a complex elaboration of postmodernism I follow Lyotard's minimal (and useful) definition of postmodernism as the questioning of all meta-narratives, including those of science.

50 My sense that feminist SF was important was confirmed by its arrogant dismissal by Sir Walter Bodmer, Director of the Human Genome Project and Chairman of the Royal Society's Committee for the Public Understanding of Science (COPUS), *Pub. Und. Sci.*, 1 (1 1992).

51 Sayre, *Rosalind Franklin and DNA*.
52 Cherki, 'Women in Physics', in H. Rose and Rose, *Radicalisation of Science*.
53 Weisstein, 'Adventures of a Woman in Science', in Ruddick and Daniels (eds), *Working it Out*.
54 Keller, 'The Anomaly of a Woman in Physics', in Ruddick and Daniels (eds), *Working it Out*. At a personal level, I remember applying for my first research post when I was living in Oxford during the early sixties. I was invited to a one-to-one interview during which I was sexually harassed; angry and humiliated, I withdrew my application. As it was the only social science research job in town, sexual harassment was a demonstrably effective strategy in keeping the number of women researchers down, at least on a short-term basis.
55 Narek, 'A Woman Scientist Speaks', in Tanner (ed.), *Voices from Women's Liberation*.
56 Rossiter, *Women Scientists in America*.
57 The pioneering work by the late Professor Daphne Jackson, engineer and physicist, in providing fellowships for married women returners has received more lip service than cash. There is some evident pressure building up among women scientists, including Conservative women scientists, and the minister responsible for science, William Waldgrave, has set up a committee to explore the wastage of women scientists and technologists. The rediscovery of the real economy forgotten during the eighties financial glitz, the change in demographics, and the anger of women scientists well placed in party politics all sustain this new-found sensitivity to the claims of women.
56 Hochschild, 'Inside the Clockwork of Male Careers', in Howe (ed.), *Women and the Power to Change*.
59 Wallsgrove, 'The Masculine Face of Science', in Brighton Women and Science Group (eds), *Alice through the Microscope*.
60 Keller, 'Making Gender Visible in Pursuit of Nature's Secrets', in DeLaurentis (ed.), *Feminist Studies/Critical Studies*.
61 Fausto Sterling, 'Building Two Way Streets: The Case of Feminism and Science', *Nat. W's Stud. Ass J.*, 4 (3 1992). Fausto Sterling is refreshingly direct in her criticism of women's studies, arts and liberal studies students' willingness to be ignorant about science.
62 Traweek, *Beamtimes and Lifetimes*.
63 Feminism was quick to make claims for methodology, but slower when it came to epistemology: cf. Roberts, *Feminist Methodology*. There are exceptions to this, some feminists were making both epistemological and methodological claims. Mies, 'Towards a Methodology for Feminist Research', in Bowles and Duelli Klein (eds), *Theories of Women's Studies*.
64 Arditti, 'Feminism and Science', in Arditti et al. (eds), *Science and Liberation*.
65 There is no systematic data source for this but UK women scientists frequently draw attention to the fact that many have children whereas they think their US sisters do not. Hilary Burrage's early study points to university women scientists not having children. *Soc. Stud. Sci.*, 13 (1983).
66 See *Women in the Scientific Professions*, New York, New York Academy of Science, 1965. Alice Rossi's paper in this otherwise unrevealing collection contains a remarkable pioneering discussion. The collection by Richter, *Women Scientists: The Road to Liberation*, contains few insights, as distinguished women scientists, unsustained by feminism, give little away in public. However in private discussion, they frequently offer observations such as that collaborating with a husband is dangerous, and more recently

that women have a different research style. Baring, 'Is there a "Female Style" in Science?', *Science*, 260, 1993.
67　A period of feminist scholarship summed up by Sheila Rowbotham's pathbreaking *Hidden from History?*
68　Watson, *The Double Helix*.
69　In a discipline singularly hostile to women, Bell's achievements have compelled recognition. She was appointed in 1991 to a chair in physics at the Open University, becoming one of that tiny handful of women science professors in the UK (currently 1 per cent).
70　Osen, *Women in Mathematics*.
71　Biographies of women in science and the autobiographies of women scientists form an immense literature which has been rediscovered and is continuously under active expansion. A number of useful bio-bibliographies of women in science survey this field: Aldrich, 'Review essay: Women in Science', *Signs*, 4 (1 1978); Hubbard et al. *Biological Woman: The Convenient Myth*; Siegel and Finley, *Women in the Scientific Search*; Scheibinger, 'History and Philosophy of Women in Science', *Signs*, 12 (2 1987).
72　Phillips, *Hidden Hands*.
73　Bourdieu and Passeron, *Reproduction in Education, Society and Culture*.
74　Brian Vickers offers an immensely scholarly and classical reading of Bacon, and yet manages to ignore his profoundly sexual imagery in his discussion of philosophy and image patterns, instead focusing on light: *Francis Bacon and Renaissance Prose*.
75　Merchant, *The Death of Nature*.
76　Easlea, *Science and Sexual Oppression*.
77　K. Jones, *Glorious Fame*.
78　Griffin, *Women and Nature*.
79　Feyerabend, *Farewell to Reason*.
80　For example: MacCormack and Strathern (eds), *Nature, Culture and Gender*; Martin, *The Woman in the Body*; Jordanova, *Sexual Visions*.
81　Hubbard et al. (eds), *Women Look at Biology Looking at Women*.
82　Goldberg, *The Inevitability of Patriarchy*.
83　Jensen, 'How Much Can We Boost IQ and Scholastic Achievement?', *Harv. Ed. Rev.*, 39 (1969); Herrnstein, *IQ in the Meritocracy*; Eysenck, *Race. Intelligence and Education*.
84　By both influential socialist feminists – Barrett and McIntosh, *The Anti-Social Family*, and Wilson, *Women and the Welfare State* – and influential radical feminists; see Lorde, 'Open letter to Mary Daly', in Moraga and Anzualdua (eds), *This Bridge Called My Back*.
85　See Collins's point concerning the standpoint of black feminism. 'The Social Construction of Black Feminist Thought', *Signs*, 14 (4 1989).
86　Science for the People, *Sociobiology as a Social Weapon*; Tobach and Rossoff (eds), *Genes and Gender*; Leibowitz, *Females, Males and Families*; Brighton Women and Science Group, *Alice through the Microscope*; Lowe and Hubbard, *Pitfalls in Genes and Gender Research*; Lowe and Hubbard, *Woman's Nature*; Sayers, *Biological Politics*; Bleier *Science and Gender*; Fausto Sterling, *Myths of Gender*; Birke, *Women, Feminism and Biology*.
87　Wilson, *On Human Nature*, p. 125.
88　Barash, *The Whispering Within*, p. 55.
89　Dorothy Nelkin cites the treatment by Benbow and Stanley of their 1980 paper on sex differences in maths: *Selling Science*, p. 29. In similar vein the British psychologist H. J. Eysenck, with strong hereditarian views on intelligence, gave an interview to a neo-Nazi magazine and was unwilling to disassociate himself from the published interview.

90 Verrall, 'Sociobiology: The Instinct in our Genes', *Spearhead*, 10 (1979); Walker, *The National Front*.
91 I do not want to be understood as suggesting that resistance through cultural struggle on its own is complete – practical resistance is equally important.
92 S. Harding, *The Science Question in Feminism*.
93 In 1993 US feminist psychiatrists were mobilizing against PMT being a classified psychiatric disorder.
94 Keller, *Reflections on Gender and Science*.
95 Piercy, *Woman on the Edge of Time*, p. 328.
96 McNeil, *Gender and Expertise*.
97 Rorty, 'Science as Solidarity', in Lawson and Appignanesi (eds), *Dismantling Truth*, p. 12.
98 Lawson, in Lawson and Appignanesi (eds), *Dismantling Truth*.
99 Jameson, 'Foreword', in Lyotard, *The Post Modern Condition*.
100 Nicholson, 'Introduction', in Nicholson (ed.), *Feminism/Postmodernism*, p. 1.
101 Walby, *Theorising Patriarchy*. p. 1.
102 'Social Science has promised certitude and self-knowledge as the result of a new rationalist quest for meaning. This promise has not been kept. Where there was certainty there was neither meaning nor self-knowledge; where there was meaning and self-knowledge, there was no certainty.' Heller, *Can Modernity Survive?*, p. 40.
103 Scott, *Gender and the Politics of History*.
104 Recent UK feminist texts show a similar move towards postmodernism. Cf. Barrett and Phillips, *Destabilising Theory*.

Chapter 2 Thinking from caring: feminism's construction of a responsible rationality

1 Pauline Bart understood this very well: 'Who really gives a damn about reading studies, particularly feminist studies, about women, their dilemmas, their problems, their attempts at solution?' 'Sexism in Social Science', *J. Marr. and Fam.*, 33 (4 1971), p. 735.
2 The concept of 'experience' was to come under intense scrutiny with the linguistic turn of the eighties; those influenced by the new current argued that experience was constructed by language, while materialists continued to argue that consciousness grows out of material reality informed by political struggle. Neither accepted the glassy mirror argument.
3 Spivak, *In other Worlds*.
4 Benston, 'The Political Economy of Women's Liberation', *Mthly R.*, 21 (4 1969), p. 13; Mainardi, 'The Politics of Housework', in Morgan (ed.), *Sisterhood is Powerful*; Lopata, *Occupation Housewife*; Oakley, *The Sociology of Housework*; Oakley, *Housewife*.
5 Anthony, 'Woman the Great Unpaid Labourer in the World' (1848), in Tanner (ed.), *Voices From Women's Liberation*.
6 Myrdal and Klein, *Women's Two Roles*.
7 Developing the concept of gender was central to this. For an early attempt to distinguish sex and nature, gender and culture, see Oakley, *Sex, Gender and Society*; as reflective of a later move within feminism to go beyond this dichotomy see Haraway, 'Gender for a Marxist Dictionary: The Sexual Politics of a Word', in Haraway, *Simians, Cyborgs and Women*.
8 Aptheker, *Tapestries of Life*.

9 Graham, 'Caring: The Labour of Love', in Finch and Groves (eds), *A Labour of Love*; Medick and Sabean, 'Interest and Emotion in Family and Kinship Studies', in Medick and Sabean (eds), *Interest and Emotion*; Lewis and Meredith, *Daughters Who Care*; Cowan, 'Genetic Technology and Reproductive Choice', in Kevles and Hood (eds), *The Code of Codes*.

10 Hochschild in *The Managed Heart* draws attention to the emotional labour of personal service by airline hostesses. The sex trade makes similar demands.

11 The Beatles' song 'A Hard Day's Night' conveyed this all too clearly, and early feminism understood and resisted it. Shulamith Firestone's wonderful analysis of 'The Bar as Microcosm' showed the psychologically demanding and sexualized nature of much of women's servicing work. When the 'sexual freshness' of the glamourous cocktail waitress fades, she is relegated to domestic service. In Tanner (ed.), *Voices From Women's Liberation*.

12 A point made strongly by both Hartsock, *Money, Sex and Power* and H. Rose, 'Hand, Brain and Heart', *Signs*, 9 (1 1983).

13 hooks, *Ain't I a Woman?*, p. 20.

14 Black Women's Group Brixton, 'Black Women and Nursing', *Race Today*, 6 (8 1974); Phizacklea, *One Way Ticket*.

15 Christine Delphy was one of the earliest to use the concept of feminist materialism. Now a number of feminist theorists in France, Canada and the UK are recognizably working in this tradition.

16 My earlier reading of marxism as arguing for a transformative knowledge imbued with an understanding of the dynamic of history – the history of humanity and the history of nature – drew on the Frankfurt school, who as thoroughgoing revisionists had a theory of nature as well as history; e.g. Adorno and Horkheimer, *Dialectic of Enlightenment*. German Greens also drew on this tradition, which was more helpful in responding to the crisis of the environment; by contrast the structuralist marxism influential in Britain and France denied the category nature even more than gender.

17 For an elegant defence of the experiment against anti-realists see Hacking, *Representing and Intervening*.

18 This is a matter of considerable difference between the US and Britain, and is perpetuated in the science parks, which give the illusion of science moving closer to industry while maintaining geographical and class distance. Massey et al., *High Tech Fantasies*.

19 As I explore within the next chapter, this problem developed as the price of the 'success' of academic feminism in the eighties.

20 D. Smith, *The Everyday World as Problematic*.

21 However, this partitioning is not fixed; see McCormack and Strathern (eds), *Nature, Culture and Gender*.

22 Shiva, *Staying Alive*.

23 Land, 'The Family Wage', *Fem. Rev.*, 6 (1980); Barrett and McIntosh, 'The Family Wage: Some Problems for Socialists and Feminists', *Cap. and Class*, 11 (1980).

24 Bradley, *Men's Work. Women's Work*.

25 Taylor, *Eve and the New Jerusalem*.

26 What Barbara Ehrenreich described for the States is also a British phenomenon: *The Hearts of Men*.

27 While this is broadly true the situation is also more complex. For an overview see Dex, 'Gender and the Labour Market', in Gallie (ed.), *Employment in Britain*.

28 Since the mid-seventies unions have addressed the issue, but most hopes were placed on the Social Chapter of the European Union – hopes which receded with the election of a third Conservative government in 1992.

29 This is discussed more fully in chapter 5.
30 The small elite that controls science includes very few women. But gains have been made, notably in the US National Institutes of Health (NIH), and the Canadian Medical Research Council (MRC), which, after long feminist campaigning, have now laid down that all studies must include women where appropriate.
31 Cockburn, 'The Gendering of Jobs' and Phizacklea, 'Gender Racism and Occupational Segregation', in Walby (ed.), *Gender Segregation at Work.*
32 Hakim, *Occupational Segregation.*
33 Braybon, *Women Workers in the First World War*; Milkman, *Gender at Work*; Braybon and Summerfield, *Out of the Cage.* For a vivid documentation see the film *Rosie the Riveter.*
34 The price of being in a 'non-traditional occupation' is high.Where women are both mothers and scientists their 'productivity' is maintained but at the price of restricting their lives to their science and their families. Cole and Zuckerman, 'Marriage, Motherhood and Research Performance in Science', in Zuckerman et al. (eds), *The Outer Circle.*
35 United Nations Decade for Women, Nairobi, 1985; for UK analyses see Gershuny et al., 'Time Budgets'. Pahl, however, suggests that unequal division of labour is constant in different classes, but that where women have a professional-level job their partners share the domestic labour more equally: *Divisions of Labour.* For the US, see Vanek, 'The Time Spent in Housework', in Amsden (ed.), *The Economics of Women and Work.*
36 Gorz, *Farewell to the Working Class.*
37 To argue this is to set aside such influential texts as Braverman, *Labour and Monopoly Capital*, and Young and Wilmott, *The Symmetrical Family.* The former argued that there was less domestic production and the latter more equality within the home. Instead it means looking at the evidence of Game and Pringle, *Gender at Work*; Cowan, *Work for Mother*; Nowotny, 'Time Structuring and Time Measurement', in Fraser and Lawrence (eds), *The Study of Time 2*; Davies, *Women and Time.*
38 The caring literature painfully underlines how carers have no time for themselves. For a general discussion of the distribution of time in both the Nordic and Italian context see Balbo and Nowotny (eds), *Time to Care in Tomorrow's Welfare Systems.*
39 See the pioneering study of Esther Boserup, *Women's Role in Economic Development.*
40 The working-class suffragist Hannah Mitchell propped up poetry books to read whilst she was washing up, and claimed that her first feminist feelings were aroused as she sat mending socks while her brothers read books.
41 Hochschild, 'The Sociology of Feeling and Emotion', in Millman and Kanter (eds), *Another Voice.* More recent work by men sociologists such as Giddens, *The Transformation of Intimacy*, has an optimism rarely shared by feminist literature.
42 Graham, 'Caring: The Labour of Love', in Finch and Groves (eds), *A Labour of Love.*
43 Gilman, *Women and Economics.*
44 Ve echoes this: 'Women's Mutual Alliances: Altruism as a Basis for Interaction', in Holter (ed.), *Patriarchy in a Welfare Society.*
45 Phillips and Taylor, 'Sex and Skill', *Fem. Rev.*, 6 (1980).
46 Elson and Pearson, 'Nimble Fingers Make Cheap Workers', *Fem. Rev.*, 7 (1981).
47 This has been fought in the US by the notion of 'comparable worth' and in Britain by 'work of equal value'.

48 Grossman, 'Women's Place in the Integrated Circuit', *S. East Asia Chron.*, 66 (1979); Mitter, *Common Fate, Common Bond.*

49 Simonen, 'Caring by the Welfare State and the Women Behind It', in Simonen (ed.), *Finnish Debates on Women's Studies.*

50 Waerness, 'On the Rationality of Care', in Sassoon (ed.), *Women and the State.*

51 Ehrenreich and English, *For Her Own Good*; McNeil *Gender and Expertise.*

52 An Open University television programme shown in 1992 on *Personality and Development* had three male researchers discussing breastfeeding, with women nursing their babies as silent illustrations.

53 Gilligan, *In a Different Voice.*

54 Clare Ungerson's work has been particularly sensitive in drawing out this tacit understanding of caring: *Policy is Personal.*

55 Ruddick, 'Maternal Thinking', in Thorne and Yalom (eds), *Rethinking the Family.*

56 Oakley, *The Captured Womb.*

57 Waerness, 'Caring as Women's Work in the Welfare State', in Holter (ed.), *Patriarchy in a Welfare Society*; K. Davies, *Women and Time.*

58 Ehrenreich and English, *For Her Own Good;* Versluysen, 'Old Wives' Tales', in C. Davies (ed.), *Rewriting Nursing History;* Donnison, *Midwives and Medical Men.*

59 Game and Pringle, *Gender at Work.*

60 Gilligan, *In a Different Voice*; Noddings, *Caring*; Belenky et al., *Women's Ways of Knowing.*

61 For a review essay of the citizenship debate see K. Jones, 'Citizenship in a Woman Friendly Polity', *Signs.*, 15 (4 1990); I am also grateful to the Women and Society Seminar at the University of Bradford during 1991–2 focusing on citizenship, where many of these themes were discussed.

62 Land and Rose, 'Compulsory Altruism for Some or an Altruistic Society for All?', in P. Bean et al. (eds), *In Defence of Welfare.*

63 Ve, 'Women's Mutual Alliances', in Holter (ed.), *Patriarchy in a Welfare Society.*

64 Hartman, 'The Family as the Locus of Gender, Class and Political Struggle', *Signs*, 6 (3 1981).

65 The 1991 General Household Survey found that men, at least as older spouses, take on a substantial level of caring labour. The difficulty with the study is that very different labours are rendered homogeneous. Currently Errollyn Bruce and I are carrying out an empirical study of care among elderly couples with both women and men carers.

66 Waerness, 'On the Rationality of Care', in Showstack Sassoon, (ed.), *Women and the State.*

67 Graham, 'Caring: The Labour of Love', in Finch and Groves (eds), *A Labour of Love.*

68 Gilligan, *In a Different Voice.*

69 Bernice Fisher and Joan Tronto in their attempt to overcome the division between labour and love always acknowledge power. 'Towards a Feminist Theory of Caring', in Abeland and Nelson (eds), *Circles of Care.*

70 Morris, *Able Lives?*

71 There are important exceptions. See Bryan et al., *Heart of the Race.* They document black women's work in the welfare state, a state which 'regarded us not as potential clients but as workers'.

72 Aptheker, *Tapestries of Life*, p. 50.

73 Aptheker, *Tapestries of Life*, p. 39.

74 Sayers, *Biological Politics.*

75 Cousins, 'Material Arguments and Feminism', *m/f*, 2 (1978).

76 Edholm et al., 'Conceptualising Women', *Crit. Anthrop*, 3 (9/10 1977).

77 O'Brien, *The Politics of Reproduction.*

78 To O'Brien's useful point I would add that what are understood as body/ nature and as culture are themselves continuously under negotiation.

79 Clutton-Brock 'How the Wild Beasts were tamed'; D. Griffin, *The Question of Mental Experience.*

80 Pateman, *The Sexual Contract.*

81 For a feminist critique of the sociobiology of violence, see Sunday and Tobach (eds), *Violence Against Women.*

82 The long prison sentences given to Sara Thornton and other women who have killed violent husbands, contrasted with the very short sentences or release given to men whose provocation for wife killing was 'nagging', have enraged feminists and their allies.

83 Fanon, *The Wretched of the Earth.*

84 The commitment of those men who do actively oppose violence is all the more precious.

85 Merchant, *The Death of Nature.*

86 Quotation from Lansbury, *The Old Brown Dog*, p. 48.

87 While the anarchist Kropotkin had a theory of co-operation in nature, this was at the level of philosophy. In the realm of practice the geneticist and marxist J. B. S. Haldane, much influenced in his life by Indian thought, developed what he called 'non-violent biology', which avoided animal experimentation.

88 Leiss, *The Domination of Nature.*

89 For feminist theology to challenge the masculinity of the conception of God is to go to the keystone of the Judaeo-Christian tradition, which frames, even for non-religious people, the culture.

90 Keller, *Reflections on Gender and Science.*

91 Mainardi, 'The Politics of Housework', in Morgan (ed.), *Sisterhood is Powerful.*

92 James and Dalla Costa, *The Power of Women and the Subversion of Community.*

93 Each has her or his own nadir; mine was P. Smith, 'Domestic Labour and Marx's Theory of Value', in Kuhn and Wolpe (eds), *Feminism and Materialism*, which restated pretty much the classic marxist position.

94 This and other cartoons illustrated Kaluzynska's survey of the domestic labour debate. 'Wiping the Floor with Theory', *Fem. Rev.*, 6 (1980).

95 Delphy, *The Main Enemy*; Delphy, *Close to Home.*

96 Hartman, 'Capitalism, Patriarchy and Job Separation by Sex', in Eisenstein (ed.), *Capitalist Patriarchy and the Case for Socialist Feminism*; Hartman, 'The Family as the Locus', *Signs*, 6 (2 1981).

97 Young, 'Beyond the Unhappy Marriage', in Sargent (ed.), *Women and Revolution.*

98 Joseph, 'The Incompatible Ménage à Trois', in Sargent (ed.), *Women and Revolution.*

99 Both unitary theory and dualism were thrown into disarray by the demand to pay attention to difference.

100 Rowbotham, 'The Trouble with Patriarchy', *New Statesman*, 21 December (1979); Barrett, *Women's Oppression Today.* The revised 1988 edition, now subtitled *The Marxist Feminist Encounter*, both accommodates the concept of patriarchy and makes overtures to postmodernism.

101 Eisenstein, 'Developing a Theory of Capitalist Patriarchy', in Eisenstein (ed.), *Capitalist Patriarchy and the Case for Socialist Feminism*.

102 Joan Smith's critique of dual systems theory knows that gender relations lie in the world capitalist system: 'Feminist Analysis of Gender: A Critique', in Lowe and Hubbard (eds), *Women's Nature*. Michèle Barrett by contrast suspects that they precede capitalism: see *Women's Oppression Today* (note 100).

103 Brown, 'Mothers, Fathers, Children', in Sargent (ed.), *Women and Revolution*; Hernes, 'Women and the Welfare State', in Holter (ed.), *Patriarchy in a Welfare Society*; Dahlrup, 'Confusing Concepts and Confusing Realities', and Borchorst and Siim, 'Women and the Advanced Welfare State', in Sassoon (ed.), *Women and the State*; Pateman, 'The Patriarchal Welfare State', in Gutman (ed.), *Democracy and the Welfare State*; Antonen, 'The Feminization of the Scandinavian Welfare State', in Simonen (ed.), *Finnish Debates on Women's Studies*; Simonen, *Contradictions of the Welfare State, Women and Caring*; Walby, *Theorising Patriarchy*.

104 Barrett and McIntosh, *The Anti-Social Family*.

105 Following Hazel Carby's searing criticism of white feminism, 'White Woman Listen!', in Centre for Contemporary Social Studies, *The Empire Strikes Back*, many white socialist feminists criticized their own race-blind analyses. E.g. Barrett and McIntosh, 'Ethnocentricism and Socialist Feminist Theory'. The contemporary white working-class family received no such recognition despite the part played by 'the family' in times of economic difficulty, such as the extended mining strike of 1984–5. Retired grandparents used the relative wealth of their state pensions to ensure that their striking adult sons and daughters and their families were fed. H. Rose, 'The Miners' Wives of Upton', *New Society*, Nov. 29 (1984).

106 Mount, *The Subversive Family*.

107 Finch and Groves, 'Community Care and the Family', *J. Soc. Pol.*, 9 (4 1980); EOC, *Who Cares for the Carers?*; Nissel and Bonnerjea, *Family Care of the Handicapped Elderly*; Finch and Groves, (eds), *A Labour of Love*; Ungerson, *Policy is Personal*.

108 Morris, *Able Lives?*

109 Briggs and Oliver, *Caring*.

110 This analysis, pioneered by Elizabeth Wilson, has had a powerful influence: *Women and the Welfare State*.

111 Siim, 'Towards A Feminist Rethinking of the Welfare State', in K. Jones and Jonasdottir (eds), *The Political Interests of Gender*.

112 Collins, 'The Social Construction of Black Feminist Thought', *Signs*, 14 (4 1989).

113 Walby, *Theorising Patriarchy*.

114 Ve, 'Women's Experience: Women's Rationality', in *Conference on the Construction of Sex/Gender*; Ve, 'Gender Differences in Rationality', and Gunnarsson and Ve, 'Technical Limited Rationality versus Responsible Rationality', in *International Conference: Ethics, Gender and Technology*.

115 Jaggar 'Love and Knowledge', in Jaggar and Bordo (eds), *Gender/Body/Knowledge*.

116 Keller, *Reflections on Gender and Science*, p. 164.

117 Emotion shadows debates between men but only the kindlier emotions are named and deplored. Anger, as a masculine emotion, is not recognized as an emotion and is thus an acceptable part of masculinist discourse. As classic expressions of rage and aggression in the philosophy of science see Lakatos and Musgrave (eds), *Criticism and the Growth of Knowledge*.

Chapter 3   Feminism and the academy: success
and incorporation

1   Two influential and linked texts set out this position: Riley, *Am I that Name?*
    and Scott, *Gender and the Politics of History.*
2   Hartsock, 'Foucalt on Power', in Nicholson (ed.), *Feminism/Postmodernism.*
3   A recent discussion of Japanese Studies noted that 80 per cent of US
    academic research on Japan is funded by Japan. The expectation built into
    the research is that the results should not be critical. *New York Times,*
    29 November 1992.
4   This joke is borrowed from Bruno Latour.
5   At the time of her request 6 million of her fellow Londoners lived in
    conditions of some distress on less than £200 a year. Women were as usual
    massively over-represented among the London poor.
6   Brown, *Medicine Men.*
7   Mitchell, 'Reflections on Twenty Years of Feminism', in Mitchell and Oakley
    (eds), *What is Feminism?*
8   Flax, 'Postmodernism and Gender Relations in Feminist Theory', *Signs,* 12
    (4 1987).
9   D. Cameron, *Feminism and Linguistic Theory.*
10   See the epigram at the head of this chapter. Gordon, 'What's New in
    Feminist Theory?', in DeLaurentis (ed.), *Feminist Studies/Critical Studies,*
    p. 10.
11   For British single parents who are financially disadvantaged women (and
    virtually all single parents are) the move from private patriarchy to public
    patriarchy has to some extent been reversed by recent socially regressive
    legislation such as that compelling absent fathers to pay towards child care.
12   While these beginnings are real, they are located within unsettled, even
    dangerous, contexts and still involve a very small number of feminists. A
    pioneering document was Center for Gender Studies, *Concept of the
    Government Programme for the Improvement of the Position of Women and the
    Family and the Protection of the Mother and Child,* Moscow; Interdisciplinary
    Research Center for the Study of the Social Problems of Gender. The
    European Left Feminist Forum had by 1992 established contacts with
    feminist groups in the former East Germany, Hungary, the former
    Yugoslavia and Russia.
13   Haug, 'Lessons from the Women's Movement in Europe', *Fem. Rev.,* 3
    (1 1989).
14   The deepening and changing objectives of feminism played a part. The
    possibility of new subjectivities became of central importance to influential
    currents within academic feminism. For example, see the conclusions of
    Code, *What Can She Know?*
15   Mies, 'Towards a Methodology for Feminist Research' and Du Bois,
    'Passionate Scholarship', in Bowles and Duelli Klein (eds), *Theories of
    Women's Studies;* Hull et al. *All the Women are White, All the Blacks are Men, But
    Some of Us are Brave.*
16   For a US criticism see Freeman, 'The Feminist Scholar', *Quest,* 5 (1979).
17   This immense commitment to radical politics was made possible by the
    fewer responsibilities of these mostly young women without dependants, a
    sort of feminist replay of the Marcusan thesis of the radical potential of the
    young and the dispossessed.
18   Rowbotham et al., *Beyond the Fragments;* Sivanandan, *Communities of Re-
    sistance.*

19 British television, perhaps the most powerful media form, has been much more resistant to admitting feminist films and programmes. Personal communication with filmmaker Midge Midgeley.
20 The market is not, however, neutral; feminist writers consistently report moves by publishers to influence their texts and titles so as to make feminism more 'marketable'.
21 See chapter 5.
22 Drafting this chapter at Uppsala, I was reminded that Foucault's doctorate was rejected here, even though he was able to use his position as director of the French studies programme, with responsibility for arranging the programme of visiting scholars, to create a wide network of significant intellectuals.
23 Bourdieu cites the 1981 *Lire* survey which asked several hundred writers, academics, students etc. to name their three intellectual heavyweights. While Lévi-Strauss, Aron and Foucault were the top of the hit parade, de Beauvoir came in fifth, and Marguerite Yourcenar sixth. Bourdieu, *Homo Academicus*.
24 Bourdieu's pessimistic neo-marxist functionalism is perhaps the key to this circularity.
25 The Danes, criticizing US cultural imperialism, have preferred the term 'feminist studies'.
26 There are very important differences between the Nordic countries but here I want to focus on the commonalities in their commitment to social democracy, particularly in how this has affected women.
27 See: Touraine, *Post Industrial Society*; Touraine, *The Voice and the Eye*; Gorz, *Farewell to the Working Class*; Laclau and Mouffe, *Hegemony and Socialist Strategy*; and indeed the entire editorial project during the late eighties of the increasingly ironically named *Marxism Today*. For an elegant dissection see Saville, 'Marxism Today: An Anatomy', in Miliband et al. (eds), *The Socialist Register 1990*.
28 See the contrast between two much-cited books: the technicist discourse of social welfare feminism in Glendining and Millar (eds), *Women and Poverty* and the social movement perspective of Piven and Cloward, *Poor People's Movements*, which includes a discussion of the largely black women's welfare rights movement.
29 There is some progress in that the social work training council has recently insisted on an anti-racist element. But the multicultural requirements that courses must meet, and that are to be found in a growing number of US universities in response to student demand and faculty concern, have not yet been demanded in Britain.
30 Like most feminist commentators I regard John Major's Opportunity 2000 as window dressing. To date, despite interesting signs of some new cross-party alliances between women, research policies which tackle the gender issues in research, as against getting more women into science, seem remote.
31 Miriam David makes the point that it is precisely this fusion of morality and politics which lies at the core of the New Right: 'The New Right in the USA and Britain', *Crit. Soc. Pol.*, 2 (1983).
32 For a discussion of how class is entwined with race see Gordon, 'The Welfare State', in Gordon (ed.), *Women, the State and Welfare*.
33 A handful of research centres has been established by such strategies in Birmingham, Lancaster, Hull and Bradford. One relatively well financed was the West Yorkshire Centre for Research on Women, established in 1985

with a 'tomb-stone' grant from West Yorkshire Metropolitan County Council, which had been abolished, along with the GLC, by Thatcher.

34 David and Land, 'Sex and Social Policy', in Glennister (ed.), *The Future of the Welfare State.*

35 This mould has been broken by the appointment of Professor Tom Blundell as Director of the Agricultural and Food Research Council, who has introduced the first equal opportunities policy.

36 The most dramatic exception to this rather cosy relationship between government and foundations was in the support given by the Wellcome Foundation (which had made a great deal of money out of AZT) to support the major social survey of sexual behaviour, estimated to cost £3 million, which had been directly blocked by the government. Sir Donald Acheson in August 1992, as the former chief scientist at the Department of Health, confirmed that this was a result of Thatcher's personal refusal to support the survey – and thus integral to her inability to think rationally and humanely about the public health hazard represented by AIDS. Reagan's response was even more irrational and he is reported to have been unwilling to say the 'A' word. This story of united opposition between a Foundation, the biomedical and survey establishments and the gay community was exceptional. However, no such happy alliance exists around breast cancer, which kills many more women each year.

37 For many years those feminists who had participated in this famous event, where the sun shone in a Scottish April, would simply say, 'Ah, Aberdeen'. This first collective declaration of a new feminist sociology was a cultural landmark.

38 Chamberlain, 'The Development of Gender Studies in the US'. As the responsible officer within Ford, Mariam Chamberlain herself has been a key player in this process. We shall have to wait for the history of the funding of US feminist research to be written before this story is clear.

39 In Britain this is purely a contested approach from within academic feminism. The National Women's Studies Conference in 1990 surveyed the extent of progress 'from margin to centre'. Aaron and Walby (eds), *Out of the Margins.*

40 The women's health movement has successfully pressed both for the inclusion of women subjects in current relevant research, such as the heart studies, and for new research to add specifically women's problems, such as cervical and breast cancers. US National Institutes of Health will now only support research where the sample, if appropriate, includes women as well as men. The implications of this policy change are not small for throughout the eighties Western governments, at the instigation of WHO, have initiated immense health promotion programmes for 'all' whose scientific justification is exclusively based on the bodies of men. The Canadian Medical Research Council has adopted a similar policy; the British Medical Research Council and Department of Health have not moved.

41 Feminist scholars Margaret Rossiter and, most recently, Evelyn Fox Keller have received prestigious and lucrative MacArthur awards.

42 Study leave is enormously varied within British universities and the former polytechnics. A generous policy has enabled some of the 1960s' 'new universities' to achieve strong research records from which feminism has also benefited. Others coming out of a college of technology or polytechnic history have had poorer conditions of research.

43 US feminism has insisted that these are women's issues too. Cf. Gordon's lead review 'Understanding Bag Ladies', *Women's Review*, 4 (1992).

44 Consciousness of race and racism have been criticizably absent from

254   Notes to pages 66–72

Scandinavian feminism; racism and neo-Nazism have developed there as elsewhere in Europe. The special issue of *Race and Class: Europe: Variations on a Theme of Racism*, 32 (3 1991) addresses the issue of growing racism in Scandinavian countries but from the perspective of ungendered beings (i.e. men): Larsson, 'Swedish Racism'; Salimi, 'Norway's National Racism'; and Quaraishy and O'Connor, 'Denmark: No Racism by Definition'.

45 For example, Uppsala University refused to accept Jewish scholars during the thirties despite the willingness of the government. K. Johannisson, *A Life of Learning*, p. 103.

46 Ruin, 'Reform Reassessment and Research Policy', in Wittrock and Elzinga (eds), *The University Research System*.

47 The danger of turning the university into primarily a centre of applied research, where research problems were defined by middle-ranking bureaucrats without the scientific capacity to know whether the problem was researchable, came from many quarters, not least the Organization for Economic Co-operation and Development in 1981.

48 Benkarts et al., *Proposal for Developing Gender Studies of Science and Technology*.

49 Jacques Delors, French socialist and EC Commissioner, speaks of his vision of a white and Christian Europe. It seems that the European citizen of the future has distressing links to the European of the imperial past. If that was not sufficient, the constitution of this new Europe expressed in the treaty of Maastricht guarantees the rights of Man internationally but the rights of women to marry and procreate only within the laws of her country.

50 'Traditional' is widely employed by policy makers and politicians to indicate an occupation currently dominated by one gender. As with much policy discourse it is saturated with a Whiggish progressivism where things are inevitably moving from the past 'traditional' to the present or future 'modern'.

51 Holter and As, *National Research Policy in Norway*.

52 Cott, *The Grounding of Modern Feminism*.

53 Le Doeuff, 'A Letter from France', *W. Philos. News.*, 8 (1992).

Chapter 4 Listening to each other: feminist voices in the theory of scientific knowledge

1 For an exploration of the extent to which women's conceptions of their normal life cycle processes have been medicalized, see: Martin, *The Woman in the Body*; Zita, 'The Premenstrual Syndrome', *Hypatia*, 3 (1 1988); for an account of the social practices of the biomedical professions which have helped bring this about, see Oakley, *The Captured Womb*.

2 'Science' disappearing within everyday talk is documented in Lambert and Rose, 'Disembodied Knowledge', in Irving and Wynne (eds), *Misunderstanding Science*.

3 The outsider status of feminist-produced knowledge of women's bodies is reflected in a remark of Donna Haraway's: 'A scientist is one authorised to name nature for industrial peoples.' 'Primatology is Politics by Other Means', in Bleier (ed.), *Feminist Approaches to Science*, p. 79.

4 Ruth Ginzberg goes further, suggesting that there has always been a gynocentric science, and seeing the conflict between gynocentric and androcentric accounts of human birth as one of conflicting paradigms. She merely hesitates about claiming the word 'science' with all its negative connotations: 'Uncovering Gynocentric Science', in Tuana (ed.), *Feminism and Science*.

5 I use the word 'truth', very much with a small 't', as signifying the most reliable account, and for that reason something feminists should not lightly abandon. Women's accounts have long been dismissed, not least in courts of law.

6 Ehrenreich and English, *For Her Own Good*.

7 Haraway also makes this point: 'Contested Bodies', in McNeil (ed.), *Gender and Expertise*, p. 72.

8 The other important series of publications, *Genes and Gender*, came out under the editorship of Ethel Tobach and Betty Rossoff.

9 This was a truly historic issue. Michelle Aldrich, Dorothy Mandelbaum, Sally Kohlstedt and Margaret Rossiter depict the social organization of science. Ann Briscoe's account is of feminist organizational resistance. Helen Lambert, Marian Lowe, Ruth Bleier, Donna Haraway and Susan Griffin expose the sexist character of Western science. Lois Magner analyses past feminist attempts to fight patriarchal science with feminist science and Adrienne Zihlman goes onto the cultural offensive, proposing a new evolutionary thesis of 'woman the gatherer'. *Signs: Journal of Women in Culture and Society*, special issue on *Women, Science and Society*, 4 (1) 1978. *Signs* was also the source for the influential collection of S. Harding and O'Barr (eds), *Sex and Scientific Inquiry*.

10 Roy Bhaskar has been influential in this debate. See *A Realist Theory of Science* and *Reclaiming Reality*.

11 Sandra Harding named 'feminist empiricism', delighting as a philosopher in the heresy of the conjunction: *The Science Question in Feminism*.

12 S. Harding and Hintikka (eds), *Discovering Reality*.

13 Collections which explore this include: Nicholson (ed.), *Feminism/Postmodernism*; Garry and Pearsall (eds), *Knowledge and Reality*.

14 My version of standpoint theory was also published that year – see 'Hand, Brain and Heart' – and further developed in a paper presented to the MIT Women's Studies Programme called 'Is a Feminist Science Possible?', and an article 'Women's Work: Women's Knowledge', in Mitchell and Oakley (eds), *What is Feminism?*

15 Chodorow, *The Reproduction of Mothering*.

16 Chodorow's work, which powerfully influenced the development of a psychology of gender, came under criticism primarily because only white, heterosexual, middle-class women and their children had been studied, but these had been given universal, ahistoric status. Other influential and connected texts, such as Dinnerstein, *The Mermaid and the Minotaur*; Gilligan, *In a Different Voice*, ran into similar criticism from which Flax's account below, because of its lack of historicity, cannot be exempted.

17 Following Engels, this signalled that women were 'oppressed' through ideology and were not exploited by men. See Mitchell, *Women's Estate*; Barrett, *Women's Oppression Today*; US marxist feminism by contrast considered domestic labour as gender exploitation: Hartman, 'The Family as the Locus' and 'The Unhappy Marriage of Marxism and Feminism', in Sargent (ed.), *Women and Revolution*.

18 Keller, 'Gender and Science' and 'The Mind's Eye', in S. Harding and Hintikka (eds), *Discovering Reality*. In her study of Barbara McClintock, Keller brilliantly uses biography to explore the complexities of the making of gender and the making of science, which goes far beyond the exploratory psychoanalytic essay: *A Feeling for the Organism*. The theoretical tension in her work is nowhere better displayed than in *Reflections on Gender and Science*: she welcomes the plurality of the voices of postmodernism, but at the same time continues to insist on the realist claims of natural science. In

her most recent book, *Secrets of Life: Secrets of Death*, Keller continues to hold
this position.

19   Flax, 'Political Philosophy and the Patriarchal Unconscious', in Harding and
     Hintikka (eds), *Discovering Reality*, p. 245.
20   Ibid., p. 251.
21   See also Bordo, 'The Cartesian Masculinization of Thought', *Signs*, 11
     (3 1986).
22   Flax, 'Political Philosophy', p. 261.
23   The constructions of masculinity and reason are also explored by Lloyd, *The
     Man of Reason*.
24   Flax's decisive move towards postmodernism, in which she refuses the
     possibility of 'a' feminist standpoint as a necessarily partial account, can be
     read following her own prescription: *Thinking Fragments*.
25   Hartsock, 'The Feminist Standpoint: Developing the Ground for a Specifi-
     cally Feminist Historical Materialism', in S. Harding and Hintikka (eds),
     Discovering Reality. Her chapter, 'Foucault on Power: A Theory for Women?',
     in Nicholson (ed.), *Feminism/Postmodernism*, makes sophisticated accommo-
     dation with postmodernism without yielding truth claims.
26   Ruddick, 'Maternal Thinking', in Thorne and Yalom (eds), *Rethinking the
     Family*. Subsequently she responded to the criticism of this as essentialist in
     Maternal Thinking.
27   It goes back to, and agrees with, Sayers's arguments for a constrained
     constructionism and limited essentialism, in *Biological Politics*.
28   D. Smith, *The Everyday World as Problematic*.
29   Hartsock, 'The Feminist Standpoint', p. 297.
30   Belenky et al., *Women's Ways of Knowing*; Noddings, *Caring*.
31   A position shared by Hubbard, 'Science, Facts and Feminism', *Hypatia*, 3
     (1 1988).
32   Millman and Kanter (eds), *Another Voice*, forms a steady reference for Sandra
     Harding from *The Science Question in Feminism* to *Whose Science? Whose
     Knowledge?*, as a classical example of feminist empiricism with its links to
     liberal feminist theory.
33   Bloor, *Knowledge and Social Imagery*; Barnes, *Interests and the Growth of
     Knowledge*; Latour and Woolgar, *Laboratory Life*.
34   H. Rose, 'Hyper-reflexivity', in Nowotny and Rose (eds), *Countermovements
     in the Sciences*.
35   I am making a distinction between sociological relativism which is crucial to
     generating accounts of a complex and diverse world, and philosophical
     relativism which abandons all truth claims.
36   S. Harding, *The Science Question*, p. 321.
37   The claims of distinctively non-Eurocentric (albeit still androcentric) stand-
     points have long been made by third-world movements. Concepts such as
     Negritude, African and panAfrican perspectives have informed movements
     of national liberation. See: Fanon, *The Wretched of the Earth*; Cabral, *Revolution
     in Guinea*.
38   S. Harding, 'Why has the Sex/Gender System Become Visible Only Now', in
     S. Harding and Hintikka (eds), *Discovering Reality*, p. 321.
39   S. Harding, *Whose Science? Whose Knowledge?*
40   Lakoff, *Language and Woman's Place*; Spender, *Man Made Language*; had
     pioneered a feminist critique of language itself. The new linguistic turn
     intensified this scrutiny of language as mediation, threatening to dissolve
     feminism's claims to name reality.
41   This commitment to remaking the sciences, discussed in chapter 1, was

widely shared. See also: Potter, *A Feminist Model of Natural Science*; Longino, *Science as Social Knowledge*.

42 Haraway, 'Animal Sociology and a Natural Economy of the Body Politic', reprinted in Haraway, *Simians, Cyborgs and Women*, p. 10.

Haraway, like many influenced by the poststructuralist currents, came to reject humanism as master discourse. For a defence of both humanism and feminist realism and a witty critical reading of postmodernism, see: Soper, 'Constructa Ergo Sum', in Soper, *Troubled Pleasures*.

43 Haraway, 'In the Beginning Was the Word', *Signs*, 6 (3 1981).

44 Lowe and Hubbard (eds), *Pitfalls in Genes and Gender*; Hubbard et al. (eds), *Women Look at Biology Looking at Women*.

45 The essay has been extensively republished.

46 Ibid., pp. 8–9.

47 Ann Oakley's action research project to increase women's happiness and the birth weights of babies embodies (literally and metaphorically) such an approach. That it was successful on both scores is a double pleasure: *Social Support and Motherhood*.

48 Hubbard, *The Politics of Women's Biology*, p. 5.

49 Ibid., p. 35.

50 See Haraway's crucial paper, 'Situated Knowledges', *Fem. Stud.*, 14 (3 1988). This refuses the one master knowledge (or even one mistress knowledge) but does admit the realist truth-claims of embodied science/knowledge in context.

51 I think feminist science studies have a good deal to thank Sandra Harding for in securing this both/and approach.

52 For example, the debates about both housework and sexuality during the seventies were markedly intolerant. Feminism's increased tolerance, which extends beyond the science debate, stems, I believe, from the greater recognition of complexity.

53 Haraway, 'Situated Knowledges'.

54 Sayers, *New Statesman*, 31 July 1987, p. 30; Zita also read *The Science Question* as 'too soft on postmodernism' and as 'abandoning feminism and socialism': 'A Feminist Question of the Science Question in Feminism', *Hypatia*, 3 (1 1988).

55 McNeil and Franklin, 'Science and Technology, Questions for Cultural Studies and Feminism', in Franklin et al., (eds), *Off Centre: Feminism and Cultural Studies*.

56 S. Harding, *The Science Question*, p. 10.

57 Ibid., p. 10.

58 Harding herself is a Quine scholar. For a discussion of holism derived from Quine see Alcoff, 'Justifying Feminist Social Science', *Hypatia*, 2 (3 1987).

59 Within Britain 'science in context' is embraced by those who are not adherents of the Edinburgh 'strong programme' and who find strong social constructionism too relativistic. Science in context has had a lively existence as a new approach to science education which, while not overtly feminist, has not been intrinsically hostile to feminism. This approach was called SISCON or science in its social context. It was a promising development in science education which was cut as part of the general Conservative attack on education. For an example of a pioneering feminist science educator finding a space within this project, see Hinton (ed.), *Women and Science*.

60 S. Harding, *The Science Question*, p. 37.

61 As Jim Watson put it in a debate on 'the limits to science' at London's Institute of Contemporary Arts in 1986, 'in the last analysis the only science is physics – everything else is social work.'

62 D. Smith, *The Everyday World as Problematic*; D. Smith, *Texts, Facts and Femininity: Exploring the Relations of Ruling*; D. Smith, *The Conceptual Practices of Power*.

63 D. Smith, *The Everyday World as Problematic*, p. 1.

64 S. Harding, *The Science Question*, p. 157.

65 Gadamer, *Philosophical Hermeneutics*; Bauman, *Hermeneutics and Social Science*; Heller, *Can Modernity Survive?* Their solutions to the problem of meaning and objectivity are substantially – if implicitly – shared by numbers of feminist social scientists and are present in the research literature, though not named as hermeneutics. Perhaps Gadamer's failure to discuss power and both Bauman's and Heller's marked androcentricity have not made them seem obvious resources for feminist epistemology. The exception is the feminist philosopher Code, *What Can She Know?*

66 The agency structure debate at its sharpest was between Althusser, *For Marx*, who sought to dissolve the subject in his structuralist project, and Thompson, who celebrated the agent to the point of voluntarism, in his account of *The Making of the English Working Classes*. See Thompson's polemic against structuralism and in particular Althusser's: *The Poverty of Theory and Other Essays*.

67 Giddens, *Central Problems in Social Theory*; Bourdieu and Passeron, *Reproduction in Education, Society and Culture*; Collins, *Theoretical Sociology*; Alexander et al., *The Macro Micro Link*.

68 Of course the whole issue of women as historical agents in their own lives has been questioned by feminist poststructuralism, most notably by Riley, *Am I that Name?* The reply from feminists cautiously sympathetic to deconstructionism may be summed up as 'Yes – sometimes but not always.' Personally I share Hartsock's scepticism concerning the attack on new historical subjects, just as women were beginning to speak and feminism was increasingly effectively challenging the universalizing discourse of masculinism. Hartsock, 'Foucault on Power', in Nicholson (ed.), *Feminism/Postmodernism.*.

69 This new concern with the subjectivity of women extends across the various strands of psychology, from Miller's pioneering *Towards a Psychology of Women* to Hollway's attempt to reconceptualize academic psychology, *Subjectivity and Method in Psychology*.

70 D. Smith, *Texts, Facts and Femininity*, pp. 33–4.

71 Haraway, *Primate Visions*.

72 See Steven Rose, 'God's Organism', *The Making of Memory*. In the neurosciences the competing advantages of octopus, chick and sea slug are passionately debated. To generate a god's-eye view of nature, getting the choice of god's organism right is correspondingly decisive.

73 Though not, as Nelly Oudshorn makes deliciously clear, the biochemistry of hormones. 'On the Making of Sex Hormones', *Soc. Stud. Sci.*, 20 (1990).

74 Jane Goodall is a defender of the chimpanzee against the depradations of the laboratory experimentalists. But this particular Western construction of eco-femininity is also deeply tied up with class; earning a living is a rather rarely mentioned item in even Haraway's account of the primatologists. A working-class woman technician with feminist politics, whose job is to kill large numbers of experimental animals, might well ask Sojourner Truth's question, 'Ain't I a woman?'

75 Think of all those millions who sat, like me, glued to the television screen watching the gorillas embrace David Attenborough.

76 It was precisely the hands-on, interventionist nature of physiology which distinguished it from mere observational zoology and gave it truly scientific

status. See Cunningham and Williams (eds), *The Laboratory Revolution in Medicine*.

77 Frances Power Cobbe published extensively on the themes of vivisection and cruelty to women: 'Wife Torture in England', *Contemp. Rev.*, 13 (1878); *The Modern Rack: Papers on Vivisection; Life of Frances Power Cobbe: by Herself*. It has been the feminist historians who have recovered this connection, lost in the mainstream account: Elston, 'Women and Anti-vivisection in Victorian England', in Rupke (ed.), *Vivisection in Historical Perspective*.

78 H. Rose, 'Gendered Reflexions on the Laboratory in Medicine', in Cunningham and Williams (eds), *The Laboratory Revolution*.

79 A medium-sized academic pharmacology laboratory would use around 100 rats a week. If toxicity testing – needed for drug production – takes place this figure would sharply escalate.

80 Within second-wave feminism this set of connections was restated by S. Griffin, *Women and Nature*; and S. Griffin, *Pornography and Silence*.

81 Spivak, 'Feminism in Decolonization', *Differences*, 3 (3 1991).

82 Because my political formation, not least about science, was fashioned in the context of the Vietnam war I am sometimes disturbed, even repelled, by her use of metaphors such as 'force fields' and 'power charging', as these are images reminiscent of the electronic battlefield, and for me always behind the disembodied language are other, uncontrollable images of real, maimed bodies.

83 The primatologist Judith Masters's review both praises the book and protests its difficult prose, which works to restrict the readership. Like her I want to have the intellectual joy of Haraway's work, but more accessibly written so that more can share it. *Women's Review*, 7 (4 1990).

84 The most unequivocal commitment to radical subjectivity for the social sciences (they assumed that objectivity reigned unproblematically in the representation of nature) was made by Liz Stanley and Sue Wise, *Breaking Out*. Subsequently they revised their position, generously explaining they had not read the standpoint literature: *Feminist Praxis* and *Breaking Out Again*.

85 In an otherwise astute reading of the literature Susan Hekman sounds merely puzzled by what to her, in her enthusiasm for the postmodernist project, is a theoretical inconsistency on the part of the feminist science critics. *Gender and Knowledge*.

86 Latour and Woolgar, *Laboratory Life*, p. 213.

87 Carr, *What is History?*

88 In the still incomplete five-year long controversy of the Baltimore case, the accuracy of an immunological paper published by Nobelist David Baltimore, President of Rockefeller University, and his co-worker Thereza Imanishi-Kari was questioned by a post-doc in Imanishi-Kari's laboratory, Margot O'Toole. O'Toole was fired and labelled a trouble-maker. Baltimore rebutted all criticism as an example of political hostility to science. However, eventually a detailed investigation mounted by the Office of Scientific Integrity of the National Institutes of Health showed evidence of 'fabrication' by Imanishi and a less than adequate part played by Baltimore. *Science and Government*, from where I draw this account, concluded by ironically noting the prizes for whistle-blowing now being offered to O'Toole after she was exiled for almost five years in the scientific wilderness.

The Gallo case concerns the identification of the AIDS virus by US scientist Robert Gallo, which was eventually shown to have been acquired from the Paris Laboratories of Luc Montagnier. Yet how does Latourian analysis work in these stories of provisional and then overturned claims to facts and authorship?

89　The crucial article was R. M. Young, 'Science is Social Relations', *Rad. Sci. J.,* 5 (1977).

90　This issue of realists and social constructionists split the British radical science movement. Young, having left his university post as a historian of science, brought the nascent social constructionist account of his discipline into the radical science movement through the *Radical Science Journal (RSJ).* The problems of strong social constructionism as philosophical relativism were both theoretical and also straightforwardly political, for in the 1970s *RSJ* spent some of its political energy not only in criticizing the racism and class supremacy in the bourgeois natural sciences but also in attacking the rather sparse numbers of left biological realists who sought to expose the racism and sexism of IQ theory and the new sociobiology as 'bad science'. Left scientists who were necessary realists were cast in *RSJ*'s analysis as part of the Leviathan. Donald MacKenzies's sophisticated analysis, which sought to move beyond the split, by and large fell on stony ground. 'Notes on the Science as Social Relations Debate', *Cap. and Class,* 14 (1981).

　　By contrast in the US, the marxist and radical tradition continued more strongly to contest sociobiology and the new genetics from a realist perspective – particularly through the Boston-based Science for the People. This grouping was able to make an alliance with the feminists active on the science question, so that key figures like Ruth Bleier, Ruth Hubbard and Marian Lowe participated in both. Within Britain this split became softened with the advent of feminist science criticism. British feminist realists such as Lynda Birke, who sought to overturn the renewed biology as destiny, were not criticized. While there are substantial theoretical differences within the British feminist critique of science, it has purposefully avoided such politically divisive debates. The crucial change of feminism seems to be in the political commitment to solidarity in working with all the critics of science.

91　Joan Smith comes to a very similar position to that of *RSJ* in 'Sociobiology and Feminism', *Philos Forum,* 13 (1982). So does Restivo, 'Modern Science as a Social Problem', *Soc. Prob.,* 35 (3 1988).

92　Haraway, *Private Visions,* p. 8.

93　See chapter 1.

94　Haraway, *Private Visions,* p. 310.

95　Haraway, *Simians, Cyborgs and Women,* p. 191.

96　Birke, *Women, Feminism and Biology;* Fausto Sterling, *Myths of Gender;* Gould, *The Mismeasure of Man;* Merchant, *The Death of Nature;* Haraway, *Primate Visions;* H. Rose and S. Rose (eds), *The Radicalisation of Science;* S. Rose et al., *Not in Our Genes.*

97　The political move towards equality of opportunity in postwar Britain saw the psychologist Cyril Burt, later knighted for his academic work, inventing a research worker and sundry statistics to shore up his conviction that heredity, and therefore class, determined IQ.

98　Collins, *Black Feminist Thought,* especially the first two chapters.

99　Fanon, *Black Skins, White Masks.*

100　Lesbian feminist literature has not directly addressed the science question, but the epistemological claims of queer theory claim both standpoint and social constructionism, and in their overall character are very close to that of Collins. See: Aptheker, *Tapestries of Life;* Zimmerman, 'Seeing, Reading, Knowing: The Lesbian Appropriation of Literature', in Hartman and Messer Davidov (eds), *(En)Gendering Knowledge: Feminists in Academe.*

101 For example: Daly, *Gyn/Ecology*; Morgan, *Sisterhood is Global*. However, not all radical feminists are essentialists; a number are materialist feminists.
102 Indeed her book richly documents the writing of black women.
103 Collins, 'Learning from the Outsider Within', *Soc. Prob.*, 33 (6 1986).
104 hooks, *Ain't I a Woman?*; Hull et al., *All the Women are White, All the Blacks are Men, But Some of Us Are Brave*; Bryan et al., *The Heart of the Race*; Narayan, 'The Project of a Feminist Epistemology', in Jaggar and Bordo (eds), *Gender/Body Knowledge*.
105 Both Sivanandam, *A Different Hunger* and Amin, *Eurocentrism* share this view that intellectual work alone cannot comprehend the viewpoints of those we oppose. Struggle is integral to making an effective sociology of the powerful.
106 hooks, *Ain't I a Woman?*, p. 39.
107 I think that the shift in my own work is discernible in 'Beyond Masculinist Realities', in Bleier (ed.), *Feminist Approaches to Science*.

Chapter 5   Gender at work in the production system of science

1 Cf. Evelyn Fox Keller's complaint that she is read as speaking of women's approach to science when what she was trying to get at was the gendered character of science – a project which is close but not interchangeable: *Secrets of Life, Secrets of Death*.
2 Snow, *The Two Cultures and the Scientific Revolution*.
3 M. F. D. Young (ed.), *Knowledge and Control*, p. 21; Young made an impassioned plea for a sociology of science education which, despite his own efforts during the seventies, faded out, to be energetically recreated in the eighties by feminists.
4 Rodney, *How Europe Underdeveloped Africa*; Goonatilake, *Aborted Discovery*; Van Sertima (ed.), *Blacks in Science Ancient and Modern*.
5 Other examples of such recovery by Western scholars include Joseph Needham's monumental history of Chinese science, *Science and Civilisation in China*, Cambridge, Cambridge University Press, continuing series of volumes, and M. Bernal, *Black Athena*.
6 Ahmad and Ahmad (eds), *Quest for a New Science*; Quarashi and Jafar, *Scientific and Technological Development in Profiles in Muslim Countries*; Ataur-Rahim, *Contribution of Muslim Scientists during the Thirteenth and Fourteenth Centuries Hijri in the Indo-Pakistan Subcontinent*.
7 Bazin, 'Their Science: Our Sciences', *Race and Class*, 34 (1993).
8 Fee, 'Critiques of Modern Science', in Bleier (ed.), *Feminist Approaches to Science*; S. Harding, *The Science Question in Feminism*.
9 Shiva, *Staying Alive*.
10 Bacon, *A Selection of his Writings*.
11 Culture in the nineteenth century included both science and the arts; for example, about to embark on his epoch-making *Beagle* voyage, and short of stowage space, Darwin left behind Haeckel's major scientific text on South American biology in order to make space for the latest novel by Jane Austen: Burkhardt and Smith (eds), *The Correspondence of Charles Darwin, Vol. 1*; Morrell and Thackray, *Gentlemen of Science*.
12 For descriptions of the Down House environment, see Raverat, *Period Piece*.
13 The writer Naomi Mitchison told me how as a young woman she routinely assisted her brother J. B. S. Haldane in the Oxford laboratory. She wanted to

become a writer but the domestic laboratory could have been a route to entering biology.

14 Levi-Montalcini, *In Praise of Imperfection*.

15 The gender-neutral term 'scientist', coined by Whewell, only gradually came into use during the mid-nineteenth century. Before then, those actively committed to scientific research were spoken of as 'cultivators of science'.

16 Haraway, *Primate Visions*, both in itself exemplifies this craft work and identifies many such women primatologists: from Jane Goodall to Jeanne Altmann and Adrienne Zihlman, the list is long and striking.

17 The statistics in this section on women's participation in the scientific labour market are heavily reliant on UNESCO data. Gender statistics do not have a high priority among OECD data.

18 In the US a third of all the working-age population has a degree, whereas in Europe only one fifth has. *Education at a Glance: OECD Indicators*, Paris, OECD, 1992.

19 Cacoullos, 'Women, Science and Politics in Greece', in Heiskanen (ed.), *Women in Science*, p. 145.

20 US federal grants require this information; UK governmental bodies, like the EU, do not. The US National Institutes of Health (NIH) also have a programme for giving additional resources to grant holders to take on extra minority research trainees. Women's share of NIH research grant money has doubled since 1980, but because they ask for less, in 1990–1 they were still only in receipt of 16 per cent of the total funds.

21 Hicks, in Heiskanen, 1991; Hawkins and Schultz, 'Women: The Academic Proletariat in West Germany and the Netherlands', in Lie and O'Leary (eds), *Storming the Tower*.

22 Hernes, *Welfare State and Woman Power*.

23 Until recently similarities between the Swedish and the German professorial system facilitated this resistance to reform.

24 Peter Scott's 1991 review of the Swedish higher education system noted how undisturbed the academy was by issues of gender justice. The reform focus was on course structure, which was controlled by the Parliament. When the Social Democrats fell in 1992 the universities were freed to structure courses in the way they wished by the incoming government, and presumably remain uninfluenced by issues of gender justice.

25 Women report, however, that firms do not like women engineers and seek to avoid using them.

26 Scandinavian data is from *The Yearbook of Nordic Statistics*; the educational year reported ranges from 1987 to 1988.

27 In addition, a tradition of anonymous competition for the design of public buildings has permitted Finnish women architects to come through in a way that has not been possible in Britain.

28 The first women's university courses in science in Russia date back to the 1860s, when they were continuously under the suspicion and close control of the Czarist censors. Sechenov, *Autobiographical Notes*.

29 Tarja Cronberg, personal communication.

30 At the International Conference on Gender Studies convened by the Moscow Institute for Management Studies, supported by UNESCO, which I attended in November 1991, the participants from global feminism were outraged (as were the handful of feminists from the former Soviet Union) by an attempt by leading academic and political men – and some women – to claim that women needed to be at home in order to develop their spirituality and save the moral welfare of the country. This is not to say that women are not exhausted with the horrendous double day associated with the past.

31 This is partly to be explained by the pre-eminence of psychology in the discussion of science education with its focus on the usually 'ungendered' 'individual' – an emphasis reproduced by teachers' legitimate concerns with real individual children, which unfortunately also carries the cost of missing their specific historical and social locations.

32 As numbers of feminists have noted, the appeal to 'biology as destiny' appears when women seek access to higher education and professional life; from the influential view of Darwin that women's brains and intellects were the inferior of men's (*The Descent of Man, and Selection in Relation to Sex*, p. 565) to Benbow and Stanley's thesis concerning the biological basis of women's inferior mathematical performance: 'Sex Difference in Mathematical Ability', *Science*, 212, 1980. This gave rise to a flurry of correspondence pointing out the similarity to the 'race and IQ' debate and again challenging the genetic reasoning: *Science*, 212, 1980 December 17, pp. 114–16. Alison Kelly made the comparative point that girls in some countries do better than boys in other countries, but that the teaching is oriented towards boys: *Girls and Science*.

33 Ruizo, 'The Intellectual Labour Market in Developed and Developing Countries', *Int. J. Sci. Ed.*, 9 (1987). Other Mediterranean countries such as Spain, Greece and southern Italy have similar statistics.

34 The immense Portuguese standing army during the long years of imperial repression in Africa also created a chance for women to enter the academic and scientific labour force.

35 While Ruizo's thesis is compelling, Finland's place as the latest and fastest to industrialize – within the Scandinavian political commitment to gender equality – needs emphasizing. For a discussion which examines the Finnish experience within this framework, see Heiskanen, 'Handmaidens of the Knowledge Class', in Heiskanen (ed.), *Women in Science*. For the immediate future the profound connections between the Finnish economy and that of the former Soviet block may prove extremely difficult to manage in this as in other respects.

36 Rather different figures are produced by looking solely at university teachers. Here UNESCO figures for 1980 (so before the Gulf War) show relatively high proportions of women teachers in some Islamic countries, such as Saudi Arabia (20 per cent) and Iran (16 per cent).

37 Acar, 'Women in Academic Science Careers in Turkey', in Heiskanen (ed.), *Women in Science*.

38 Philippa Ingram, *Times Higher Educational Supplement*, March 3, 1972, p. 1.

39 Westergaard and Resler, *Class in a Capitalist Society*.

40 Sadler, 'Patterns of Discrimination and Discouragement', in *Report of the New York City Commission on Human Rights*.

41 Roby, 'Institutional Barriers to Women Students in Higher Education', in Rossi and Calderwood (eds), *Academic Woman on the Move*, pp. 39–40.

42 Evelyn Keller remarks that in 1956, almost a century after admitting its first woman student, Ellen Swallow Richards, MIT decided to exclude women students. 'The Wo/Man Scientist', in Zuckerman et al. (eds), *The Outer Circle*, p. 230.

43 In January 1970, Sadler, on behalf of the Women's Equity Action League, filed against the colleges on sex discrimination grounds that as federal contractors they were legally obliged to be non-discriminatory. Subsequently the 1964 Civil Rights legislation was amended to cover equal pay. See chapter 7, Chamberlain, *Women in Academe*. The power of class action suits in the US to compel reform compares very interestingly with the British

tradition of passive non-compliance, for example with the 1919 legislation discussed in the next chapter.

44 See Chamberlain, *Women in Academe*.

45 Farley, 'Women Professors in the USA: Where are They?', and D. Davies and Astin, 'Life Cycle, Career Patterns and Gender Stratification in Academe', in Lie and O' Leary, *Storming the Tower*.

46 The American Association of University Professors issued a report in November 1992 showing that women were over-represented among part-time and temporary faculty, a conclusion also made by a 1989 report from the UK Association of University Teachers.

47 The category of 'minorities' includes African-Americans, Native American, Hispanics, Latins and Asian Americans. The situation of these and other, smaller groups is by no means identical; however, women are under-represented among the already under-represented, except amongst African-Americans.

48 In the context of fee rises and diminishing grant support, 'diversity' increasingly means minorities and women, while class, including how that locks with race and gender, is ignored. Protests against wealth being openly considered as an admission factor have taken place at a number of elite institutions including Columbia, Smith, Brown and Wesleyan. *International Herald Tribune*, 27 April 1992.

49 There is an acute problem for specific minorities, not least Native Americans.

50 *National Research Council Doctoral Recipients for US Universities*, National Academic Press, 1980 and 1986. Data reported in *Science* in an issue devoted to Women in Science, 260, 16 April 1993, p. 409, reported still improving percentages in the PhD figures.

51 The current picture for the 'semi professions' is contradictory. Recent changes introduced by the Conservative government mean that while nursing becomes a college-based education, social work returns to an employer-controlled apprenticeship model, and college education for school teachers has been under sharp ideological attack.

52 Engineering and scientific laboratories during the eighties radiated their contempt for women in their abundant display of highly sexist 'funny' official safety posters. Workshops under the control of male technicians not infrequently displayed pornographic posters.

53 Aziz, 'Women in UK Universities', in Lie and Leary, *Storming the Tower*. Cambridge University has 33 per cent contract staff on fixed-term appointments; these form 50 per cent of all university staff. *Proposals for the Waldegrave White Paper on Science and Technology from Cambridge Women Scientists, Mathematicians and Engineers*.

54 The most recent initiative was William Waldegrave's. As the minister responsible for science, Waldegrave set up a committee of women natural scientists, chaired by Bill Stewart, the chief government science adviser, to explore ways of using women scientists' talents fully. The position of Nancy Lane as a distinguished cell biologist, as a senior member of this committee but who has only an honorary post at Cambridge, speaks to the severity of the problem.

55 Women were a very evident part of the computer science labour force when computing was a new development during World War Two. With the formalization of the area came also an increasing mathematization of the subject itself, the two together serving to exclude women. See: EOC, *Information Technology and Gender*; EOC, *Girls and Information Technology*, Lovejoy and Hall, 'Where Have All the Girls Gone?', *Univ. Comput*, 9 (1987);

Turkle, 'Computational Reticence: Why Women Fear the Intimate Machine', in Kramarae (ed.), *Technology and Women's Voices*; Griffiths, 'Strong Feelings about Computers', *W's Stud. Int. Forum*, 11 (1988); Dain, 'Women and Computing', *W's Stud. Int. Forum*, 14 (1991).

56 Similar pressures had developed in Oxbridge; the 1972 opening of the men's colleges to women students weakened the recruiting capacity of the women's colleges. In response Girton became a mixed college, and during 1992, Somerville, despite considerable student opposition, voted to become mixed, leaving the two remaining women's colleges increasingly isolated. The irony of the loss of the women's colleges at the same time that the feminist sociology of education was pointing to the advantages of single-sex education was not missed. By the early nineties, what were originally the men's colleges had for almost twenty years been recruiting women students, so that in Cambridge 34 per cent of all students were women. None the less, the former men's colleges showed few signs of appointing women academic staff in appropriate proportions – in 1991 there were only 7 per cent women staff. Reform at the student level shored up rather than diluted the men's academic control.

57 During the eighties, when the influential *Marxism Today* was celebrating the move to post-Fordism in industrial management, the university system was being forced into a classically Fordist model, withdrawing resources from the primary purposes of teaching and research and increasing accounting, reporting and other bureaucratic functions. Hierarchical structures like this, as had been seen earlier in the cases of nursing and social services organizational reforms, facilitated the preservation/rise of the male manager.

58 Norwegian research suggests that formal commitment and organizational equal opportunity structures do not secure much change in the gender pattern within hierarchical organizations. Kvande and Rasmussen, 'Gender Technology and Organisation in Postmodern Times', *Gender, Technology and Ethics*.

59 In 1992 the Association of University Teachers pointed out that women professors were systematically paid less than men professors. Alas, the British unlike the US do not have 'class action'.

60 The debate concerning a suitable education for girls and women continued from the pre-war arguments. The Newsom Report with its recommendations for gender-specific education represented the political low point.

61 Swann Report. The discussion of science education in a multicultural society has been much slower to develop: Brandt et al. (eds), *Science Education in a Multi-cultural Society*; Nott and Watts go much further and criticize the notion that there is 'one world science': 'Towards a Multi-cultural and Anti-racist Science Education Policy', *Ed. Sci.*, 121 (1985).

62 S. Acker, 'No Woman's Hand: British Sociology of Education 1960–79', *Soc. Rev.*, 29 (2 1981).

63 Althusser, *Lenin and Philosophy and Other Essays*, especially 'Ideology and Ideological State Apparatuses', pp. 121–73; Bernstein, 'Education Cannot Compensate for Society', in Cosin et al. (eds), *School and Society*; Bourdieu and Passeron, *Reproduction in Education, Society and Culture*; Apple (ed.), *Cultural and Economic Reproduction in Education*.

64 Bowles and Gintis, *Schooling in Capitalist America*.

65 See: Brighton Women and Science Group (eds), *Alice through the Microscope*; BSSRS, *Science for People*, for an example of this conditioning argument.

66 Whyte, *Girls into Science and Technology*.

67 See Traweek, *Beam times and Life times*.

68   Rosser, *Female Friendly Science.*
69   J. Harding (ed.), *Perspectives on Science and Gender.*
70   Kelly, 'The Construction of Masculine Science', *B. J. Soc. Ed.*, 6 (2 1985);
     Kelly, *Science For Girls,*p. 2.
71   Thomas, *Gender and Subject in Higher Education;* her findings echo those of
     Sharon Traweek.
72   Willis, *Learning to Labour.*
73   McDonald, 'Schooling and the Reproduction of Class and Gender Rela-
     tions', in Barton et al. (eds), *Schooling, Ideology and the Curriculum;* Arnot,
     'Male Hegemony', Social Class and Women's Education', *J. Ed.*, 164 (1982).
74   Spender, *Men's Studies Modified;* Spender, *Invisible Women;* Spender, *For the
     Record.*
75   H. Rose, 'Nothing Less than Half the Labs', in Finch and Rustin (eds),
     Agenda For Higher Education.
76   Yet other strands of feminism, which stress spirituality and closeness or
     even identity with an essentialist concept of nature, seek to have little truck
     with science as occupation and do not see it as susceptible to feminist
     criticism and change. This anti-science mood, which is currently quite
     strong within feminism, is neither new nor limited to feminism; it has for
     example strong links with the animal rights movement. Here I want to leave
     such matters to one side not because they are uninteresting or unimportant,
     as they speak both of the revulsion of many women at science and also of
     diverse and changing constructions of femininity, but because I want to
     focus on the over-representation of men within the organized production of
     scientific and technological knowledge.

Chapter 6   Joining the procession: (man)aging the entry of women into the Royal Society

1   Margaret Cavendish not only was the first woman scientist and philosopher
    who tried to participate in the proceedings of the Royal Society in its
    founding years, but also on being rejected wrote a first feminist SF/utopia:
    The Description of a New World Called the Blazing-World; K. Jones, *Glorious
    Fame.*
2   Sanctification has long been a means of dealing with the acclaimed translator
    of *The Mechanism of the Heavens.* Marina Benjamin contrasts the public image
    and acceptablity of Somerville's selflessness and womanliness to the
    gentlemen of science with her letters and diaries, which show a lively
    competitiveness with her friends and peers. 'Elbow Room: Woman Writers
    on Science 1790–1840', in Benjamin (ed.), *Science and Sensibility.*
3   In saying this I do not want to erase the achievement of Kathleen Lonsdale,
    who was a vice-president at the time of the tercentenary celebrations in
    1960. However, it is the geneticist Anne McLaren who became the first
    working officer, as Foreign Secretary in 1991. Officers hold considerable
    social power marked by the fact that only their papers are kept by the
    Society regardless of the scientific or other distinction of non-officer
    members. Lonsdale's are held by University College, London.
4   Thatcher was, however, not given the traditional academic accolade, an
    honorary degree from Oxford. While this insult was richly merited for her
    government's onslaught against both education and research, it was difficult
    to escape the feeling that it was more easily brought off against a woman
    rather than a man prime minister. Despite such qualms, there was
    considerable rejoicing even in women's studies circles.

5   Home, 'A World-wide Scientific Network and Patronage System', in Home and Kohlsted (eds), *International Science and National Scientific Identity Between Australia, Britain and America*; Stearns, 'Colonial Fellows of the Royal Society of London 1661–1788', *Notes Rec. Roy. Soc.*, 8 (1951). These accounts are primarily concerned with white British colonialists and their descendants in first the empire and then the Commonwealth; however, Home does note an Indian shipbuilder, Ardaseer Curstejee, with an amateur interest in science elected as early as 1841 – that is, before the 1847 reform (p. 157). A number of Indian men mathematicians and physicists have been elected this century.

6   J. D. Bernal was Hodgkin's supervisor and central both in the social relations of science movement and in crystallography. I interviewed Dorothy Hodgkin (D. C. H.) in March 1991 as one of the key actors in this story who is still alive. Dorothy (she is called Dorothy by everyone, which is quite extraordinary in view of her eminence, but perhaps provides a gendered means of managing eminence) said how much she would have liked to share the Nobel Prize with Bernal, and how disappointed she was when she learned that it had been awarded to her alone.

7   D. C. H., letter, 25 January 1991.

8   Within the Royal Society there was some discussion of the view that she would make a richly qualified president, but this would have meant forgetting her gender, and the patriarchal institution settled for an unmarked scientist.

9   In Royal Society parlance these candidacy papers are spoken of as a certificate. The certificate, once properly drawn up by a Fellow and signed by an appropriate number of Fellows, is then registered and suspended while the candidate is considered for election. The candidate has officially no active part in the process, although informally candidates may help their advocates prepare the case.

10 'By marriage, the very being or legal existence of a woman is suspended, or at least it is incorporated or consolidated into that of her husband, under whose wing, protection and care she performs everything, and she is therefore called a femme couvert.' Blackstone, *Commentaries on the Laws of England*.

11 Internationally the feminist movement had been challenging this conception from the 1860s. Within Britain by the turn of the century there was access to a number of university colleges and some professions, etc. A number of professional societies admitted women as fellows long before the Royal Society. Thus Agnes Arber (FRS 1946) had been a Fellow of the Linnean Society since 1908.

12 Mason cites both this letter from Lord Huggins and that from his wife. 'Hertha Ayrton 1854–1923 and the Admission of Women to the Royal Society of London', *Notes Rec. Roy. Soc.*, 45 (1991).

13 Barbara Bodichon (Leigh Smith) was an active feminist and used her money to support 'the cause'. She was also the inspiration for George Eliot's *Romola*. Both Eliot and Leigh Smith were students at Bedford College, London.

14 Such extramural courses were key means for women to acquire a scientific education at this time.

15 Matilde Ayrton's story was part of that struggle for a medical education described by: Jex-Blake, *Medical Women*; Blackwell, *Pioneer Work in Opening the Medical Profession to Women*.

16 Eliot wrote to Harriet Beecher Stowe of the resistance towards her sympathetic treatment of Jewish culture. She saw anti-semitism as an aspect of that typical English superiority 'towards all oriental peoples' and 'a national disgrace'. Barbara Hardy, 'Introduction', to Eliot, *Daniel Deronda*.

17   See Mason, 'Hertha Ayrton'.

18   See E. Curie, *Madame Curie*.

19   The Women's Engineering Society was established in 1919 in response to women engineers losing their jobs when men soldiers returned from the 1914–18 war. It is still extant.

20   The potential implications for a number of similarly incorporated bodies, i.e. colleges and universities, was spotted by the Society, but the lawyers, scarcely surprisingly, refused to be drawn to comment on this. If this lay but not unsophisticated reading was correct the legality of the preservation of single-sex access to many prestigious higher education institutions and fee-paying and state schools until the 1960s and later looks dubious.

21   *Royal Society*, NLB, 67,493.

22   While the feminist movement knew of the 1922 legal opinion, there was some confusion as to whether Ayrton was then finally admitted. Strachey, *The Cause*, p. 377, gets it wrong.

23   *Proceedings of the Royal Society*, A. Vol., 102 (1923).

24   The speed at which Sir Henry Dale (H. H. D.) was able (see below) to find a second candidate once Lonsdale had been proposed should not be forgotten.

25   D. C. H., letter, 25 January 1991.

26   Rossiter, *Women Scientists in America*.

27   This sense of being both in and out of culture has been long discussed by feminists – from Virginia Woolf to Dorothy Smith and Patricia Hill Collins.

28   The Society's legal advisers had been changed the previous year; serendipity or design?

29   H. H. D., 'Memorandum on Action required by Presentation of Certificates for Women Candidates', 26 November 1943.

30   H. H. D., 'Memorandum'.

31   Lonsdale steadily published with Smith, a practice characteristic of the more egalitarian laboratories.

32   This left commitment led to many British scientists being regarded with suspicion, verging on paranoia, by the CIA. See chapter 7 for an account of Dorothy Hodgkin's proscription.

33   Hodgkin, *Kathleen Lonsdale*. The handwriting in which Lonsdale's certificate is written appears to be Astbury's.

34   P. M. S. Blackett, letter to W. T. Astbury, 9 February 1945. Blackett shows his sympathy in citing Bernal's view that 'if she had been a man, she probably would have been elected four or five years ago. Is this also your view?'

35   W. T. Astbury, letter to P. M. S. Blackett, 12 February 1945. Astbury's last sentence indicates the extent to which Dale had framed the terms of the discussion of the two women nominees. 'Another important point to my mind is this, that I believe the biologists are putting up Marjory Stephenson, and it would be much more satisfactory if two women got in together the first time any woman was elected.'

36   Hodgkin, *Kathleen Lonsdale*, p. 454.

37   H. H. D., letter to Paul Fildes, 15 November 1943.

38   Fildes enjoyed claiming a first in biological warfare. He believed he had helped cause the death of Heydrich, the Nazi Gauleiter of occupied Czechoslovakia, through septicaemia by introducing a pathogen into a grenade. The Nazi response was the massacre of the village of Lidice. Harris and Paxman, *A Higher Form of Killing*, pp. 88–94.

39   Fildes was proud of his great-grandmother Marjory Fildes. She had been

president of the Female Radical Reformers of Manchester, which had taken a leading part in the demonstration put down by the massacre of Peterloo.

40 H. H. D., 15 November 1943 (93 HHD).

41 H. H. D., letter to J. B. S. Haldane, 22 November 1945.

42 Fildes letter to H. H. D., 16 November 1943 (93 HHD).

43 Interview, D. C. H., March 1991.

44 H. H. D., 'Memorandum'.

45 Indeed the plaint was not unheeded; in 1946 the Society expanded its numbers by twenty-five. See note 62.

46 Zuckerman, *Scientific Elite*.

47 This section draws extensively on Hodgkin, *Kathleen Lonsdale*. Most of the women's biographies are written by other women. Even among eminent women scientists, who so often tell us that they have no specially gendered experience, a quiet network of mutual support, in life and in writing the obituary record, seems to operate.

48 Annan, *The Disintegration of an Old Culture*.

49 Hodgkin, *Kathleen Lonsdale*.

50 Lonsdale, 'Women in Science', *Impact*, 20 (1970).

51 The women's colleges were very skilled at supporting their gifted women research students – particularly in the sciences.

52 This was by no means an unusual arrangement. The laboratory replaced the married woman's proper contribution to family life.

53 Hodgkin notes, 'for the sake of women's liberation', that Lonsdale romanticized the merits of the child-induced breaks as being helpful to research. This is incidentally the one moment when Hodgkin speaks of feminist politics directly. The memoir was written in 1971, when the women's liberation movement exploded into existence in Britain, with the first huge meeting in Oxford, where Hodgkin had her laboratories. Hodgkin, *Kathleen Lonsdale*, p. 473.

54 Julian, 'Women in Crystallography', in Kass-Simon and Farnes (eds), *Women of Science*.

55 In the early eighties this still prevailed, so that while there were an estimated 2 per cent women in physics generally, there were 14 per cent in crystallography. Julian, 'Women in Crystallography', p. 336.

56 Hodgkin, *Kathleen Lonsdale*, p. 457.

57 Robertson, *Marjory Stephenson*.

58 The biographies of women scientists rarely claim feminism, but as against men's exceptional acknowledgement they usually acknowledge the contribution of their mothers to their scientific success.

59 Brittain, *Testament of Youth*.

60 This publication marked Stephenson's biochemical achievement and was the kind of contribution which would have led to a man scientist being proposed. The crucial point is that no man scientist was quite able to bring himself to propose her, including the biological leftists and liberals who were well established in the Royal Society – until Astbury proposed Lonsdale. That Stephenson was a socialist and a member of the Cambridge Scientists Anti-war Group indicates the extent of the left scientist's refusal to accept the claims of women colleagues and comrades.

61 Robertson, *Marjory Stephenson*.

62 Since their admission forty-five years before, fifty-two women Fellows had been elected, with an additional five as foreign Members.

63 The situation has deteriorated, to some extent masked by the increase in the annual limit of scientists to be admitted to the Fellowship: from fifteen each year – set in 1847, the year of reform against aristocratic degeneracy – to

seventeen in 1930 and twenty in 1937. In 1946 (perhaps because of fear of the avalanche of women who would deny men their places) the limit was raised to twenty-five. This limit has continued to rise without more women being elected, but both men and women Fellows remain convinced that gender has no bearing on election, and only scientific contribution counts.

Chapter 7   Nine decades, nine women, ten Nobel Prizes: gender politics at the apex of science

1   The award of the 1992 medicine and physiology Nobel Prize to two men biochemists in their seventies, E. Krebs and M. Fischer, caused considerable surprise and concern for just this reason.
2   Harriet Zuckerman's pioneering study covered five women Laureates. She notes that the interval between the work and its recognition is longer for women than men, yet those women Nobelists whose awards followed her study were older and the gap greater. Her study draws attention to the achievements of Jewish scientists who are strongly represented among the US Nobel Laureates. Both Zuckerman and her mentor Robert K. Merton, the pioneering sociologist of science, are Jewish and sensitive to the history of institutionalized anti-semitism in US academic life. Like many of his generation, Merton had found it necessary to change his name to one sounding more acceptably Anglo. Zuckerman, *Scientific Elite*.
3   Levi-Montalcini, *Le Prix Nobel*, 1986, p. 276 (hereafter as *LPN* plus date).
4   Zuckerman, *Scientific Elite*, p. 82.
5   For a sustained examination, see H. Rose and Rose, 'The Incorporation of Science', in H. Rose and Rose (eds), *Political Economy of Science*.
6   However, the papers of individuals not infrequently have fifty-year restrictions placed on them.
7   The biographies range from popular accounts, such as Opfell, *The Lady Laureates*; Phillips, *The Scientific Lady*, to scientific and biographical accounts written by feminist scientists and historians as part of a project to write women back into the 'his'tory of science, to bibliographic guides: Kohlstedt, 'In from the Periphery', *Signs*, 4 (1 1978); Scheibinger, 'The History and Philosophy of Women in Science', *Signs*, 12 (2 1987); Kass-Simon and Farnes (eds), *Women of Science*; Alic, *Hypatia's Heritage*; Amir-Am and Outram (eds), *Uneasy Careers and Intimate Lives*; Scheibinger *The Mind has no Sex*; Searing (ed.), *The History of Women in Science, Technology and Medicine*; Ogilvie, *Women in Science*; Herzenberg, *Women Scientists from Antiquity to the Present*; Siegel and Finley, *Women in the Scientific Search*.
8   His stuffed horse and bloody clothes are displayed in the historical museum in Stockholm.
9   Nuclear physics itself was still young enough to be open to women.
10   Author's translation: *LPN*, 1903, p. 2. Note the division of recognition in both the words and the portrait size: Becquerel a half, Marie Curie and Pierre a quarter each.
11   E. Curie, *Madame Curie*.
12   The book was published in many countries and was inspirational for young women scientists. See Rosalyn Yalow's autobiographical note, *LNP*, 1977.
13   E. Crawford, *The Beginnings of the Nobel Institution*. I am indebted to this study for the material on Marie Curie's two prizes. See also Giroud, *Marie Curie: A Life*.
14   Initially the groups and individuals consulted were very narrowly drawn,

primarily the national academies and existing Nobel Laureates. Then as now the personal international connections of Swedish Nobel committee members were influential. Today the consultations are much wider, but with little effect so far as recognizing women scientists is concerned.

15   E. Crawford, *Beginnings of the Nobel Institution*, p. 112.
16   Koblitz, *A Convergence of Lives*; Margaret Rossiter reports that 'a Swedish mathematician' (Leffler?) wrote to Henrietta Leavitt, the Harvard astronomer, in 1925, saying that he wanted to nominate her. She was, alas, already dead. Rossiter, *Women Scientists in America*.
17   E. Crawford, *Beginnings of the Nobel Institution*, p. 194.
18   According to H. J. Mozan (John Zahm), *Woman in Science*, Ann Carlotta Leffler also published a study of the admired mathematician: *Sophia Kovalevskaia*.
19   Senta Troemel-Ploetz draws attention to a little-known biography by Desamka Trbuhovic Gjuric, herself a mathematician acquainted with the Swiss milieu where the Einsteins lived and worked. This has been republished, but rather heavily edited, in German. 'Mileva Einstein Marić: The Woman who did Einstein's Mathematics', *W's Stud. Int. Forum*, 13 (5 1990).
20   *Ibid.*, p. 418.
21   Walker, 'Did Einstein Espouse his Spouse's ideas?', *Physics Today*, February (1989). However, more disturbingly, John Hackel, editor of *The Collected Papers of Albert Einstein*, Vols I and II, ignores this evidence. Despite my feeling that historians of science have recently been more willing to accept the contribution of women scientist, it seems that in the case of Einstein the myth of the unaided male genius must be preserved.
22   E. Crawford, *Beginnings of the Nobel Institution*, p. 148
23   Then as now we have to be impressed by the physical effort – it took 6,000 kg of pitchblende to produce 0.1 g of the new element.
24   It was on this visit that Marie Curie met Hertha Ayrton.
25   See Clarke, *Working Life of Women in the Seventeenth Century*, for a similar picture for the widows of brewers and opticians – sometimes the widow or a surviving daughter was able to inherit a 'male' occupation. Ivy Pinchbeck, for a slightly later period, shows widows and even wives taking part in their husbands' skilled trades: *Women Workers and the Industrial Revolution 1750–1850*. A. D. Morrison Lowe makes a similar argument for scientific instrument makers: 'Women in the Nineteenth Century Instrument Trade', in Benjamin (ed.), *Science and Sensibility*. The argument made here has to be understood against this general pattern of women and highly skilled occupations and activities.
26   While Eve Curie's biography dismisses the story of Langevin and Curie with outrage, others, including feminist historians, have accepted it as fact. I prefer Robert Reid's sober conclusion that there is no real way of knowing what happened between Langevin and Curie, and that it is irrelevant, the critical point being that a woman scientist could not, without comment, spend leisure time in the company of a man scientist unrelated to her. At the time the right-wing press wallowed in the sexual innuendo and attacked mixing anti-semitism and nationalism, while the left and liberal press defended her. She had every need to accept Hertha Ayrton's invitation to be an incognito guest in England. Reid, *Marie Curie*.
27   The scientific community intensely debated the issue. *Nature* editorialized, 'we have confidence that the doors of science will eventually be open to women on equal terms with men.' 12 January 1911.

28 There was some criticism of Joliot's opportunism in claiming the Curie name.

29 The other women in office with some claims to be scientists have been Thatcher, Ceauşescu and Zhukova.

30 Deborah Crawford wrote a lively, if somewhat inventive, biography for children published the year after Meitner's death: *Lise Meitner: Atomic Pioneer*.

31 Hahn, *My Life*, p. 88.

32 Johannisson, *A Life of Learning*, p. 103.

33 Frisch (ed.), *Trends in Atomic Physics*.

34 Haberer, *Politics and the Community of Science*.

35 Beyerchen, *Scientists Under Hitler*.

36 Steven Rose and I named this 'the Physicists' War' in H. Rose and Rose, *Science and Society*.

37 By contrast with these rather unsatisfactory interpretations Herbert Mehrtens has produced a much more nuanced account of the mathematicians: 'Mathematics in the Third Reich', in Vigger et al. (eds), *New Trends in the History of Science*.

38 The Royal Society recognized her very slowly, more than ten years after women were admitted to the Society and ten years after Hahn became a Laureate.

39 Miller, 'Women in Chemistry', in Kass-Simon and Farnes (eds), *Women of Science*, p. 315.

40 Zuckerman, *Scientific Elite*, p. 192.

41 Aaron Klug, a crystallographer, a Nobel Laureate, and also Jewish, consistently wrote Franklin's contribution back into the scientific record.

42 Franklin was not, however, the only gifted woman in the Randall laboratory, a fact used by Maurice Wilkins to suggest that the laboratory was not hostile to women.

43 Perutz was complex. Though this action was no help to Franklin, he also lobbied the influential Swede Gunner Haag for Dorothy Hodgkin as a candidate for the Nobel Prize. Personal communication, C. Haag.

44 Biologist Richard Lewontin entertainingly deconstructs 'Honest Jim Watson's Big Thriller about DNA', in Stent (ed.), *James D. Watson*.

45 Maria Goeppert Mayer, *LPN*, 1963, p. 98.

46 'Science for fun' was another way of acknowledging that there was then little or no paid employment for women scientists in either the US or the UK. In Cambridge, England, a number of married women scientists who were eventually to become very eminent biochemists were either unpaid or where they had children, only paid their child care costs. (Personal communication, Dorothy Needham, FRS.)

47 J. E. Mayer, 'My Wife's Secret: The Atomic Bomb', in *The Nobel Prize*.

48 Ibid., p. 25.

49 L. M. Jones, 'Intellectual Contributions of Women in Physics', in Kass-Simon and Farnes (eds), *Women of Science*, p. 200.

50 Zuckerman, *Scientific Elite*.

51 Goeppert Mayer, *LPN*, 1963, p. 151.

52 As mentioned in chapter 1, Feynman describes theory in his Nobel speech as 'at first as an elegant woman you love and marry, then as she ages at best a good mother to her children'. *LPN*, 1966.

53 Men Laureates at this time do sometimes talk about the importance of their mothers; more rarely do they insist on their independent status as persons.

54 Dorothy Crowfoot Hodgkin, *LPN*, 1964, p. 84.

55 Dorothy Crowfoot Hodgkin, interview, April 1991.

56 Hudson, 'Unfathering the Thinkable', in Benjamin (ed.), *Science and Sensibility*. In an otherwise sensitive analysis Hudson unfortunately conflates pacificism and anti-militarism. While they were close, not to say muddled up within particular organizations and people, in interview Hodgkin explicitly refused the label pacificist and described herself as anti-militarist. The Women's International League of Peace and Freedom (WILPF), which Hudson discusses, was quintessentially a bourgeois feminist organization, and while socialist and communist anti-militarists frequently made alliances with WILPF, the political differences between them were considerable. However, Hodgkin has what I would call an 'aesthetic pacificism', a feeling shared by many anti-militarist women who could not call themselves pacificist.

57 Max Perutz even praises her as a 'wife and mother'. Just how many festschrifts of men scientists honour their status as 'a husband and father'? 'Forty Years Friendship with Dorothy', in Dodson et al. (eds), *Structural Studies in Molecules of Biological Interest*.

58 The nepotism rule continued on both sides of the Atlantic for many years, often solved by married women working as unpaid research associates. In 1970 the primatologist Jeanne Altmann was an honorary associate, her husband Stuart having the post. In the late sixties the biochemist Clare Woodward was refused a research post on a named grant at Minnesota because her geneticist husband had a teaching post. At the London School of Economics in the sixties the sociologist Ruth Glass was advised against applying because her demographer husband held a sociology chair. At the same period sociologist Olive Banks' husband and fellow sociologist Jo was invited to apply for a Cambridge chair. On learning the nepotism rule, he declined, and the Banks went first to Liverpool and then to Leicester, where they held joint chairs. To me as a sociologist, the Banks' feminist principles and practices have long been a source of personal inspiration.

59 Yalow, *LPN*, 1977, p. 243.

60 *The Women's Biography* confuses this situation, claiming that Berson, who was by then dead, shared the prize with Yalow: Uglow and Hinton, *The Women's Biography*.

61 Yalow, *LPN*, 1977, p. 243.

62 Ibid., p. 245.

63 Ibid., p. 248.

64 Keller, *A Feeling for the Organism*.

65 From Florence Nightingale to Virginia Woolf the plea for privacy is constant. Even today few women have a space within a family home which is exclusively theirs.

66 Balbo, 'Crazy Quilts', in Sassoon (ed.), *Women and the State*.

67 Barbara McClintock, *LPN*, 1983, p. 192.

68 Her invisible college celebrated her achievements in the last year of her life. Fedora and Botstein (eds), *The Dynamic Genome*.

69 Levi-Montalcini in Knudsin (ed.), *Successful Women in the Sciences*.

70 Levi-Montalcini, *LPN*, 1986, p. 277.

71 Levi-Montalcini, *In Praise of Imperfection*.

72 The culture of biomedical research in Italy has long been relatively unregulated by ethical debate. Even today 'smart' drugs are tested on Alzheimer patients by pharmacologists with industrial connections without reference to ethical committees or controls. Personal communication Steven Rose.

73 Because the Italian highly educated elite is relatively small, women's participation in science has been acceptable as a cultural activity for upper-

class women. Thus campaigning for a woman scientist created fewer problems than in other countries, although it unquestionably required much patience.

74 Levi-Montalcini, *LPN*, 1986, p. 283.
75 This is the only Nobel Laureate, caught perhaps by constructions of feminity, who does not give a birthdate.
76 Gertrude Elion is the least well documented of the women Nobel Prize-winners. As an industrial scientist she has a less public profile; for example she has published no books, unlike all the other women Laureates. The *LPN* 1988 publication is in this case particularly important.
77 There is a problem about the 'capturing' of an institution like the Nobel Prize by a discipline or even by a school, which then acts as a self-recruiting oligarchy. This latter is displayed to absurdity in the case of the Nobel Prize for Economics and Chicago Economics.

Chapter 8   Feminism and the genetic turn: challenging reproductive technoscience

1 Petchesky, 'Reproductive Freedom', in Stimpson and Person (eds), *Women, Sex and Sexuality*; Rothman, *The Tentative Pregnancy*; Rowland, 'The Meanings of Choice in Reproductive Technology', in Arditti et al. (eds), *Test Tube Women*; Cowan makes an attempt to rescue choice in 'Genetic Technology and Reproductive Choice', in Kevles and Hood (eds), *The Code of Codes*.
2 Feminist work on the one-child family policy in China expresses this tension. The problem of both how many people the land can bear without widespread starvation and also the horror of female infanticide, to which the policy led, cannot be evaded by a simple appeal to individual 'rights'. Croll et al. (eds), *China's One Child Family*.
3 I have avoided the term 'population', which seems to slide too quickly into a discourse of control: whether to increase, stabilize or decrease 'population' becomes a societal abstraction, managed and dominated by masculinist and usually Western others.
4 This is not to ignore the rich feminist literature exploring gender and technology both within the home and within employment.
5 Nordic feminism has made a significant contribution in articulating a vision of *The New Everyday Life*, for it is here, in that intermediate location between the 'private' and the 'public', where much of women's activity takes place.
6 Yoxen, *The Gene Business*; Duster, *Backdoor to Eugenics*.
7 Marx, *Capital*, vol. I, p. 177.
8 The irony is that the post-cold-war militarism of the present British government is likely to sustain this commitment even while the funding problems of civil society, including science, increase.
9 Suzuki and Knudtson, *Genethics*.
10 The media has freely speculated as to whether Watson resigned over patenting sequences, where he was known to disagree strongly with Bernadine Healy, the Director of NIH, or whether she forced his resignation because of his commercial interests in biotechnology. Healy's feminism and Watson's anti-feminism have been left unacknowledged.
11 Marx, *Capital*, vol. I, pp. 352–65.
12 Oakley, *The Captured Womb*; Adams, *The Womb Revisited*.
13 This point is directed solely towards feminist discussion of reproductive

technology; that of technology in general is theoretically framed by both the science and technology debates.

14 Ellul, *The Technological Society*.

15 Noble, *Forces of Production*, p. xi; also Winner, *Autonomous Technology*.

16 A point made by S. Harding, *Whose Science? Whose Knowledge?*, and Proctor, *Value Free Science?*.

17 In vitro fertilization and the impending successful implant in humans were extensively discussed in scientific journals in the late sixties and early seventies. See Handler (ed.), *Biology and the Future of Man*; the British Society for Social Responsibility in Science (BSSRS) convened an international meeting on the Biological Revolution' in 1970, published as Fuller (ed.), *The Social Impact of Modern Biology*. This included a paper by R. G. Edwards reporting his work on human reproduction and the associated social and ethical issues. To me, as a BSSRS activist and slowly dawning feminist, it was clear this had immense implications for women. Some of these anxieties are discussed in H. Rose and Hanmer 'Women's Liberation, Reproduction and the Technological Fix', in Allen and Barker (eds), *Sexual Divisions and Society*.

18 Such powers are not new but were not located within the rhetoric of 'helping the infertile'. The history of the involvement of the medical profession in compulsory sterilization programmes in the twentieth century is grim reading: from the thirties in Nazi Germany and the US, population control programmes in the third world, sterilization programmes during the State of Emergency in India, US and British sterilization of poor black women. Muller Hill, *Murderous Science*; Proctor, *Racial Hygiene*; Ludmerer, *Genetics and American Society*; Mass, *Population Target*; National Welfare Rights Organisation, 'Forced Sterilisation of the Poor', *Welfare Fighter*, 4 (1 1974); Rodriguez-Trias, 'Sterilisation Abuse', in Hubbard et al. (eds), *Biological Woman*; Bryan et al. *The Heart of the Race*. pp. 100–7.

19 Stanworth, 'Reproductive Technologies and the Deconstruction of Motherhood', in Stanworth (ed.), *Reproductive Technologies*.

20 The social and political context of use is also important, particularly for biomedicine, where legitimacy is always an issue. See Petchesky, 'Reproductive Freedom', in Stimpson and Person (eds), *Women, Sex and Sexuality*.

21 Etzioni, *Genetic Fix*.

22 The impetus for the new developments in reproductive engineering come in part from their relevance not to human but to animal engineering – the search for innovation and profit in agriculture. Yet one of the keys to the new biological understanding is the similarity between the biologies of non-human animals and humans. Thus, what is done on animals achieves almost automatic, willy-nilly relevance to the prospect of human intervention. Despite the current unease about animal experimentation, for most experimental biologists, working on animals remains a conscious ethical choice, which has served to insulate embryological and genetic advances in other species from ethical and political debate. What was unthinkable about human beings became unthought. Science and scientists – as in the case of IVF – were thus freed from the irritation of popular debate with all its misunderstandings – and understandings – and they escaped, not for the first time, popular and democratic accountability and control. In an interesting way the advent of the animal rights movement could – on an optimistic reading of its impact – help to close the ethical gap through which science escapes. None the less, however much discounted by the scientific establishment, from the 1920s onwards, the new reproductive technology began to be considered in terms of its transferability to human beings.

23 R. G. Edwards had taken part in the 1970 BSSRS debate.

24 It was only in the 1980s that the Commmittee for the Public Understanding of Science (COPUS) was established by the Royal Society, chaired by Walter Bodmer (also president of the Human Genome Organization). COPUS believes that if the public understood science better, science would be better supported. For research contesting this milkjug theory of understanding see Irving and Wynne (eds), *Misunderstanding Science*.

25 Technology assessment, which was the earlier US governmental response to anxiety about the nuclear industry, was never adopted by the British government and somewhat nominally introduced in the EU with STOA (Science and Technology Assessment group).

26 The Nuffield Foundation and the Hastings Institute have played an active part in extending the role of ethics as guidance for biomedicine. 'Nuffield Collections' of 400 volumes have been recently established in Warsaw, Prague, and Budapest, with shorter 'Hastings/Nuffield Collections' to be set up in Krakow, Brno and Pecs. The list contains only one feminist book (Gilligan, *In a Different Voice*, despite feminism's passionate debate on reproductive medicine. *Bull. Med. Ethics*, 75, January/February, 1992.

27 Mainstream biomedical ethics have moved in in strength to colonize the new field: Austin, *Reproductive Technologies*; Bartels et al., *Beyond Baby M*; Cameron, *Embryos and Ethics*; Chadwick, *Ethics. Reproduction and Genetic Control*; Ciba Foundation, *Human Embryo Research*; Dyson and Harris, *Experiments on Embryos*; Ford, *When Did I Begin?*; Glover et al., *Fertility and the Family*; Hull, *Ethical Issues in the New Reproductive Technologies*; Shannon, *Surrogate Motherhood*; Sutton, *Prenatal Diagnosis*; Warnock, *A Question of Life*.

28 Nowhere is this clearer than in the contribution of the philosopher and ethicist Peter Singer. Having established his international reputation with his defence of *Animal Liberation*, Singer, in collaboration first with the obstetrician William Walters and then with D. Wells, produced an apologia for IVF, curiously able to set aside the fact that its existence was predicated on extensive animal research. Singer and Wells, *The Reproductive Revolution*; Walters and Singer, *Test Tube Babies*.

29 Despite the belief of many ecofeminists that genetic engineering in plants and animals is especially relevant to the lives of women, the 1988 Advisory Committee for Genetic Manipulation and Planned Release contained no women members and one woman assessor from the Ministry of Agriculture. This is the more usual picture in the British politics of science.

30 McLaren had served on the Royal College of Obstetricians and Gynaecologists (RCOG) in their study of *Artificial Insemination*. Donors were to have the following attributes: 'intelligence . . . pleasing personality . . . better educated' (p.64). The RCOG had also specifically opposed donors who had parents in prison or in a psychiatric hospital. It sees 'the use of university students as "donors" as entirely appropriate'. Snowden and Mitchell, *The Artificial Family*.

31 Pfeffer and Woolett, *The Experience of Infertility*.

32 This was subsequently modified in the guidance for fertility treatment required by sections 25 and 26 of the Human Fertilization and Embryology Act 1990. The new Human Fertilization and Embryology Authority now requires that clinics advise patients of pain and risk involved in collecting gametes.

33 This is not quite such an unproblematic gain, as the 1992 Child Support Act aims to reduce a mother's dependency on the state and transfer it to the biological father – a move from public patriarchy to private patriarchy.

34 Hansard, House of Lords, 'Human Fertilization and Embryology Bill', 20 March 1990, cols 209–10.
35 The public debate on AI began with the optimism of Lord Brabazon's speech: 'It is our duty . . . to know the problems that are about to face us [in the use of AI] and in our wisdom to do the best that in us lies, so as to direct those new forces that they will result in bringing happiness and good into the world.' Hansard, House of Lords, 1943, vol. 128, col. 823. The debate rumbled through the *British Medical Journal*, with the church, through the Archbishop of Canterbury's committee, appointing itself as the moral custodians, in its 1948 report, approving AIH (artificial insemination by husband) and condemming AID (artificial insemination by donor). Donors other than husbands were constructed as 'megalomaniac'. The later Faversham Committee of 1958 was precipitated in the context of a high point in the moral panic around the AI child within marriage. Despite this panic of the bishops and the lords, AI escaped regulation and was simply left to medical practice, with the result that while most clinics limited treatment to married couples some were, by the seventies, helping lesbian couples and single women. Warnock's recommendation and the 1990 legislation stopped such developing practices and introduced penalties for those who assisted – the target being the professionals. Sweden also legislated, insisting on known donors to guarantee fathers for AID children, with the result that donors dramatically declined.
36 Self-help insemination, which has frequently turned to gay men as donors, has had to rethink in the context of the AIDS crisis. Donors now seem to be gay men with tested negative HIV status and heterosexual men with low risk behaviour and/or with tested negative status.
37 In the context of a massive attack on the health and welfare professions as welfare state parasites, this was a notable exception.
38 Warnock Report: *Report of the Committee of Inquiry into Human Fertilisation Embryology*, Cmnd. 9314, HMSO, para. 2.12, also published as Warnock, *A Question of Life*.
39 Known anti-abortionists, like the then Labour science spokesperson Jeremy Bray MP, were very quick to attack the fourteen-day experimental period. Others such as Enoch Powell MP saw the admission of the unacceptability of research after fourteen days as some measure of personhood and thus as support for his bill to protect the 'unborn child'. Many British feminists shared Powell's view, though not his conclusions, and saw the fourteen-day marker as dangerously admitting personhood.
40 A ruling in a Tennesee court, 1 June 1992, concluded that an eight-cell embryo was a 'pre-embryo' and thus not a person to be protected by the law. In writing this moral economy, Justice Daughtrey drew on the American Fertility Society's report, *Ethical Considerations of the New Reproductive Technologies*, and rejected the testimony of geneticist Jerome Lejeune that life and therefore personhood began at conception. Daughtrey rejected the pre-embryo both as a person and as a pure commodity. The embryo 'occupies an interim category that entitles them to special respect because of their potential for life.' *Nature*, 357 (6378), 1992, pp. 425–6.
41 Sarah Franklin 'The Changing Cultural Construction of Reproduction in the Context of New Reproductive Technologies: Redefining Reproductive Choice', paper delivered at the British Sociological Association/Political Science Association Meeting on Political Rights and Reproduction, London, 1990.
42 The feminist obstetrician and gynaecologist Wendy Savage speaks movingly

of this, a point sometimes under-estimated within non-biomedical feminist discourse.

43 Gallagher, 'Eggs, Embryos and Foetuses', in Stanworth (ed.), *Reproductive Technologies*.

44 The proposal also left open the possibility of regulated embryological research, and McLaren is a developmental biologist. As research by Bent Brandth and Agnes Bolsø indicates, while 'women's values' appear to be different from 'men's values', the values of women and men biologists are quite close to one another. 'Men and Women on Biotechnology', in *Gender, Technology and Ethics*.

45 Hansard, House of Lords, 'Human Fertilization and Embryology Bill', 20 March 1990.

46 Firestone, *The Dialectic of Sex*.

47 The Lords' debate over the homophobic legislation of Clause 28 and the discourse of 'pretend family forms', when known homosexual peers voted with the government, speaks of the power of this institution.

48 Dworkin, *Right Wing Women*, pp. 181–8. Dworkin almost undoes the work of feminist historians who have patiently removed the intense moralizing around prostitution, showing it as an economic activity for working-class women denied other means of supporting themselves and their children. Zipper and Sevenhuijsen, 'Surrogacy and Feminist Notions of Motherhood', in Stanworth (ed.), *Reproduction Technologies*.

49 Thus an undocumented story is reported that a woman had conceived purely to provide foetal material to treat her father's Parkinsonism. In Akhter et al. (eds), *Declaration of Comilla*, p. 168.

50 Within India sex selection has precipitated a tremendous outcry among feminists and has been extensively documented in the feminist journal *Manushi* and even by mainstream media such as *The Times of India*. What is at stake is the universalization of 'femicide' from the specific and very negative experience of one country. Dworkin 1983, *Right Wing Women*, p. 188.

51 What is interesting is how this feminist fundamentalism picks up the metaphors of the patriarchs they oppose. E.g. Postgate, 'Bat's Chance in Hell', *New Sci.*, 5 April (1983), overtly links the three political concerns of race, class and sex and sees 'leaping to breed male' especially in the third world as resolving all eugenicist and population concerns. He speculates about farming models for future breeding. See Dworkin, *Right Wing Women*; also Corea, 'The Reproductive Brothel', in Arditti et al. (eds), *Man Made Women*, 1985; Corea, 'How the New Reproductive Technologies Could be Used to Apply the Brothel Model of Social Control Over Women', *W's Stud. Int. Forum*, 8 (4 1985).

52 Hanmer and Allen, 'Reproductive Engineering', in Brighton Women and Science Group, *Alice through the Microscope*, p. 208.

53 Hanmer, 'Transforming Consciousness', in Arditti et al. (eds), *Man Made Women*.

54 Robin Rowland holds out this possibility purely on the basis of an unconfirmed quotation from one scientist: *Living Laboratories*.

55 Klein, 'The Crucial Role of In Vitro Fertilisation as a Means of the Social Control of Women', in *Documentation*, European Parliament (1986).

56 Raymond, 'Preface' in Arditti et al., (eds), *Man Made Woman*.

57 O'Brien, *The Politics of Reproduction*.

58 The presence of such strong ecofeminists as Vandana Shiva contributed to this shift.

59 Mies, 'What Unites, What Divides Women from the South and from the North in the Field of Reproductive Technologies', in Akhter et al. (eds),

Declaration of Comilla. p. 37. As Anne Donchin observes an argument 'never successfully established'. *Hypatia,* 4 (3 1989).

60 At this point most of British socialist feminism was socially determinist, a position Steven Rose and I criticized in S. Rose, *Against Biological Determinism.* See also: Sayers, *Biological Politics.* Birke, *Women, Feminism and Biology;* Benton, 'Biology and Social Science', *Soc.,* 25 (1 1991).

61 Some success has been achieved in mobilizing public opinion, but new legislation to prevent misuse of prenatal testing is widely seen as unenforceable.

62 Swinbank, 'Japanese Gynaecology', *Nature,* 321 (1986), p. 720. While feminists will enjoy the suggestion that 'gender' rather than sex may now be selected, this news item raised intense public criticism, particularly when the Sugiyama Clinic revealed that they too were developing sex-selection techniques, were part of a world-wide network involving 770 groups, and had achieved a 90 per cent success rate in promoting the conception of girl babies, although in a relatively small group. Sugiyama's statement that some women had sought the technique not to avoid hereditary disease but purely out of desire for a daughter inflamed debate. What is interesting in the account is the high passions that these reports aroused, compared with the relatively muted response to male sex selection. Sex selection causes considerable difficulties among feminists and has led to some countries such as Norway, where abortion is on demand up to the first trimester forbidding the giving of such information to pregnant women until the first trimester has passed. See Vines, 'The Hidden Cost of Sex Selection', *New Sci.,* 1 May (1993).

63 Amongst the nineteen clauses, the woman agreed that the doctor had sole power to determine how many embryos shall be transferred. Clinic practice is varied: some expect women to sign the agreement at the interview, others permit some time for reflection, and some still accept consent. Pfeffer *Bulletin of Medical Ethics,* 69, June 1991, pp. 28–31. The subsequent legislation limited number of embryos to three and there is evidence from Frances Price's research that at least one clinic now only transfers two because of women's distress at triplets.This distress of women and their partners at high birth orders arising from infertility treatment speaks to the inadequacy of counselling and informed consent. Botting et al., *Three, Four and More.* In the US the number of transfers may still be as high as five or six.

64 Adams (ed.), *The Wellborn Science;* also 'Science and the Future', Haldane Centenary Conference London, 10–11 April 1992.

65 The slippage between research from the other to the human animal is a good if pragmatic reason for feminists being interested in political arguments for animal welfare or liberation.

66 Preserving a sense of Gramscian optimism in the context of 1990s' postmodernist scepticism and political defeat is crucial for a feminism which seeks to change rather than merely comment on events.

67 D. Wilkinson, 'For Whose Benefit? Politics and Sickle Cells', *The Black Scholar,* 5(8), 1974.

68 Certainly the self-help organization the Family Heart Association is committed to this view, while by and large UK lipidologists are cautious about mass screening.

69 Any gardener will understand this point; genetically identical seeds do very differently in different soils.

70 The thalassaemia story is often described as a success story of screening, but within Greek Cyprus the story began negatively and is still associated with a loss of human rights and more than a slight push towards eugenicism.

The first screening programme led to identified carriers becoming social outcasts, and it was only in the second phase, after careful work with religious and community leaders (predominantly powerful men), that the abortion of affected foetuses after pregnancy screening became a socially accepted outcome. Stigma moved from the carrier to the affected foetus, a by no means positive development from the perspective of disability politics. In Cyprus today, a religiously led state has fused marriage and parenthood so that screening for carrier status is obligatory for people getting married. Suzuki and Knudtson, *Genethics*.

71 One major objective has been to search for the genes 'predisposing' their carriers towards mental disorders such as schizophrenia and manic depression. A goal of eugenic thinking for the major part of this century, it seemed to have been achieved with the triumphalist announcement in *Nature* in 1988 in the chromosomal localization of marker genes for these conditions in, for schizophrenia, an Icelandic and an English family ('pedigree' to use the geneticists' term) and, for manic depression, a population of the US religious community, the Old Amish, in which the disorder is apparently particularly prevalent. The editor of *Nature*, John Maddox, greeted these claims with an editorial announcing that the era of genetic therapy for mental disorder was now at hand. Within two years both claims had been discreetly withdrawn as their original authors proved unable to confirm the findings with large samples. Although the 'non-replications' were also reported in *Nature*, Maddox remained silent on the implications of this debacle for the next five years, only the most recent in a long line of such failures of the genetic approach to mental disease. For a discussion of this and similar evidence see Rose et al., *Not in our Genes*, and Metcalfe (ed.), *Disorder, Disease and Degeneration*.

72 Hubbard and Henifin, 'Genetic Screening of Prospective Parents and of Workers', in Humber and Almeder (eds), *Biomedical Ethics Reviews*. also Hubbard, 'Eugenics and Prenatal Testing', *Int. Health Ser. J.*, 16 (2 1986).

73 Rapp, 'The Story of XYLO', in Arditti et al., (eds), *Test Tube Women*; Rothman, *The Tentative Pregnancy*.

74 Farrant, 'Who's for Amniocentesis?', in Homans (ed.), *The Sexual Politics of Reproduction*.

75 Murdrey and Slack, *B. Med. J.*, 291 (1985).

76 See for example the lack of sensitivity to pregnant women's emotions in the evidence given to the Black Report: *Report by the Working Group on Screening for Neural Tube Defect*. The Royal College of Physicians in its guidance to 'purchasers of clinical genetics' makes its arguments primarily in economic terms of the 'burden' of caring for parents and the state.

77 Taylor et al., *Mental Handicap*.

78 Clinical geneticists and molecular biologists constantly express two conflicting concerns: whether the routine screening they take on drives out research and whether the levels of accuracy of commercial and other non-research laboratories is satisfactory. Both AFP (alpha feto protein) testing and more recently DNA fingerprinting have been criticized on this score. For a discussion of the accuracy and related problems, see Hubbard and Wald, *Exploding the Gene Myth*.

79 For a social view of screening, which looks to the 'total effect on women and their families', see Kings Fund, Consensus Forum, *Screening for Foetal and Genetic Abnormality*, p. 2.

80 Cooper and Schmidtke, 'Diagnosis of Genetic Disease using Recombinant DNA', *Hum. Gen.*, 73 (1986). With the move in Britain to Health Trusts and an emphasis on purchasing services, the advantages of screening and

termination to save the burden of care for both the parents and the state have been openly built into the medical rhetoric. As reduction of the births of the genetic 'abnormal' was hard to demonstrate, clinical genetics defended its expanding services with the claim that it was increasing intending parents' choice.

81 Counsellors in clinical genetics debate whether non-directive screening is possible or desirable.

82 McKusick, *Mendelian Inheritance in Man*.

83 Wingerson, *Mapping Our Genes*; Bishop and Waldholz, *Genome*.

84 Wexler contributed to the NIH Report: *Report of the Working Group on the Ethical, Legal and Social Issues Related to Mapping and Sequencing the Human Genome*. Her colleague James Gusella is also committed to pre- and post-counselling: 'Accuracy of Testing Huntington's Disease', *Nature*, 323 (1986) 118.

85 A Cardiff study of the 'non-disclosing' test which maintains individual choice further points to complexity. Of a group of would-be parents with a background risk of Huntington's, almost all said that with only a 50:50 chance of the foetus being unaffected they would choose abortion. However, of the nine couples where the woman became pregnant, two refused a test, for two tests were not possible, three were told the foetus was unaffected and the remaining three were told that they had only a 50:50 chance. All these women had abortions: Quarrell et al., 'Exclusion Testing for Huntington's Disease in Pregnancy with a Closely Linked DNA Marker', *Lancet*, 8545 (1987). For a useful overview of the field see J. Green, *Calming or Harming?*.

86 Krimsky, *Biotechnics and Society*.

87 Krimsky, on the basis that the US is 'at peace', sees biotechnology as more about industrial regeneration. Susan Wright, by contrast, has maintained a long vigilance over biological warfare development: 'The Military and the New Biology'; Wright (ed.), *Preventing a Biological Arms Race*.

88 In reproducing this reductionist claim it is also important to acknowledge that many non-reductionist biologists flatly refuse it. 'Genes do not reproduce and DNA does not replicate itself, as they are sometimes said to do. Their reproduction or replication happens as part of the metabolism of living cells': see Hubbard, *The Politics of Women's Biology*, p. 81; also Hubbard and Wald, *Exploding the Gene Myth*.

89 Daniel Cohen, the Director of Généthon, claims to have mapped and sequenced some 28 per cent of the genome and almost the whole of chromosome 21. The next stage plans to move towards industrial partnership as Génépole in order to turn gene-mapping technology into products. *Nature*, 357 (1992), p. 527.

90 Lewontin, 'The Dream of the Human Genome', *NY Rev Bks*, 39 (10 1992).

91 This claim to 'truth' has been eroded by a combination of critical biologists and lawyers shrewd enough to see what sociology would call the actor network and what the participants more evocatively speak of as the 'forensic mafia'. Technically incompetent work in the US commercial labs played a part.

92 *Nature*, 358, 9 July 1992, p. 95.

93 Gilbert, in Kevles and Hood (eds), *The Code of Codes*.

94 Adams, *The Well Born Science*.

95 Kevles, *In the Name of Eugenics*. Regrettably Kevles's critical view of the slippery path is reserved for past genetics, for him current genetics is immune.

96 Dulbecco, 'A Turning Point in Cancer Research', *Science*, 231 (1986).

97 Epstein, *The Politics of Cancer*.

98 Koshland, 'Editorial', *Science*, 246 (1989) p. 189.
99 Luria, 'Letter', *Science*, 246 (1989), p. 873.
100 Jon Beckwith, Ruth Hubbard, Jonathan King, Sheldon Krimsky, Richard Lewontin and Susan Wright are among the members. The Council for Responsible Genetics in part developed from Science for the People, which had played an earlier and significant role in contesting sociobiology. Cf. *Sociobiology as a Social Weapon*. The Council's publication *Genewatch* is an invaluable resource.
101 Davis, *Mapping the Code*.
102 McKeown, *The Role of Medicine*, is the single most influential text.
103 US critics of Genome have pointed out that tackling the shocking infant mortality figures for low-income Americans by modifying those environmental factors known to be causal would be much more effective. Even the Bush administration was forced in an election year to announce a prenatal care programme. The UK politics of health during the eighties were framed by the Inequalities in Health debate arising from the Black Report (also published as Townsend and Davidson, *Inequalities in Health*, with the consequence that the Conservative government has already successfully moved on this same acute issue – although many others are neglected.
104 Milio, *Promoting Health through Public Policy*.
105 The objectives of the new public health and Health For All are, however, like much of the epidemiological research that underpins them, androcentric. See H. Rose, 'Gender Politics in the New Public Health', in Draper (ed.), *Health through Public Policy*.
106 Not for nothing do Dorothy Nelkin and Lawrence Tancredi call their book *Dangerous Diagnostics: The Social Power of Biological Information*.
107 Keller 'Nature, Nurture and the Human Genome Project', in Kevles and Hood (eds), *The Code of Codes*.
108 These observations are derived from interviewing and listening to women biologists working in this area.
109 Benno Muller Hill, a Genome participant and author of an important critique of Nazi genetics (*Murderous Science*), sees the eugenicist dangers of the new genetics and argues for social justice policies.
110 A number of the US contributors to Kevles and Hood (eds), *The Code of Codes*, make loud, impersonal claims for Genome, matched only by their silences about their personal financial stakes in it.
111 Patients with the genetic disorder of famial hypercholesterolaemia whom we interviewed were indifferent to the rhetoric of the new genetics and its claims. Instead they draw on a folk language heredity – 'Our family doesn't make old bones' – and were primarily concerned with critically gathering knowledge about how to manage the disorder through life-style adaptation and medication, both of which they studied with considerable energy. Lambert and Rose, 'Disembodied Knowledge', in Irving and Wynne (eds), *Misunderstanding Science*.

Chapter 9 Dreaming the future: other wor(l)ds

1 This is particularly true in Britain; in the USA and in France SF has been treated more seriously.
2 Donna Haraway is one of the exceptions and uses science fiction as a lens through which to read primatology: *Primate Visions*; the philosopher Mary Midgley makes a brilliant, albeit ungendered, analysis of scientists' futuristic visions: *Science as Salvation*.

3   J. D. Bernal, *The World, the Flesh and the Devil*; J. D. Bernal, *The Social Function of Science*.

4   C. Haldane, *Man's World*; J. B. S. Haldane, *Daedalus: Or Science and the Future*. The feminist Naomi Mitchison (*Memoirs of a Spacewoman*) is yet another member of the Haldane family: her brother was J. B. S.

5   Burdekin, *Swastika Night*; Russell, *Hypatia*: Or *Woman and Knowledge*.

6   See Armitt, 'Introduction', in Armitt (ed.), *Where No Man Has Gone Before*.

7   Russ, 'What Can a Heroine Do?', in Glasgow and Ingram (eds), *Courage and Tools*.

8   Cavendish, *The Description of the New World Called the Blazing-World*.

9   Merchant, *The Death of Nature*, pp. 270–2.

10   K. Jones, *Glorious Fame*.

11   Shelley, *Frankenstein*.

12   Mary Wollstonecraft Godwin was the only child of Mary Wollstonecraft and William Godwin, her mother dying a few days after the birth in 1797. At 17 she met Percy Shelley and eloped with him. She had a series of rapid pregancies; one miscarried, one baby girl died, and another small son was killed in an accident. When Mary was still only 25 Percy was drowned in a boating accident, leaving her with one remaining son to provide for. Arguably Percy's own Promethean tendencies made her understand the price of the separation of nurturant love from creativity.

13   Such naturalism is expressed in Pearson, 'Coming Home', in Barr (ed.), *Future Females*.

14   Adrienne Rich says this much better. 'Revision . . ., the act of looking back, of seeing again with fresh eyes, of entering an old text from a new critical direction: 'When We Dead Waken: Writing as Re-vision', in Gelpi and Gelpi (eds), *Adrienne Rich's Poetry*.

15   Le Guin, *The Language of the Night*.

16   Piercy, *Woman on the Edge of Time*.

17   Mattapoisett is Piercy's Utopia; Whileaway is from Joanna Russ, *The Female Man*, and Sally Gearhart's utopia is *The Wanderground*.

18   Piercy, *Woman on the Edge*, p. 228.

19   I am indeed indebted to Inge Lise Paulser for clarifying Gearhart in this way, as, like Paulsen, I was interested in what Gearhart wrote about even while her literary style gave me difficulties: 'Can Women Fly?', *W's Stud. Int. Forum*, 7 (1989).

20   Josephine Saxton writes with a bitter wit of the experience of being classified as an SF writer 'Goodbye to All That . . .', in Armitt (ed.), *Where No Man Has Gone Before*.

21   Lessing, *Colonized Planet 5 Shikasta*; *The Marriages between Zones Three, Four and Five*; *The Sirian Experiments*. Lessing is preoccupied with both individual and societal change in these utopian and dystopian novels. 'Things change: that is all we may be sure of' (*Colonized Planet*, p. 3).

22   I have not discussed time within feminist SF here, but it is not constructed within either the clock or the clockwork of male careers. To enter the feminist SF novel is to abandon chronology.

23   For a critical account of the passive and beautiful good heroine and the active, ugly and bad stepmother/witch see Lieberman, 'Some Day My Prince will Come', in Zipes (ed.), *Don't Bet on the Prince*.

24   Sargent (ed.), *Women of Wonder*, cited by Barr in Palumbo (ed.), *Erotic Universe*.

25   The geneticist Herman J. Muller proposed this in *Out of the Night*.

26   Bernal, *The World, the Flesh and the Devil*.

27   Ibid., p. 46.

28 Bogdanov, *Red Star: The First Bolshevik Utopia*.
29 Graham discusses this in his introduction to Bogdanov, *Red Star*.
30 Bogdanov, *Red Star*, p. 119.
31 Ibid., p. 119.
32 Mead, quoted in Rohrlich and Hoffman, *Women in Search of Utopia*.
33 Earlier utopias, even those written in the same century by Gilman and Mead, were initially erased by second-wave feminism. Shulamith Firestone wrote: 'We haven't even a literary image of this future society: there is not even a utopian feminist literature yet in existence' (*The Dialectic of Sex*, p. 135).
34 Gilman, *Herland*.
35 If Vandyck was a sociologist of science then maybe we should take heart from Gilman, but the current political and educational task is hard going.
36 Gilman, *The Yellow Wallpaper*.
37 Reported in Kevles, *In the Name of Eugenics*.
38 Firestone, *Dialectic of Sex*.
39 Le Guin, *The Left Hand of Darkness*.
40 Lefanu, *In the Chinks of the World Machine*, p. 132.
41 In *Dancing at the Edge of the World*, Le Guin makes her own feminist self-criticism of *The Left Hand of Darkness*.
42 Le Guin, *The Dispossessed*.
43 Ibid., p. 276.
44 Piercy, *Woman on the Edge*; Russ, *The Female Man*; Gearhart, *The Wanderground*.
45 Nuttall, *Bomb Culture*.
46 In that it is Janet from Whileaway who asks if there is an alternative, Russ invites the reader to share both feminism's and humanism's reactions.
47 Role reversal is fun for a cartoon, an interlude as here, or even a short story. It palls as a full-length novel. E.g. Brantenberg, *The Daughters of Egalia*.
48 The creation of Dave enables Russ to play out the role reversal joke, while his synthetic reminds us that only a robot can be the suitable recipient of this sort of heterosexist nonsense.
49 The dream of a common language has been echoed by feminists and others from Cavendish to Rich (*The Dream of a Common Language*). Gearhart's intense ecological concerns mean that her linguistic community bridges the natural/cultural divide. By contrast Wittig's *Les Guérilleres* wage war not just against men but against language. Donna Haraway offers 'an Ironic Dream of a Common Language for Women in the Integrated Circuit': 'A Manifesto for Cyborgs', *Soc. Rev.*, 80 (1985).
50 Gearhart, 'Future Visions', in Rohrlich and Hoffman (eds), *Woman in Search of Utopia*, p. 308.
51 This has been a continuing silence within feminist SF. While Sara Lefanu's celebration of feminist SF, *In the Chinks of the World Machine*, includes Octavia Butler the contributors to the Armitt collection, *Where No Man Has Gone Before*, are silent on 'race' and racism.
52 See June Jordan's poem, 'Song of Sojourner Truth', in Rohrlich and Hoffman (eds), *Women in Search of Utopia*; Walker, *The Color Purple*; Bambara, *The Salt Eaters*; Anzaldua, 'Towards a Construction of El Mundo Zurdo', in Moraga and Anzaldua (eds), *This Bridge called my Back*.
53 Butler, *Wild Seed: Xenogenesis I*, Butler, *Dawn: Xenogenes II*; Butler, *Adulthood Rites: Xenogenesis III*.
54 The figure of Lilith also evokes a recent origin story from biology of an African Eve proposed the universal mother of humanity.
55 Haraway, 'Manifesto for Cyborgs'.

56 Haraway, *Primate Visions*, p. 377.
57 Joanna Russ teases this feminist enthusiasm for non-human reproduction: 'The Clichés from Outer Space', *W's Stud. Int. Forum*, 7 (2 1984).

Epilogue: Women's work is never done

1 For a criticism of the erasing practices of mainstream science studies see H. Rose, 'Rhetoric, Feminism and Scientific Knowledge', in Roberts and Good (eds), *The Recovery of Rhetoric*.

2 Joan Rothschild, on the basis of her survey of feminist science, technology and society courses, argues that interdisciplinarity is a strength: *Teaching Technology from a Feminist Perspective*.

3 MacIntosh, 'Interactive Phases of Curricular Revision', in Spanier et al., (eds), *Toward a Balanced Curriculum*.

4 Innovators include Oakley, *Social Support and Motherhood*. Oakley fuses so-called 'soft' approaches – women's happiness and social support – with 'hard' quantitative approaches, crucially birth weight. In this she moves beyond that equation of feminist methodology with phenomenological approaches, a position which has dominated UK feminist social research in its resistance to the mathematization of reality. In a very different but again very male-dominated field, Sherry Turkle and Seymour Papert also argue for epistemological space. 'Epistemological Pluralism: Styles and Voices within the Computer Culture', *Signs*, 16 (1 1990).

5 Biology and Gender Study group, 'The Importance of Feminist Critique for Contemporary Cell Biology', in Tuana (ed.), *Feminism and Science*.

6 Teaching or showing evolutionary theory is still problematic for many US school districts and science museums, because of the continued strength of Christian religious fundamentalism in the US.

7 Desmond and Moore, *Darwin*.

8 While the moral agony is clear the moral prescriptions are not. Thus while Josephine Donovan's brilliant survey endorses Peter Singer's opposition to animal experimentation, her own proposal for a feminist ethics of animal rights is more restricted. She opposes beauty and cleaning product tests, the notorious LD-50 test and the more vicious experiments such as those of Harlow's primate lab. My purpose is not to score points off Donovan, rather to underline the difficulty – which I share – of deciding what to support and what to oppose. 'Animal Rights and Feminist Theory', *Signs*, 15 (2 1990).

9 See the special supplement 'Women in Science: Gender and the Culture of Science', *Science*, 260, 1993.

10 Merchant, *Radical Economy*.

11 Biehl, *Rethinking Eco Feminist Politics*.

12 Bordo, 'The Cartesian Masculinization of Thought', *Signs*, 11 (3 1986).

Bibliography

Aaron, Jane and Sylvia Walby, eds. *Out of the Margins: Women's Studies in the Nineties*. National Women's Studies Conference. Sussex: Falmer, 1991.

Abeland, Emily and Margaret Nelson, eds. *Circles of Care: Towards a Feminist Theory of Caring*. Albany: New York State University Press, 1990.

Acar, Feride. 'Women in Academic Science Careers in Turkey.' In *Women in Science: Token Women or Gender Equality*, ed. Veronica Stolte Heiskanen. Oxford: Berg, 1991.

Acker, Sandra. 'No Woman's Hand: British Sociology of Education 1960–79.' *Sociological Review* 29 (2 1981): 77–104.

Adams, Alice. *The Womb Revisited: Birth and Human Design in Science. Feminism and Literature*. (forthcoming).

Adams, Mark, ed. *The Wellborn Science: Eugenics in Germany. France. Brazil and Russia*. Oxford: Oxford University Press, 1990.

Adorno, Theodor and Max Horkheimer. *Dialectic of Enlightenment*. London, Verso, 1979 (1972).

Ahmad, Raia and S. Naseen Ahmad, eds. *Quest for a New Science*. Aligarh, India: International Printing Press, 1986.

Akhter, Ferida, Wilma van Berkel and Natasha Ahmad, eds. *Declaration of Comilla*. Dhaka: Finrrage and Ubinig, 1992.

Alcoff, Linda. 'Justifying Feminist Social Science.' *Hypatia: Journal of Feminist Philosophy*, 2 (3 1987): 107–27.

Aldrich, Michele. 'Review Essay: Women in Science.' *Signs: Journal of Women in Culture and Society* 4 (1 1978): 126–35.

Alexander, Jeffrey et al. *The Macro Micro Link*. Berkeley, CA: University of California Press, 1987.

Alic, Margaret. *Hypatia's Heritage: A History of Women in Science from Antiquity through the Nineteenth Century*. Beacon, 1986.

Allen, Sheila and Diana Leonard Barker, eds *Sexual Divisions and Society: Process and Change*. London: Tavistock, 1976.

Althusser, Louis. *Lenin and Philosophy and Other Essays*. London: New Left Books, 1971.

Althusser, Louis. *For Marx*. London: New Left Books, 1975.

Amin, Samir. *Eurocentrism*. London: Zed Books, 1989.

Amir-Am, Pnina and Dorinda Outram, eds. *Uneasy Careers and Intimate Lives: Women in Science 1789–1989*. New Brunswick: Rutgers, 1987.

Amsden, Alice, ed. *The Economics of Women and Work*. Harmondsworth: Penguin, 1980.

Andersen, Margaret and Patricia Hill Collins, eds. *Race. Class and Gender: An Anthology*. Belmont, CA.: Wadsworth, 1992.

Anderson, Sam. 'Science, Technology and Black Liberation.' In *Radicalisation of Science*, eds Hilary Rose and Steven Rose. London: Macmillan, 1976.

Annan, Noel. *The Disintegration of an Old Culture*. Oxford: Clarendon, 1966.

Anthony, Susan B. 'Woman the Great Unpaid Labourer in the World.' (1848) In *Voices from Women's Liberation*, ed. Leslie Tanner. New York: Signet, 1970.

Antonen, Anneli. 'The Feminization of the Scandinavian Welfare State.' In *Finnish Debates on Women's Studies*, ed. Leila Simonen. Tampere: University of Tampere, 1990.

Anzaldua, Gloria. 'Towards a Construction of El Mundo Zurdo.' In *This Bridge Called My Back*, eds. Cherie Moraga and Gloria Anzaldua. Watertown, MA: Persephone, 1981.

Apple, M., ed. *Cultural and Economic Reproduction in Education*. London: Routledge and Kegan Paul, 1982.

Aptheker, Bettina. *Tapestries of Life: Women's Work: Women's Consciousness and the Meaning of Daily Experience*. Amherst, MA: University of Massachusetts Press, 1989.

Arditti, Rita. 'Feminism and Science.' In *Science and Liberation*, eds. Rita Arditti, Pat Brennan, and Steve Cavrack. Boston, MA: South End Press, 1980.

Armitt, Lucie. 'Introduction.' In *Where No Man Has Gone Before*, ed. Lucie Armitt. London and New York: Routledge, 1990.

Arnot, Madeline. 'Male Hegemony, Social Class and Women's Education.' *Journal of Education* 164 (1982): 64–898.

Arnot, Madeleine and Gaby Weiner, eds. *Gender and the Politics of Schooling*. Milton Keynes: Open University Press, 1987.

Ashmore, Michael. 'The Life and Opinions of a Replication Claim: Reflexivity and Symmetry in the Sociology of Scientific Knowledge.' In *Knowledge and Reflexivity*, ed. Steven Woolgar. London: Sage, 1988.

Ataur-Rahim, Mohammed. *Contribution of Muslim Scientists during the Thirteenth and Fourteenth Hijri in the Indo-Pakistan Subcontinent*. Islamabad: Research Unit for Science in Islamic Polity, Islamabad Natural Science Research Council, 1986.

Austin, C. R. *Reproductive Technologies*. Oxford: Oxford University Press, 1989.

Aziz, Adrienne. 'Women in UK Universities: The Road to Casualization.' In *Storming the Tower*, eds. Suzanne Lie and Virginia O'Leary. London: Kogan Page, 1990.

Bacon, Francis. *A Selection of His Writings: Proemium to the Great Instauration*. New York: Odyssey, 1965.

Balbo, Laura. 'Crazy Quilts: Women's Perspectives on the Welfare State Crisis.' In *Women and the State*, ed. Ann Showstack Sassoon. London: Hutchinson, 1987.

Balbo, Laura and Helga Nowotny, eds. *Time to Care in Tomorrow's Welfare Systems*. Vienna: European Centre for Social Welfare Training and Research, 1986.

Bambera, Toni Cade. *The Salt Eaters*. London: Women's Press, 1983.

Barash, D. P. *Sociobiology and Behaviour*. Amsterdam: Elsevier, 1977.

Barash, D. *The Whispering Within*. New York: Harper and Row, 1979.

Baring, Marcia. 'Is There a "Female Style" in Science?' Special Issue on Women in Science. *Science* 260 (1993): 383–91.

Barnes, Barry. *Interests and the Growth of Knowledge*. London and Boston, MA: Routledge and Kegan Paul, 1977.

Barr, Marlene, ed. *Future Females: A Critical Anthology*. Ohio: Bowling Green, 1981.

Barrett, Michèle. *Women's Oppression Today: Problems in Marxist Feminist Analysis*. London: Verso, 1980.

Barrett, Michèle and Mary McIntosh. 'The Family Wage: Some Problems for Socialists and Feminists.' *Capital and Class* 11 (1980): 51–72.

Barrett, Michèle. *Women's Oppression Today; The Marxist Feminist Encounter*. London: Verso, 1980.

Barrett, Michèle and Mary McIntosh. *The Anti-Social Family*. London: Verso, 1982.

Barrett, Michèle and Mary McIntosh. 'Ethnocentricism and Socialist Feminist Theory.' *Feminist Review* 20 (1985): 23–47.

Barrett, Michèle and Anne Phillips, eds. *Destabilizing Theory: Contemporary Feminist Debates*. Cambridge: Polity, 1992.

Bart, Pauline. 'Sexism in Social Science: From the Gilded Cage to the Iron Cage.' *Journal of Marriage and the Family* 33 (4 1971): 734–45.

Bartels, D. M., R. Priester, D. E. Vawter, and A. L. Capian. *Beyond Baby M: Ethical Issues in New Reproductive Technologies*. Clifton, NJ: Humana Press, 1990.

Barton, L., R. Meighan, and S. Walker, eds. *Schooling. Ideology and the Curriculum*. Brighton: Falmer, 1980.

Bauman, Zygmunt. *Hermeneutics and Social Science*. London: Hutchinson, 1978.

Bazin, Maurice. 'Their Science: Our Sciences.' *Race and Class* 34 (1993): 35–46.

Belenky, Mary Field, B. M. Clinchy, N. R. Goldberger, and J. M. Tarule. *Women's Ways of Knowing: The Development of Self, Voice and Mind*. New York: Basic Books, 1986.

Benbow, Camilla and Julian Stanley. 'Sex Difference in Mathematical Ability: Fact or Artifact?' *Science* 211 (12 December 1980): 1262.

Benedict, Ruth. *Race, Science and Politics*. New York: Modern Age Books, 1940.

Benjamin, Marina. 'Elbow Room: Woman Writers on Science 1790–1840.' In *Science and Sensibility: Gender and Scientific Enquiry 1780–1945*, ed. Marina Benjamin. Oxford: Blackwell, 1991.

Benjamin, Marina, ed. *Science and Sensibility: Gender and Scientific Enquiry 1780–1945*. Oxford: Blackwell, 1991.

Benkarts, Sylvia, Boel Berner, Hilary Rose et al. *Proposal for Developing Gender Studies of Science and Technology*. Stockholm: Swedish Planning and Research Council, mimeo, 1992.

Benston, Margaret. 'The Political Economy of Women's Liberation.' *Monthly Review* 21 (4 1969).

Benton, T. ed. 'Biology and Social Science: Why the Return of the Repressed should be given a (Cautious) Welcome,' *Sociology*, 25 (1 1991): 1–29.

Bernal, J. D. *The World, the Flesh and the Devil*. London: Cape, 1929.

Bernal, J. D. *The Social Function of Science*. London: Routledge and Kegan Paul, 1939.

Bernal, J. D. *Science in History*. London: Watts, 1954.

Bernal, Martin. *Black Athena*. London: Free Association Books, 1988.

Bernstein, Basil. 'Education Cannot Compensate for Society.' eds. B. R. Cosin et al. London: Routledge and Kegan Paul, New Haven, CT: 1977.

Beyerchen, Alan. *Scientists Under Hitler*. New Haven, CT: Yale University Press, 1977.

Bhaskar, Roy. *A Realist Theory of Science*. Leeds: Leeds Books, 1975.

Bhaskar, Roy. *Reclaiming Reality*. London: Verso, 1989.

Biehl, Janet, *Rethinking Eco-Feminist Politics*. Boston, MA: South End Press, 1991.

Billig, Michael. *Arguing and Thinking: A Rhetorical Approach to Social Psychology*. Cambridge: Cambridge University Press, 1987.

Biology and Gender Study Group. 'The Importance of Feminist Critique for Contemporary Cell Biology.' In *Feminism and Science*, ed. Nancy Tuana. Bloomington: Indiana University Press, 1989.

Birke, Lynda. *Women, Feminism and Biology: The Feminist Challenge*. Brighton: Harvester Wheatsheaf, 1986.

Bishop, Jerry and Michael Waldholz. *Genome*. New York: Simon and Schuster, 1990.

Black Report. *Report by the Working Group on Screening for Neural Tube Defect*. London: DHSS, 1979.

Black Women's Group Brixton. 'Black Women and Nursing: A Job like Any Other.' *Race Today* 6 (8 1974): 226–30.

Blackstone, William. *Commentaries on the Laws of England*. 1765.

Blackwell, Elizabeth. *Pioneer Work in Opening the Medical Profession to Women*. London: 1885.

Bleier, Ruth. *Science and Gender: A Critique of Biology and its Theories on Women*. Oxford: Pergamon, 1984.

Bleier, Ruth, ed. *Feminist Approaches to Science*. Oxford: Pergamon, 1986.

Bloor, David. *Knowledge and Social Imagery*. London: Routledge and Kegan Paul, 1977.

Bodmer, Walter. *Public Understanding of Science*. 1 (1 1992): 7–10.

Bogdanov, Alexander. *Red Star: The First Bolshevik Utopia*. Bloomington: Indiana University Press, 1984 (1915).

Bohme, Gernot, Wolfgang van den Daele and Wolfgang Krohn. Finalization of Science', *Social Science Information* 15 (1976): 307–30.

Bois, Barbara Du, 'Passionate Scholarship: Notes on Values, Knowing and Method in Feminist Social Science.' In *Theories of Women's Studies: Sociological Theology*, eds Gloria Bowles and Renate Duelli Klein. London: Routledge and Kegan Paul, 1983.

Borchorst, Annette and Birte Siim. 'Women and the Advanced Welfare State: A New Kind of Patriarchal Power.' In *Women and the State*, ed. Anne Showstack Sassoon. London: Hutchinson, 1987.

Bordo, Susan. 'The Cartesian Masculinization of Thought.' *Signs: Journal of Women in Culture and Society* 11 (3 1986): 439–56.

Boserup, Esther. *Women's Role in Economic Development*. New York: St Martin's Press, 1970.

Boston Women and Health Collective. *Our Bodies, Ourselves*. New York: Simon and Schuster, 1969.

Botting, Beverley, Alison Macfarlane, and Frances Price. *Three, Four and More: A Study of Triplet and Higher Order Births*. London: HMSO, 1990.

Bourdieu, Pierre, *Homo Academicus*. Stanford: University of California Press, 1988.

Bourdieu, Pierre. 'Social Space and Symbolic Power.' *Sociological Theory* 7 (1 1989): 14–25.

Bourdieu, Pierre and J. C. Passeron. *Reproduction in Education. Society and Culture*. London: Sage, 1977.

Bowles, Gloria and Renate Duelli-Klein, eds *Theories of Women's Studies*. London: Routledge and Kegan Paul, 1983.

Bowles, S. and H. Gintis, *Schooling in Capitalist America*. London: Routledge and Kegan Paul, 1976.

Bradley, Harriet. *Men's Work, Women's Work: A Sociological History of the Sexual Division of Labour in Employment*. Cambridge: Polity, 1989.

Brandt, G., S. Turner, and T. Turner, eds *Science Education in a Multi-cultural Society*. London: Institute of Education, 1985.

Brandth, Bent and Agnes Bolsø, 'Men and Women on Biotechnology.' In

International Conference on Gender Technology and Ethics, Luleå: University of Luleå, 1992.

Brantenberg, Gerd. *The Daughters of Egalia.* London: Journeyman Press, 1985.

Braverman, Harry. *Labour and Monopoly Capital: The Degradation of Work in the Twentieth Century.* New York: Monthly Review Press, 1974.

Braybon, Gail. *Women Workers in the First World War.* London: Croom Helm 1981.

Braybon, Gail and Penny Summerfield. *Out of the Cage.* London: Pandora, 1987.

Briggs, Anna and Judith Oliver, eds. *Caring: Experiences of Looking After Disabled Relatives.* London: Routledge and Kegan Paul, 1985.

Brighton Women and Science Group, eds *Alice through the Microscope: The Power of Science over Women's Lives.* London: Virago, 1980.

Brittain, Vera. *Testament of Youth: An Autobiographical Study of the Years 1900–1925.* London: Gollancz, 1933.

Brody, Elaine. 'Parent Care as Normative Family Stress.' *The Gerontologist* 25 (1 1985): 19–27.

Brown, Carol. 'Mothers, Fathers, Children: from Private to Public Patriarchy.' In *Women and Revolution*, ed. Lydia Sargent. Boston, MA: South End Press, 1981.

Brown, Richard. *Medicine Men: Medicine and Capitalism in America.* Berkeley, University of California Press, 1979.

Bryan, Beverley, Stella Dadzie, and Suzanne Scarfe. *The Heart of the Race: Black Women's Lives in Britain.* London: Virago, 1985.

BSSRS (British Society for Social Responsibility in Science). *Science for People: Women and Science Issue.* 1974.

Bukharin, Nicholai et al. *Science at the Cross Roads.* London: Cass, 1973 (1931).

Burdekin, Katherine. *Swastika Night.* London: Lawrence and Wishart, 1985 (1937).

Burkhardt, Frederick and Sydney Smith, eds. *The Correspondence of Charles Darwin.* Vol. 1. Cambridge: Cambridge University Press, 1985.

Burrage, Hilary. 'Women University Teachers of Natural Science', *Social Studies of Science* 13, (1983) 147–60.

Butler, Octavia. *Wild Seed: Xenogenesis I.* New York: Doubleday, 1980.

Butler, Octavia. *Dawn: Xenogenesis II.* New York: Warner, 1987.

Butler, Octavia. *Adulthood Rites: Xenogenesis III.* New York: Warner, 1988.

Butler, Octavia. *Imago: Ectogenesis 3.* London: Gollancz, 1989.

Cabral, Amilcar. *Revolution in Guinea: An African People's Struggles.* London: Stage One, 1974.

Cacoullos, Ann 'Women, Science and Politics in Greece: Three is a Crowd.' In *Women in Science: Token Women or Gender Equality*, ed. Veronica Stolte Heiskanen. Oxford: Berg, 1991.

Cameron, Deborah, *Feminism and Linguistic Theory.* London: Macmillan, 1985.

Cameron, N., ed. *Embryos and Ethics.* Edinburgh: Rutherford Horne, 1987.

Caplan, Arthur. *Beyond Baby M: Ethical Issues in New Reproductive Technologies.* Clifton, NJ: Humana Press, 1990.

Carby, Hazel. 'White Woman Listen! Black Feminism and the Boundaries of Sisterhood.' *In the Empire Strikes Back*, Centre for Contemporary Cultural Studies. London: Hutchinson, 1982.

Card, Claudia, ed. *Feminist Ethics.* Lawrence, Kan., Kansas University Press, 1991.

Carr, E. H. *What is History?* Basingstoke: Macmillan, 1986 (1961).

Carson, Rachel. *Silent Spring.* Boston, MA: Houghton Mifflin, 1962.

Cavendish, Margaret. *The Description of the New World Called the Blazing-World.* London: Pickering and Chatto, 1992 (1688).

Centre for Contemporary Cultural Studies. *The Empire Strikes Back.* London: Hutchinson, 1982.

Centre for Gender Studies. *Concept of the Government Programme for the Improvement of the Position of Women and the Family and the Protection of the Mother and Child.*

Moscow: mimeo, Interdisciplinary Research Center for the Study of the Social Problems of Gender, 1990.

Chadwick, R., ed. *Ethics. Reproduction and Genetic Control.* London: Routledge, 1990.

Chamberlain, Mariam. *Women in Academe.* New York: Russell Sage, 1988.

Chamberlain, Mariam. 'The Development of Gender Studies in the US.' In *International Seminar: Gender Studies: Issues and Comparative Perspectives.* May. Moscow, 1992.

Chamberllaine, Prue. 'The Mothers' Manifesto and Disputes over Mutterlichkeit.' *Feminist Review*, 35, (1990): 9–23.

Cherki, Monique Couture. 'Women in Physics.' In *The Radicalisation of Science*, eds Hilary Rose and Steven Rose. London: Macmillan, 1976.

Chodorow, Nancy. *The Reproduction of Mothering.* Berkeley. CA: University of California Press, 1978.

Ciba Foundation. *Human Embryo Research: Yes or No.* London: Tavistock, 1986.

Cicotti, Giovanni, Marcello Cini, and Michelangelo de Maria. 'The Production of Science in Advanced Capitalist Countries.' In *The Political Economy of Science*, eds Hilary Rose and Steven Rose. London: Macmillan, 1976.

Clarke, Alice. *Working Life of Women in the Seventeenth Century.* London: Routledge, 1919.

Clutton-Brock, Juliet. 'How the Wild Beasts were Tamed.' *New Scientist* (15 February 1992): 41–3.

Cobbe, Frances Power. 'Wife Torture in England.' *Contemporary Review* 13 (1878): 55–87.

Cobbe, Frances Power. *The Modern Rack: Papers on Vivisection.* London: Sonnenstein, 1889.

Cobbe, Frances Power. *Life of Frances Power Cobbe: By Herself.* Boston, MA: Houghton Mifflin, 1904.

Cockburn, Cynthia. 'The Gendering of Jobs: Workplace Relations and the Reproduction of Sex Segregation.' In *Gender Segregation at Work*, ed. Sylvia Walby. Milton Keynes: Open University Press, 1988.

Code, Lorraine. *What Can She Know? Feminist Theory and the Construction of Knowledge.* Ithaca, NY: Cornell University Press, 1991.

Cole, Jonathan and Harriet Zuckeman. 'Marriage, Motherhood and Research Performance in Science.' In *The Outer Circle: Women in the Scientific Community*, eds Harriet Zuckerman, Jonathan Cole and John Bruer. New York: Norton, 1992.

Collins, Harry. 'Perspective.' *Times Higher Educational Supplement*, 21 October 1988, 14.

Collins, Patricia Hill. 'Learning from the Outsider Within: The Sociological Significance of Black Feminist Thought.' *Social Problems* 33 (6 1986): 514–32.

Collins, Patricia Hill. 'The Social Construction of Black Feminist Thought.' *Signs: Journal of Women in Culture and Society* 14 (4 1989): 745–73.

Collins, Patricia Hill. *Black Feminist Thought.* Boston, MA: Unwin Hyman, 1990.

Collins, Randall. *Theoretical Sociology.* San Diego: Harcourt Brace, 1988.

Cooley, Mike. *Architect or Bee?* Slough: Hand and Brain Publications, 1979.

Cooper, D. N. and J. K. Schmidtke. 'Diagnosis of Genetic Disease using Recombinant DNA.' *Human Genetics* 73 (1986): 1–11.

Corea, Gena. 'How the New Reproductive Technologies Could be Used to Apply the Brothel Model of Social Control over Women.' *Women's Studies International Forum* 8 (4 1985): 294–305.

Corea, Gena. *The Mother Machine.* New York: Harper and Row, 1985.

Corea, Gena. 'The Reproductive Brothel.' In *Man Made Women: How the New*

Reproductive Technologies Affect Women, eds Rita Arditti, Gena Corea, and Renate Duelli Klein. London: Hutchinson, 1985.

Cosin, B. R. and et al. *School and Society: A Sociological Reader*. London: Routledge and Kegan Paul, 1977.

Cott, Nancy. *The Grounding of Modern Feminism*. New Haven, CT: Yale University Press, 1987.

Cousins, Mark. 'Material Argument and Feminism.' *m/f* 2 (1978): 62–70.

Cowan, Ruth Schwartz. *More Work for Mother: The Ironies of Household Technology from the Open Hearth to the Microwave*. New York: Basic Books, 1983.

Cowan, Ruth Schwartz. 'Genetic Technology and Reproductive Choice: An Ethics for Autonomy.' In *The Code of Codes: Scientific and Social Issues in the Human Genome Project*, eds Daniel Kevles and LeRoy Hood. Cambridge, MA: Harvard University Press, 1992.

Crawford, Deborah. *Lise Meitner: Atomic Pioneer*. New York: Crown, 1969.

Crawford, Elizabeth. *The Beginnings of the Nobel Institution: 1901–15*. Cambridge: Cambridge University Press, 1984.

Croll, Elizabeth, Delia Davin and Penny Kane, eds. *China's One Child Family*. London: Macmillan, 1985.

Cunningham, Andrew and Perry Williams, eds. *The Laboratory Revolution in Medicine*. Cambridge: Cambridge University Press, 1992.

Curie, Eve. *Madame Curie*. London: Heinemann, 1945.

Daele, Walter Van den. 'The Social Construction of Knowledge.' In *The Social Production of Scientific Knowledge*, eds. E. Mendelsohn, P. Weingart, and R. Whitley. Dordrecht: Reidel, 1977.

Dahlrup, Drude. 'Confusing Concepts and Confusing Realities: A Discussion of the Patriarchal State.' In *Women and the State*, ed. Anne Showstack Sassoon. London: Hutchinson, 1987.

Dain, Julia. 'Women and Computing: Some Responses to Falling Numbers in Higher Education.' *Women's Studies International Forum* 14 (3 1991): 217–25.

Daly, Mary. *Gyn/Ecology: the Meta-ethics of Radical Feminism*. London: Women's Press, 1978.

Darwin, Charles. *The Descent of Man, and Selection in Relation to Sex*. London: John Murray, 1888.

Daughtrey, Justice. 'Ethical Considerations of the New Reproductive Technologies.' *Nature* 357 (6378, 1992): 425–6.

David, Miriam. 'The New Right in the USA and Britain: A New Anti Feminist Moral Economy.' *Critical Social Policy* 2 (1983): 31–45.

David, Miriam and Hilary Land. 'Sex and Social Policy.' In *The Future of the Welfare State*, ed. Howard Glennister. London: Heinemann, 1983.

Davies, Celia, ed. *Rewriting Nursing History*. London: Croom Helm, 1980.

Davies, Diane and Helen Astin. 'Life Cycle, Career Patterns and Gender Stratification in Academe.' In *Storming the Tower*, eds Suzanne Lie and Virginia O'Leary. London: Kogan Page, 1990.

Davies, Karen. *Women and Time: Weaving the Strands of Everyday Life*. London: Coronet, 1989.

Davin, Delia and Penny Kane, eds. *China's One Child Family*. London: Macmillan, 1985.

Davis, Angela. *Women, Race and Class*. New York: Random House, 1981.

Davis, Joel. *Mapping the Code: The Human Genome Proiect and the Choices of Modern Science*. New York: John Wiley, 1990.

Delamont, Sara. 'Three Blind Spots? A Comment on the Sociology of Science by a Puzzled Outsider.' *Social Studies of Science* 17 (1987): 163–70.

DeLaurentis, Teresa. *Feminist Studies/Critical Studies*. Bloomington: Indiana University Press, 1986.

Delphy, Christine. *The Main Enemy: A Materialist Analysis of Women's Oppression.* London: Women's Research and Resources Centre, 1977.

Delphy, Christine. *Close to Home: A Materialist Analysis of Women's Oppression.* London: Hutchinson, 1984.

Desmond, Adrian and James Moore. *Darwin.* Harmondsworth: Penguin, 1991.

Devault, Marjorie. 'Talking and Listening from Women's Standpoint.' *Social Problems* 37 (1 1990): 96–116.

Dex, Shirley. 'Gender and the Labour Market.' In *Employment in Britain* ed. Duncan Gallie. Oxford: Blackwell, 1988.

Dinnerstein, Dorothy. *The Mermaid and the Minotaur: Sexual Arrangements and Human Malaise.* New York: Harper, 1976.

Dodson, G., J. Glusker, and D. Sayre, eds. *Structural Studies in Molecules of Biological Interest.* Oxford: Clarendon, 1981.

Doeuff, Michelle le. 'A Letter from France.' *Women in Philosophy Newsletter* 8 (1992): 13–19.

Donchin, Anne. 'The Growing Feminist Debate Over the New Reproductive Technologies.' *Hypatia: Journal of Feminist Philosophy* 4 (3 1989), 136–49.

Donnison, Jean. *Midwives and Medical Men: A History of Inter-professional Rivalries and Women's Rights.* London: Heinemann, 1977.

Donovan, Josephine. 'Animal Rights and Feminist Theory.' *Signs: Journal of Women in Culture and Society* 15 (2 1990): 350–75.

Doyal, Lesley. 'Infertility a Life Sentence? Women and the National Health Service.' In *Reproductive Technologies: Gender, Motherhood and Medicine*, ed. Michelle Stanworth. Cambridge: Polity, 1987.

Draper, Peter, ed. *Health Through Public Policy: The Greening of Public Health.* London: Merlin, 1991.

Dulbecco, Renato. 'A Turning Point in Cancer Research: Sequencing the Human Genome.' *Science* (231 1986): 1055–6.

Duster, Troy. *Backdoor to Eugenics.* New York: Routledge, 1990.

Dworkin, Anthea. *Right Wing Women: The Politics of Domesticated Females.* London: Women's Press, 1983.

Dyson, Anthony and John Harris, eds. *Experiments on Embryos.* London: Routledge, 1990.

Easlea, Brian. *Science and Sexual Oppression.* London: Weidenfeld and Nicolson, 1981.

Edholm, Felicity, Olivia Harris, and Kate Young. 'Conceptualising Women.' *Critique of Anthropology* 3 (9 and 10 1977): 103–30.

Ehrenreich, Barbara. *The Hearts of Men: American Dreams and the Flight from Commitment.* London: Pluto, 1983.

Ehrenreich, Barbara and Deirdre English. *For Her Own Good: 150 Years of the Experts' Advice to Women.* London: Pluto, 1979.

Eisenstein, Zillah. 'Developing a Theory of Capitalist Patriarchy.' In *Capitalist Patriarchy and the Case for Socialist Feminism*, ed. Zillah Eisenstein. New York: Monthy Review Press, 1978.

Elion, Gertrude. *Le Prix Nobel.* Stockholm: Nobel Foundation, 1986.

Eliot, George. *Romola.* Harmondsworth: Penguin, 1980 (1863).

Ellul, Jacques. *The Technological Society.* London: Cape, 1965.

Elson, Diane and Ruth Pearson. 'Nimble Fingers Make Cheap Workers: An Analysis of Women's Employment in Third World Export Manufacture.' *Feminist Review* 7 (Spring 1981): 87–107.

Elston, Mary Ann. 'Women and Anti-vivisection in Victorian England.' In *Vivisection in Historical Perspective*, ed. Nicholaas Rupke. London: Routledge, 1990.

EOC (Equal Opportunities Commission). *Who Cares for the Carers? Opportunities for those Caring for the Elderly and Handicapped*. Manchester: EOC, 1982.

EOC (Equal Opportunities Commission). *Girls and Information Technology*. Manchester: EOC, 1985.

EOC (Equal Opportunities Commission). *Information Technology and Gender: An Overview*. Manchester: EOC, 1985.

Epstein, Samuel. *The Politics of Cancer*. San Francisco: Sierra Club Books 1979.

Etzioni, Amitai. *Genetic Fix: The Next Technological Revolution*. New York: Harper, 1973.

Eysenck, Hans. *Race, Intelligence and Education*. London: Temple Smith, 1971.

Fanon, Frantz. *The Wretched of the Earth*. London: McGibbon Kee, 1965.

Fanon, Frantz. *Black Skins, White Masks*. New York: Grove, 1967.

Farley, Jenny. 'Women Professors in the USA: Where are They?' In *Storming the Tower*, eds Suzanne Lie and Virginia O'Leary. London: Kogan Page, 1990.

Farrant, Wendy. 'Who's for Amniocentesis? The Politics of Pre-natal Screening.' In *The Sexual Politics of Reproduction*, ed. Hilary Homans. London: Gower, 1985.

Fausto-Sterling, Anne. *Myths of Gender: Biological Theories about Women and Men*. New York: Basic Books, 1992 (1985).

Fausto-Sterling, Anne. 'Building Two Way Streets: The Case of Feminism and Science. *National Women's Studies Association Journal* 4 (3 1992): 336–49.

Fedora, Nina and David Botstein, eds. *The Dynamic Genome: Barbara McClintock's Achievements in the Century of Genetics*. Cold Spring Harbor: Cold Spring Harbor Press, 1992.

Fee, Elizabeth. 'Is Feminism a Threat to Scientific Objectivity?' *International Journal of Women's Studies* 4 (1981): 378–92.

Fee, Elizabeth. 'Critiques of Modern Science: The Relationship of Feminism to other Radical Epistemologies.' In *Feminist Approaches to Science*, ed. Ruth Bleier. Oxford: Pergamon, 1986.

Feyerabend, Paul. *Farewell to Reason*. London: Verso, 1987.

Feynman, Richard. *Le Prix Nobel*. Stockholm: Nobel Foundation, 1963.

Finch, Janet and Dulcie Groves. 'Community Care and the Family: A Case for Equal Opportunities.' *Journal of Social Policy* 9 (4 1980) 487–511.

Finch, Janet and Dulcie Groves, eds. *A Labour of Love: Women, Work and Caring*. London: Routledge and Kegan Paul, 1983.

Firestone, Shulamith. 'The Bar as Microcosm.' In *Voices from Women's Liberation*, ed. Leslie Tanner. New York: Signet, 1970.

Firestone, Shulamith. *The Dialectic of Sex: The Case for Feminist Revolution*. London: Cape, 1971 (1970).

Fisher, Bernice and Joan Tronto. 'Towards a Feminist Theory of Caring.' In *Circles of Care: Work and Identity in Women's Lives*, eds Emily Abeland and Margaret Nelson. Albany: New York State University Press, 1990.

Flax, Jane. 'Political Philosophy and the Patriarchal Unconscious.' In *Discovering Realty: Feminist Perspectives on Epistemology, Methodology and Philosophy of Science*, eds Sandra Harding and Merrill Hintikka. Dordrecht: Reidel, 1983.

Flax, Jane. 'Postmodernism and Gender Relations in Feminist Theory.' *Signs: Journal of Women in Culture and Society* 12 (4 1987): 621–43.

Flax, Jane. *Thinking Fragments: Psychoanalysis. Feminism and Postmodernism in the Contemporary West*. Berkeley, CA: University of California Press, 1990.

Foden, Myra, ed. *Second X and Women's Health*. New York: Gordian Press, 1983.

Ford, Norman. *When Did I Begin? Conception of the Human Individual in History, Philosophy and Science*. Cambridge: Cambridge University Press, 1988.

Fraser, Elizabeth, Jennifer Hornsby, and Sabina Lovibond, eds. *Ethics: A Feminist Reader*. Oxford: Blackwell, 1992.

Frazer, J. T. and N. Lawrence, eds. *The Study of Time II: Second Conference of the International Society for the Study of Time.* New York: Springer Verlag, 1975.

Frazer, Nancy and Linda Nicholson. 'Social Criticism without Philosophy: An Encounter between Feminism and Postmodernism.' In Linda Nicholson (ed.) *Feminism/Postmodernism,* New York: Routledge 1990.

Freeman, Jo. 'The Feminist Scholar.' *Quest* 5 (1 1979): 26–36.

French, Anderson W. 'Prospects for Gene Therapy.' *Science* 226 (24 October 1984): 401–409.

Frisch, Otto, ed. *Trends in Atomic Physics: Essays Dedicated to Lise Meitner. Otto Hahn and Fritz van Laue on the Occasion of their 80th Birthday.* New York: Interscience Publishers, 1957.

Frobel, Folker, Jurgen Heinrichs, and Otto Krege. *The New International Division of Labour: Structural Unemployment in Industrialized Countries and Industrialization in Developing Countries.* Cambridge: Cambridge University Press 1980.

Fuller, Watson, ed. *The Social Impact of Modern Biology.* London: Routledge, 1971.

Gadamer, H. G. *Philosophical Hermeneutics.* Berkeley, CA: University of California Press, 1977.

Gallagher, Janet. 'Eggs, Embryos and Foetuses: Anxiety and the Law.' In *Reproductive Technologies: Gender, Motherhood and Medicine,* ed. Michelle Stanworth. Cambridge: Polity, 1987.

Gallie, Duncan, ed. *Employment in Britain.* Oxford: Blackwell, 1988.

Game, Ann and Rosemary Pringle. *Gender at Work.* London: Pluto, 1984.

Garry, Ann and Marilyn Pearsall, eds. *Knowledge and Reality.* London: Unwin Hyman, 1989.

Gearhart, Sally. 'Future Visions: Today's Politics: Feminist Utopias in Review.' In *Women in Search of Utopia,* eds E. Rohrlich and B. Hoffman. New York: Schocken, 1984.

Gearhart, Sally. *The Wanderground: Stories of the Hill Women.* London: Women's Press, 1985.

Gershuny, Jay et al. 'Time Budgets: Preliminary Analyses of a National Survey.' *Quarterly Journal of Social Affairs* 2 (1 1986): 13–39.

Giddens, Anthony. *Central Problems in Social Theory: Action, Structure and Contradiction in Social Analysis.* Berkeley, CA: University of California Press, 1976.

Giddens, Anthony. *The Transformation of Intimacy: Love, Sexuality and Eroticism in Modern Societies.* Cambridge: Polity, 1992.

Gilligan, Carol. *In a Different Voice: Psychological Theory and Women's Development.* Cambridge, MA.: Harvard University Press, 1982.

Gilman, Charlotte Perkins. *Women and Economics.* Boston, MA: Small Maynard, 1900.

Gilman, Charlotte Perkins. *The Yellow Wallpaper.* Old Westbury: Feminist Press, 1973.

Gilman, Charlotte Perkins. *Herland.* London: Women's Press, 1979 (1915).

Ginzberg, Ruth. 'Uncovering Gynocentric Science.' In *Feminism and Science,* ed. Nancy Tuana, Bloomington: Indiana University Press, 1989.

Giroud, Francoise. *Marie Curie: A Life.* Trans. Lydia Davis. New York: Holmes and Meier, 1986.

Glasgow, Joanne and Angela Ingram, eds. *Courage and Tools: The Florence Howe Award for Feminist Scholarship 1974–89.* New York: Modern Language Association of America, 1990.

Glendining, Carol and Jane Millar, eds. *Women and Poverty.* Brighton: Harvester Wheatsheaf, 1987.

Glover, Jonathan. *What Sort of People Should There Be?* Harmondsworth: Penguin, 1984.

Glover, Jonathan et al. *Fertility and the Family: The Glover Report on Reproductive Technologies to the European Commission*. London: Fourth Estate, 1989.

Goldberg, Steven. *The Inevitability of Patriarchy*. New York: Morrow, 1975.

Golden, Stephanie. *The Women Outside: Meaning and Myths of Homelessness*. Berkeley, CA: University of California Press, 1992.

Goonatilake, Susantha. *Aborted Discovery: Science and Creativity in the Third World*. London: Zed Books, 1984.

Gordon, Linda. 'What's New in Feminist History?' In *Feminist Studies/Critical Studies*, ed. Teresa DeLaurentis. Bloomington: Indiana University Press, 1986.

Gordon, Linda. 'The Welfare State: Towards a Socialist Feminist Perspective.' In *Women, the State and Welfare*, ed. Linda Gordon. Madison: University of Wisconsin, 1990.

Gordon, Linda. 'Understanding Bag Ladies'. *Women's Review of Books* 4 (November 1992): 1–4.

Gorz, André. *Farewell to the Working Class*. London: Pluto, 1980.

Gould, Stephen Jay. *The Mismeasure of Man*. New York: Norton, 1981.

Graham, Hilary. 'Caring: The Labour of Love.' In *A Labour of Love: Women, Work and Caring*, ed. Janet Finch and Dulcie Groves. London: Routledge and Kegan Paul, 1983.

Green, Josephine. *Calming or Harming? A Critical Overview of Psychological Effects of Foetal Diagnosis on Pregnant Women*. Vol. 2. London: Galton Institute Occasional Papers, 1990.

Griffin, Donald. *The Question of Mental Experience. Animal Awareness: Evolutionary Continuity*. New York: Rockefeller University Press, 1981.

Griffin, Susan. *Women and Nature: The Roaring Inside Her*. New York: Harper and Row, 1978.

Griffin, Susan. *Made from this Earth: Selections from Her Writings*. New York: Harper and Row, 1981.

Griffin, Susan. *Pornography and Silence*. New York: Harper and Row, 1981.

Griffiths, Morwenna. 'Strong Feelings about Computers.' *Women's Studies International Forum* 11 (1988): 145–54.

Grossman, Rachel. 'Women's Place in the Integrated Circuit.' *South East Asia Chronicle* 66 (1979): 2–17.

Gunnarsson, Eva and Hildur Ve. 'Technical Limited Rationality versus Responsible Rationality.' In International Conference: Ethics, Gender and Technology, Luleå, University of Luleå, 1992.

Gusella, James. 'Accuracy of Testing Huntington's Disease.' *Nature* 323 (1986): 118.

Gutman, Amy, ed. *Democracy and the Welfare State*. Princeton, NJ: Princeton University Press, 1988.

Haberer, Joseph. *Politics and the Community of Science*. New York: Van Nostrand Reinhold, 1969.

Hackel, John, ed. *The Collected Papers of Albert Einstein*. Vols. I and II. Princeton, NJ: Princeton University Press, 1987, 1989.

Hacking, Ian. *Representing and Intervening: Introductory Topics in the Philosophy of Natural Science*. Cambridge: Cambridge University Press, 1983.

Hahn, Otto. *My Life*. London: Macdonald, 1968.

Hakim, Catherine. *Occupational Segregation*. Research Paper No. 9. London: Department of Employment, 1979.

Haldane, Charlotte. *Man's World*. New York: Doran, 1927.

Haldane, J. B. S. *Daedelus: Or Science and the Future*. London: Cape, 1924.

Hall, Catherine. 'Review of J. W. Scott, Gender and the Politics of History.' *Gender and History* 3 (2 1991): 204–10.

Handler, Philip, ed. *Biology and the Future of Man.* Oxford: Oxford University Press, 1970.

Hanmer, Jalna. 'Transforming Consciousness: Women and the New Reproductive Technologies.' In *Man Made Women: How the New Reproductive Technologies Affect Women,* eds Rita Arditti, Gena Corea, and Renate Duelli Klein. London: Hutchinson, 1985.

Hanmer, Jalna and Pat Allen. 'Reproductive Engineering: The Final Solution.' In *Alice through the Microscope,* ed. Brighton Women and Science Group. London: Virago, 1980.

Haraway, Donna. 'Animal Sociology and a Natural Economy of the Body Politic.' *Signs: Journal of Women in Culture and Society* 4 (1978): 21–36.

Haraway, Donna. 'The Biological Enterprise: Sex, Mind and Profit from Human Engineering to Sociobiology.' *The Radical History Review* 20 (1979): 206–37.

Haraway, Donna. 'In the Beginning was the Word: The Genesis of Biological Thought.' *Signs: Journal of Women in Culture and Society* 6 (3 1981): 469–81.

Haraway, Donna. 'A Manifesto for Cyborgs: Science, Technology and Socialist Feminism in the 1980s.' *Socialist Review* (80 1985): 65–107.

Haraway, Donna. 'Primatology is Politics by Other Means.' In *Feminist Approaches to Science,* ed. Ruth Bleier. Oxford: Pergamon, 1986.

Haraway, Donna. 'Contested Bodies.' In *Gender and Expertise,* ed. Maureen McNeil. London: Free Association Books, 1987.

Haraway, Donna. 'Situated Knowledges: The Science Question in Feminism and the Privilege of Partial Perspective.' *Feminist Studies* 14 (3 1988): 575–99.

Haraway, Donna. *Primate Visions: Gender. Race and Nature in the World of Modern Science.* New York: Routledge, 1989.

Haraway, Donna. *Simians. Cyborgs and Women: The Reinvention of Nature.* London: Free Association Books, 1991.

Harding, Jan. *Switched Off: The Science Education of Girls.* York: Longman for the Schools Council, 1983.

Harding, Jan, ed. *Perspectives on Science and Gender.* Brighton: Falmer, 1986.

Harding, Sandra. 'Why has the Sex/Gender System Become Visible Only Now?' In *Discovering Reality: Feminist Perspectives on Epistemology, Metaphysics, Methodology and Philosophy of Science,* eds Sandra Harding and Merrill Hintikka. Dodrecht: Reidel, 1983.

Harding, Sandra. 'The Instability of the Analytical Categories of Feminist Theory.' *Signs: Journal of Women in Culture and Society* 11 (4 Summer 1986): 645–64.

Harding, Sandra. *The Science Question in Feminism.* Milton Keynes: Open University Press, 1986.

Harding, Sandra. *Whose Science? Whose Knowledge?: Thinking from Women's Lives.* Milton Keynes: Open University Press, 1991.

Harding, Sandra and Merrill Hintikka, eds. *Discovering Reality: Feminist Perspectives on Epistemology, Metaphysics, Methodology and Philosophy of Science.* Dordrecht: Reidel, 1983.

Harding, Sandra and Jean O'Barr, eds. *Sex and Scientific Inquiry.* Chicago: University of Chicago Press, 1987.

Harris, Robert and Jeremy Paxman. *A Higher Form of Killing: The Secret Story of Gas and Germ Warfare.* London: Paladin, 1982.

Hartman, Heidi. 'Capitalism, Patriarchy and Job Separation by Sex.' In *Capitalist Patriarchy and the Case for Socialist Feminism,* ed. Zillah Eisenstein. New York: Monthly Review Press, 1979.

Hartman, Heidi. 'The Family as the Locus of Gender, Class and Political Struggle The Example of Housework.' *Signs: Journal of Women in Culture and Society* 6 (3 1981): 366–94.

Hartman, Heidi. 'The Unhappy Marriage of Marxism and Feminism.' In *Women and Revolution*, ed. Lydia Sargent. Boston, MA: South End Press, 1981.

Hartman, Heidi and Diana Pearce. *High Skills and Low Pay: The Economics of Child Care Work*. Washington, DC: Institute of Women's Policy Research, 1989.

Hartsock, Nancy. 'The Feminist Standpoint: Developing the Ground for a Specifically Feminist Historical Materialism.' In *Discovering Reality: Feminist Perspectives on Metaphysics, Epistemology, Methodology and Philosophy of Science*, eds Sandra Harding and Merrill Hintikka. Dordrecht: Reidel, 1983.

Hartsock, Nancy. *Money, Sex and Power: Towards a Feminist Historical Materialism*. Amherst, MA: University of Masachusetts Press, 1984.

Hartsock, Nancy. 'Foucalt on Power: A Theory for Women?' In *Feminism/Postmodernism*, ed. Linda Nicholson. London and New York: Routledge, 1990.

Haug, Frigga. 'Lessons from the Women's Movement in Europe.' *Feminist Review* 3 (1 1989): 107–16.

Hawkins, Anne and Dagmar Schulte. 'Women, the Academic Proletariat in West Germany.' In Susanne Lie and Virginia O'Leary, eds. *Storming the Tower*. London: Kogan Page, 1987.

Heiskanen, Veronica Stolte. 'Handmaidens of the Knowledge Class.' In *Women in Science: Token Women or Gender Equality*, ed. Heiskanen. Oxford: Berg, 1991.

Heiskanen, Veronica Stolte, ed. *Women in Science: Token Women or Gender Equality*. Oxford: Berg, 1991.

Hekman, Susan. *Gender and Knowledge*. Cambridge: Polity, 1990.

Heller, Agnes. *Can Modernity Survive*? Cambridge: Polity, 1990.

Hernes, Helga. 'Women and the Welfare State: The Transition from Private to Public Dependence.' In *Patriarchy in a Welfare Society*, ed. Hamet Holter. Bergen: University of Bergen Press, 1984.

Hernes, Helga. *Welfare State and Woman Power: Essays in State Feminism*. Oslo: University of Oslo Press, 1987.

Herrnstein, Richard. *IQ in the Meritocracy*. Boston, MA: Little, Brown, 1971.

Herzenberg, Caroline. *Women Scientists from Antiquity to the Present: An International Reference History and Biographical Directory of Some Notable Women Scientists from Ancient to Modern Times*. West Cornwall: Locust Hill Press, 1986.

Hessen, B. 'The Social and Economic Roots of Newton's Principia.' In Nicholai Bukharin et al., *Science at the Crossroads*, London: Cass, 1971.

Hicks, Esther. In *Women in Science*, ed. Veronica Stolte Heiskanen. Oxford: Berg, 1991.

Hinton, Kate, ed. *Women and Science*. SISCON Project. Manchester: Manchester University Press, 1976.

Hochschild, Airlie. 'Inside the Clockwork of Male Careers.' In *Women and the Power to Change: Essays Sponsored by the Carnegie Commission on Higher Education*, ed. Florence Howe. New York: McGraw-Hill, 1975.

Hochschild, Airlie. 'The Sociology of Feeling and Emotion: Selected Possibilities.' In *Another Voice*, eds Marcia Millman and Rosbeth Moss Kanter. New York: Anchor, 1975.

Hochschild, Airlie. *The Managed Heart: The Commercialisation of Human Feeling*. Berkeley, CA: University of California Press, 1983.

Hodgkin, Dorothy Crowfoot. *Le Prix Nobel*. Stockholm: Nobel Foundation, 1964.

Hodgkin, Dorothy Crowfoot. *Kathleen Lonsdale: A Biographical Memoir*. London: Royal Society, 1971.

Hollway, Wendy. *Subjectivity and Method in Psychology*. London: Sage, 1989.

Holter, Harriet, ed. *Patriarchy in a Welfare Society*. Bergen: University of Bergen Press, 1984.

Holter, Harriet and Gro Hanne As. *National Research Policy in Norway: A Feminist Perspective*. Mimeo. Oslo: 1991.

Homans, Hilary, ed. *The Sexual Politics of Reproduction.* London: Gower, 1985.

Home, R. W. 'A World-wide Scientific Network and Patronage System.' In *International Science and National Scientific Identity Between Australia, Britain and America,* eds R. W. Home and Sally Kohlsted. Dordrecht: Kluwer, 1991.

hooks, bell. *Ain't I a Woman?: Black Women and Feminism.* Boston, MA: South End Press, 1981.

hooks, bell. 'Talking Back: Thinking Feminist – Thinking Black.' London: Sheba, 1989.

Hrdy, Sarah. *The Woman that Never Evolved.* Cambridge, MA: Harvard University Press, 1981.

Hubbard, Ruth. 'Eugenics and Pre-natal Testing.' *International Journal of Health Services* 16 (2 1986): 227–42.

Hubbard, Ruth. 'Science, Facts and Feminism.' *Hypatia* 3 (1 1988): 5–17.

Hubbard, Ruth, *The Politics of Women's Biology.* New Brunswick: Rutgers, 1990.

Hubbard, Ruth and Mary Sue Henefin. 'Genetic Screening of Prospective Parents and of Workers.' In *Biomedical Ethics Reviews* eds James Humber and Robert Almeder. Clifton, NJ: Humana Press, 1984.

Hubbard, Ruth and Elijah Wald. *Exploding the Gene Myth.* Boston, MA: Beacon, 1993.

Hubbard, Ruth, Mary Sue Henefin, and Barbara Fried, eds. *Women Look at Biology Looking at Women.* Cambridge, MA: Schenkman, 1979.

Hubbard, Ruth, Mary Sue Henifin, and Barbara Fried, eds. *Biological Woman: The Convenient Myth.* Cambridge, MA: Schenkman, 1982.

Hudson, Gill. 'Unfathering the Thinkable: Gender Science and Pacificism in the 1930s.' *Science and Sensibility,* ed. Marina Benjamin. Oxford: Blackwell, 1991.

Hull, Gloria, Patricia Scott, and Barbara Smith. *All the Women are White. All the Blacks are Men. But Some of Us are Brave: Black Women's Studies.* New York: Feminist Press, 1982.

Hull, R., ed. *Ethical Issues in the New Reproductive Technologies.* Belmont, CA: Wadsworth, 1990.

Humber, James and Robert Almeder, eds. *Biomedical Ethics Reviews.* Clifton, NJ: Humana, 1984.

Hunter, Anne, ed. *On War, Peace and Gender: A Challenge to Genetic Explanations.* New York: Feminist Press, 1991.

Irving, Alan and Brian Wynne, eds. *Misunderstanding Science: Making Sense of Science and Technology in Everyday Life.* Cambridge: Cambridge University Press (forthcoming).

Jaggar, Alison. 'Love and Knowledge: Emotion in Feminist Epistemology.' In *Gender/Body/Knowledge: Feminist Reconstructions of Being and Knowing,* eds Alison Jaggar and Susan Bordo. New Brunswick: Rutgers, 1989.

James, Selma and Maria Rosa Dalla Costa. *The Power of Women and the Subversion of the Community.* Bristol: Falling Wall Press, 1972.

Jameson, Frederic. 'Foreword.' In *The Post Modern Condition,* Jean Francois Lyotard. Minneapolis: University of Minnesota Press, 1984.

Jensen, Arthur. 'How Much Can we Boost IQ and Scholastic Achievement?' *Harvard Educational Review* 39 (1969): 1–123.

Jex-Blake, Sophia. *Medical Women: A Thesis and a History.* Edinburgh: Oliphant Anderson and Ferrier, 1886.

Johannisson, Karin. *A Life of Learning: Uppsala University during Five Centuries.* Uppsala: Uppsala University Press, 1989.

Jones, Kathleen. *A Glorious Fame: The Life of Margaret Cavendish, Duchess of Newcastle 1623–1673.* London: Bloomsbury, 1988.

Jones, Kathleen B. 'Citizenship in a Woman Friendly Polity.' *Signs: Journal of Women in Culture and Society* 15 (4 1990): 781–812.

Jones, L. M. 'Intellectual Contributions of Women in Physics.' In *Women of Science: Righting the Record*, eds G. Kass-Simon and Patricia Farnes. Bloomington: Indiana University Press, 1990.

Jordan, June. 'Song of Sojourner Truth.' In *Women in Search of Utopia*, eds R. Rohrlich and E. Hoffman. New York: Schocken, 1984.

Jordan, June. *Moving Towards Home: Political Essays*. London: Virago, 1981 (1989).

Jordanova, Ludmilla. 'Natural Facts: A Historical Perspective on Science and Sexuality.' In *Nature, Culture and Gender*, eds C. MacCormack and M. Strathern. Cambridge: Cambridge University Press, 1980.

Jordanova, Ludmilla. *Sexual Visions: Images of Gender in Science and Medicine between the Eighteenth and Twentieth Centuries*. Brighton: Harvester Wheatsheaf, 1989.

Joseph, Gloria. 'The Incompatible Ménage à Trois: Marxism, Feminism and Racism.' In *Women and Revolution*, ed. Lydia Sargent. Boston, MA: South End Press, 1981.

Julian, Maureen. 'Women in Crystallography.' In *Women of Science: Righting the Record*, eds G. Kass-Simon and Patricia Farnes. Bloomington: Indiana University Press, 1990.

Kaluzynska, Eva. 'Wiping the Floor with Theory: A Survey of Writings on Housework.' *Feminist Review* 6 (1980): 27–54.

Kass-Simon, G. and Patricia Farnes, eds. *Women of Science: Righting the Record*. Bloomington: Indiana University Press, 1990.

Keller, Evelyn Fox. 'The Anomaly of a Woman in Physics.' In *Working it Out: Twenty-three Writers, Scientists and Scholars Talk About Their Lives*, ed. Sara Ruddick and Pamela Daniels. New York: Pantheon, 1977.

Keller, Evelyn Fox. 'Feminism and Science.' *Signs: Journal of Women in Culture and Society* 7 (3 1982): 589–602.

Keller, Evelyn Fox. *A Feeling for the Organism: The Life and Work of Barbara McClintock*. San Francisco: Freeman, 1983.

Keller, Evelyn Fox. 'Gender and Science.' In *Discovering Reality: Feminist Perspectives on Epistemology. Metaphysics, Methodology and Philosophy of Science*, eds Sandra Harding and Merrill Hintikka. Dordrecht: Reidel, 1983.

Keller, Evelyn Fox. 'The Mind's Eye.' In *Discovering Reality: Feminist Perspectives on Epistemology, Metaphysics, Methodology and Philosohy of Science*, eds Sandra Harding and Merrill Hintikka. Dordrecht: Reidel, 1983.

Keller, Evelyn Fox. *Reflections on Gender and Science*. New Haven, CT: Yale University Press, 1985.

Keller, Evelyn Fox. 'Making Gender Visible in Pursuit of Nature's Secrets'. In *Feminist Studies/Critical Studies*, ed. Teresa DeLaureutis. Bloomington: Indiana University Press, 1986.

Keller, Evelyn Fox. 'The Wo/Man Scientist.' In *The Outer Circle: Women in the Scientific Community*, eds Harriet Zuckerman, Jonathan Cole and John Bruer. New York: Norton, 1991.

Keller, Evelyn Fox. 'Nature, Nurture and the Human Genome Project.' In *The Code of Codes: Scientific and Social Issues in the Human Genome Project*, ed. Daniel Kevles and LeRoy Hood. Cambridge, MA: Harvard University Press, 1992.

Keller, Evelyn Fox. *Secrets of Life: Secrets of Death*. New York: Routledge, 1992.

Kelly, Alison. *Girls and Science: International Study of Sex Differences in School Science*. Stockholm: Almquist and Wiksell, 1978.

Kelly, Alison. 'The Construction of Masculine Science.' *British Journal of the Sociology of Education* 6 (2 1985): 133–54.

Kelly, Alison. *Science for Girls*. Milton Keynes: Open University Press, 1987.

Kevles, Daniel. *In the Name of Eugenics: Genetics and the Uses of Human Heridity*. New York: Knopf, 1985.

Kevles, Daniel and Leroy Hood, eds. *The Code of Codes: Scientific and Social Issues in the Human Genome Project*, Cambridge MA: Harvard, 1992.

Kings Fund Consensus Forum. *Screening for Foetal and Genetic Abnormality*. London: Kings Fund, 1987.

Klein, Renate Duelli. 'The Crucial Role of In Vitro Fertilisation as a Means of the Social Control of Women.' In *Documentation: Women's Hearing on Genetic and Reproductive Technology*. Brussels: European Parliament, 1986.

Koblitz, Ann Hibner. *A Convergence of Lives: Sofia Kovalevskaia. Scientist, Writer, Revolutionary*. Boston, MA: Birkhaüser, 1984.

Kohlstedt, Sally. 'In from the Periphery: American Women in Science 1830–1880.' *Signs: Journal of Woman in Culture and Society* 4 (1 1978): 81–96.

Koshland, D. 'Editorial.' *Science* 246 (13 October 1989): 189.

Kramarae, Cheris, ed. *Technology and Women's Voices: Keeping in Touch*. London: Routledge, 1988.

Krimsky, Sheldon. *Biotechnics and Society: The Rise of Industrial Genetics* New York: Praeger, 1991.

Kuhn, Thomas. *The Structure of Scientific Revolutions*. Chicago: University of Chicago Press, 1962.

Kuper, Leo, ed. *Race, Science and Society*. Paris: UNESCO, 1975.

Kvande, Elin and Bente Rasmussen. 'Gender, Technology and Organisations in Post-Modern Times.' Conference on *Gender, Technology and Ethics*. Luleå: University of Luleå, 1992.

Laclau, Ernest and Chantal Mouffe. *Hegemony and Socialist Strategy: Towards a Radical Democratic Politics*. London: Verso, 1983.

Lakatos, Imrie and Alan Musgrave, eds. *Criticism and the Growth of Knowledge*. Cambridge: Cambridge University Press, 1970.

Lakoff, Robin. *Language and Woman's Place*. New York: Harper and Row, 1975

Lambert, Helen and Hilary Rose. 'Disembodied Knowledge: Making Sense of Biomedical Science.' In *Misunderstanding Science*, eds Alan Irving and Brian Wynne. Cambridge: Cambridge University Press (forthcoming).

Land, Hilary. 'The Family Wage.' *Feminist Review* 6 (1980): 55–77.

Land, Hilary and Hilary, Rose. 'Compulsory Altruism for Some or an Altruistic Society for All?' In *In Defence of Welfare*, eds P. Bean, J. Ferris and D. Whynes, London: Tavistock, 1985.

Lansbury, Carol. *The Old Brown Dog: Women, Workers and Anti-vivisection in Edwardian England*. Madison: University of Wisconsin Press, 1985.

Larsson, Stieg. 'Swedish Racism: The Democratic Way.' *Race and Class* 32 (3 1991): 102–110.

Latour, Bruno. 'Give Me a Laboratory and I Will Raise the World.' In *Science Observed: Perspectives on the Social Study of Science*, eds Karin Knorr and Michael Mulkay. London: Sage, 1983.

Latour, Bruno. *Science in Action*. Milton Keynes: Open University Press, 1987.

Latour, Bruno and Steve Woolgar. *Laboratory Life: The Social Construction of Scientific Facts*. Beverley Hills: Sage, 1979.

Lawson, Hilary and Lisa Appignanesi, eds *Dismantling Truth: Reality in the Post Modern World*. London: Weidenfeld and Nicolson, 1989.

Lefanu, Sara. *In the Chinks of the World Machine: Feminism and Science Fiction*. London: Women's Press, 1988.

Le Guin, Ursula. *The Dispossessed*. London: Gollancz, 1974.

Le Guin, Ursula. *The Language of the Night*. New York: Perigree, 1979.

Le Guin, Ursula. *Dancing at the Edge of the World*. London: Gollancz, 1989.

Le Guin, Ursula. *The Left Hand of Darkness*. London: Orbit, 1992.

Lecourt, Dominic. *Proletarian Science? The Case of Lysenko*. London: New Left Books, 1977.

Leffler, Ann Carlotta. *Sophia Kovalevskaia: Her Recollection of Childhood with a Biography*. New York: 1885.

Leibowitz, Leila. *Females, Males and Families: A Biosocial Approach*. North Scituate, MA: Duxbury Press, 1978.

Leiss, William. *The Domination of Nature*. New York: Brazillier, 1972.

Lessing, Doris. *Colonized Planet Shikasta 5*. London: Cape, 1979.

Lessing, Doris. *The Marriages between Zones Three, Four and Five*. London: Cape, 1980.

Lessing, Doris. *The Sirian Experiments: The Report by Ambien II of the Fire*. London:Cape, 1981.

Levidow, Les. 'A Marxist Critique of the IQ Debate.' *Radical Science Journal* 6/7 1978: 13–72.

Levi-Montalcini, Rita. In *Successful Women in the Sciences*, ed. Ruth Kudsin. New York: New York Academy of Sciences, 1973.

Levi-Montalcini, Rita. *Le Prix Nobel*. Stockholm: Nobel Foundation, 1986.

Levi-Montalcini, Rita. *In Praise of Imperfection: My Life and Work*. New York: Basic Books, 1988.

Lewis, Jane and Barbara Meredith. *Daughters Who Care: Daughters Caring for Mothers at Home*. London: Routledge, 1988.

Lewontin, Richard. 'Honest Jim Watson's Big Thriller about DNA.' In *James D. Watson: The Double Helix*, ed. Gunther Stent. London: Weidenfeld and Nicolson, 1981.

Lewontin, Richard. 'The Dream of the Human Genome.' In *New York Review of Books*, 39 (10 1992): 30–40.

Lewontin, Richard and Richard Levins. 'The Problem of Lysenkoism.' In *The Radicalisation of Science*, eds Hilary Rose and Steven Rose. London: Macmillan, 1976.

Lie, Suzanne Stiver and Virginia O'Leary eds. *Storming the Tower*, London: Kogan Page, 1990.

Lieberman, Marcia K. 'Some Day My Prince Will Come: Female Acculturation through the Fairy Tale.' In *Don't Bet on the Prince: Contemporary Feminist Fairy Tales in North America and England*, ed. Jack Zipes. Aldershot: Gower, 1986.

Lin, Pam. 'Gender Stereotypes, Technology Stereotypes.' In *Gender and Expertise*, ed. Maureen McNeil. London: Free Association Books, 1987.

Lloyd, Genevieve. *The Man of Reason: 'Male' and 'Female' in Western Philosophy*. London: Methuen, 1984.

Longino, Helen. *Science as Social Knowledge: Values and Objectivity in Scientific Enquiry*. Princeton, NJ: Princeton University Press, 1990.

Longino, Helen and Ruth Doell. 'Body Bias and Behaviour.' *Signs: Journal of Women in Culture and Society* 9 (2 1983): 206–27.

Lonsdale, Kathleen. 'Women in Science: Reminiscences and Reflections.' *Impact of Science on Society* 20 (1970): 45–59.

Lopata, Helen. *Occupation Housewife*. Oxford: Oxford University Press, 1971.

Lorde, Audre. 'Open Letter to Mary Daly.' In *This Bridge Called My Back*, eds Cherie Moraga and Gloria Anzaldua. Watertown, MA: Persephone, 1981.

Lowe, A.D. Morrison. 'Women in the Nineteenth Century Instrument Trade.' In *Science and Sensibility: Gender and Scientific Enquiry 1780–1945*, ed. Marina Benjamin. Oxford: Blackwell, 1991.

Lowe, Marian and Ruth Hubbard. eds. *Pitfalls in Genes and Gender Research*. New York: Gordian Press, 1979.

Lowe, Marian and Ruth Hubbard. eds. *Woman's Nature: Rationalisation of Women's Inequality*. Oxford: Pergamon, 1983.

Lovejoy, Gillian and Wendy Hall. 'Where Have All the Girls Gone?' *University Computing* 9 (1987): 207–10.

Ludmerer, K. M. *Genetics and American Society*. Baltimore: Johns Hopkins University Press, 1972.

Luria, Salvador. 'Letter,' *Science* 246 (17 November, 1989): 873.

MacCormack, Carol and Marilyn Strathern, eds. *Nature, Culture and Gender*. Cambridge: Cambridge University Press, 1980.

MacIntosh, Peggy. 'Interactive Phases of Curricular Revision,' In *Towards a Balanced Curriculum*, eds Bonnie Spanier, Alexander Bloom and Darlene Boroviak. Rochester: Schenkman, 1984.

Mackenzie, Donald. 'Notes on the Science as Social Relations Debate', *Capital and Class*, 14 (1981) 47–60.

Mainardi, Pat. 'The Politics of Housework.' In *Sisterhood is Powerful*, ed. Robin Morgan. New York: Vintage, 1970.

Martin, Emily. *The Woman in the Body: A Cultural Analysis of Reproduction*. Boston, MA: Beacon, 1987.

Marx, Karl. *Capital*. Vols 1 and 2. London: Allen and Unwin, 1938 (1889).

Mason, Joan. 'Hertha Ayrton 1854–1923 and the Admission of Women to the Royal Society of London.' *Notes and Records of the Royal Society of London* (45 1991): 201–20.

Mass, Bonnie. *Population Target: The Political Economy of Population in Latin America*. Brampton, Ontario: Charters, 1976.

Massey, Doreen, Paul Quintas, and David Wield. *High Tech Fantasies: Science Parks in Society, Science and Space*. London: Routledge, 1991.

Masters, Judith. 'Natural Selection, Cultural Construction', *Women's Review of Books*, 7 (4 1990) 18–19.

Mayer, Joseph E. 'My Wife's Secret: The Atomic Bomb.' In *The Nobel Prize*, Stockholm: Nobel Institute, 1963.

Mayer, Maria Goeppert. *Le Prix Nobel*. Stockholm: Nobel Foundation, 1963.

McClintock, Barbara. *Le Prix Nobel*, Stockholm, Nobel Foundation, 1983.

McDonald, Madeleine. 'Schooling and the Reproduction of Class and Gender Relations.' In *Schooling, Ideology and the Curriculum*, eds L. Barton, R. Meighan and S. Walker. Brighton: Falmer, 1980.

McKeown, Thomas. *The Role of Medicine: Dream, Mirage or Nemesis*. Oxford: Blackwell, 1979.

McKusick, Victor A. *Mendelian Inheritance in Man: Catalogs of Autosomal Dominant, Autosomal Recessive and X-linked Phenotypes*. Baltimore: Johns Hopkins University Press, 1992 (1966).

McNeil, Maureen, ed. *Gender and Expertise*. London: Free Association Books, 1987.

McNeil, Maureen and Sara Franklin. In *Off Centre: Feminism and Cultural Studies*, eds. Sara Franklin, Celia Long and Jackie Stacey. London: HarperCollins, 1991.

McNeil, Maureen and Sara Franklin. 'Science and Technology: Questions for Cultural Studies and Feminism.'

Medick, H. and D. W. Sabean. 'Interest and Emotion in Family and Kinship Studies.' In *Interest and Emotion: Essays on the Study of Family and Kinship*, eds H. Medick and D. W. Sabean. Cambridge: Cambridge University Press, 1984.

Mehrtens, Herbert. 'Mathematics in the Third Reich: Resistance, Adaptations and Collaboration.' In *New Trends in the History of Science*, eds R. W. Vigger et al. Amsterdam: Rodopi, 1989.

Mendelsohn, E., P. Weingart, and R. Whitley. *The Social Production of Scientific Knowledge*. Dordrecht: Reidel, 1977.

Merchant, Carolyn. *The Death of Nature: Women, Ecology and the Scientific Revolution*. London: Wilwood, 1980.

Merchant, Carolyn. ' "Isis' Consciousness Raised?' *Isis* 73 (268 1982): 398–409.

Merchant, Carolyn. *Radical Economy: The Search for a Livable World*, New York: Routledge, 1992.

Metcalfe, Judith, ed. *Disorder, Disease and Degeneration*. Milton Keynes: Open University Press, 1992.

Midgley, Mary. *Science as Salvation: A Modern Myth and its Meaning*. London and New York: Routledge, 1992.

Mies, Maria. 'Towards a Methodology for Feminist Research.' In *Theories of Women's Studies*, eds Gloria Bowles and Renate Duelli Klein. London: Routledge and Kegan Paul, 1983 (1978).

Mies, Maria. 'What Unites, What Divides Women from the South and from the North in the Field of Reproductive Technologies.' In *Declaration of Comilla*, eds Ferida Akhter, Wilma van Berkel and Natasha Ahmad. Dhaka: Finrrage and Ubinig, 1992.

Milio, Nancy. *Promoting Health through Public Policy*. Philadelphia: Davis, 1981.

Milkman, Ruth. *Gender at Work: The Dynamics of Gender Segregation by Sex During World War II*. Chicago: University of Illinois Press, 1987.

Miller, Jane. 'Women in Chemistry.' In *Women of Science: Righting the Record*, eds G. Kass-Simon and Patricia Farnes. Bloomington: Indiana University Press, 1990.

Miller, Jean Baker. *Towards a Psychology of Women*. Boston, MA: Beacon, 1976.

Millman, Marcia and Rosbeth Moss Kanter, eds. *Another Voice: Feminist Perspectives on Social Life and Social Science*. New York: Anchor, 1975.

Mitchell, Juliet. *Women's Estate*. Harmondsworth: Penguin, 1971.

Mitchell, Juliet. 'Reflections on Twenty Years of Feminism.' In *What is Feminism?*, eds Juliet Mitchell and Ann Oakley. Oxford: Blackwell, 1986.

Mitchison, Naomi. *Memoirs of a Spacewoman*. London: Gollancz, 1962.

Mitter, Swasti. *Common Fate, Common Bond: Women in the Global Economy*. London: Pluto, 1986.

Mitter, Swasti. *Computer Aided Manufacturing and Women's Employment: The Clothing Industry in Four EC Countries*. London and New York: Springer Verlag, 1992.

Montagu, Ashley. *Race, Science and Humanity*. Princeton, NJ: Van Nostrand, 1963.

Moraga, Cherie and Gloria Anzaldua, eds *This Bridge Called My Back*. Watertown, MA: Persephone, 1981.

Morgan, Robin, ed. *Sisterhood is Powerful*. New York: Vintage, 1970.

Morgan, Robin, ed. *Sisterhood is Global*. New York: Doubleday, 1984.

Morrell, Jack and Arnold Thackray. *Gentlemen of Science: Early Years of the British Association for the Advancement of Science*. Oxford, Clarendon, 1981.

Morris, Jenny. *Able Lives? Women's Experience of Paralysis*. London: Women's Press, 1989.

Mount, Ferdinand. *The Subversive Family: An Alternative History of Love and Marriage*. London: Cape, 1982.

Mozan, H. J. (John Zahm). *Woman in Science*. New York and London: Appleton, 1913.

Muller, Herman J. *Out of the Night*. London: Gollancz, 1936.

Muller Hill, Benno. *Murderous Science: Elimination by Scientific Selection of Jews, Gypsies and Others. Germany 1933–1945*. Oxford: Oxford University Press, 1988.

Murdrey, V. and J. Slack. *British Medical Journal* 291 (1985): 1315–18.

Myrdal, Alva and Viola Klein. *Woman's Two Roles: Home and Work*. London: Routledge and Kegan Paul, 1956.

Narayan, Uma. 'The Project of a Feminist Epistemology: Perspectives from a Non Western Feminist.' In *Gender/Body Knowledge: Feminist Reconstructions of Being and Knowing*, eds Alison Jaggar and Susan Bordo. New Brunswick: Rutgers, 1989.

Narek, Diane. 'A Woman Scientist Speaks.' In *Voices from Women's Liberation*, ed. Leslie Tanner. New York: Signet, 1970.

Needham, J. 'Foreword.' *Science at the Crossroads*, Nicholai Bukharin et al., London: Cass, 1973 (1931).

Needham, Joseph. 'History and Human Values: A Chinese Perspective for World Science and Technology.' In *The Radicalisation of Science*, eds Hilary Rose and Steven Rose. London: Macmillan, 1976.

Nelkin, Dorothy. *Selling Science*. New York: Freeman, 1987.

Nelkin, Dorothy and Lawrence Tancredi. *Dangerous Diagnostics: The Social Power of Biological Information*. New York: Basic Books, 1989.

Nicholson, Linda, ed. *Feminism/Postmodernism*. London: Routledge, 1990.

Nielsen, Joyce McCarl. *Feminist Research Methods*. San Francisco and London: Westview Press, 1990.

NIH Report. *Report of the Working Group on the Ethical, Legal and Social Issues Related to Mapping and Sequencing the Human Genome*. Bethesda: National Institutes of Health, 1989.

Nissel, Muriel and Lucy Bonnerjea. *Family Care of the Handicapped Elderly: Who Pays?* London: Policy Studies Institute, 1982.

Noble, David. *America by Design*. Cambridge, MA: MIT Press, 1979.

Noble, David. *Forces of Production: A Social History of Industrial Automation*. New York: Knopf, 1984.

Noddings, Nell. *Caring: A Feminine Approach to Ethics and Moral Education*. Berkeley, CA: University of California Press, 1984.

Nott, M. and M. Watts. 'Towards a Multi-Cultural and Anti-Racist Science Education Policy.' *Education in Science* 121 (1987): 37–8.

Nowotny, Helga. 'Time Structuring and Time Measurement: On the Interrelation Between Time Keepers and Social Time.' In *The Study of Time II: Second Conference of the International Society for the Study of Time*, eds J. T. Fraser and N. Lawrence. New York: Springer Verlag, 1975.

Nowotny, Helga and Hilary Rose, eds. *Countermovements in the Sciences: Alternatives to Big Science*. Dordrecht: Reidel, 1978.

Nuttall, Jeff. *Bomb Culture*. London: MacGibbon Kee 1968.

O'Brien, Mary. *The Politics of Reproduction*. Boston, MA, and London: Routledge and Kegan Paul, 1981.

Oakley, Ann. *Sex, Gender and Society*. London: Temple Smith, 1972.

Oakley, Ann. *Housewife*. Harmondsworth: Allen Lane, 1974; Penguin, 1990.

Oakley, Ann. *The Sociology of Housework*. Bath: Martin Robertson, 1974.

Oakley, Ann. *The Captured Womb: A History of the Medical Care of Pregnant Women* Oxford: Blackwell, 1984.

Oakley, Ann. *Social Support and Motherhood: The Natural History of a Research Project*. Oxford: Blackwell, 1993.

Ogilvie, Marilyn Bailey. *Women in Science: Antiquity through the Nineteenth Century: A Biographical Dictionary with Annotated Bibliography*. Cambridge, MA.: 1986.

Opfell, Olga. *The Lady Laureates: Women who have Won the Nobel Prize*. Mutuchen, NJ: Scarecrow Press, 1986.

Osen, Laura. *Women in Mathematics*. Cambridge, MA: MIT Press, 1974.

Oudshorn, Nelly. 'On the Making of Sex Hormones', *Social Studies of Science* 20 (1990): 5–33.

Pahl, Ray. *Divisions of Labour*. Oxford: Blackwell, 1984.

Palumbo, D., ed. *Erotic Universe: Sexuality and Fantastic Literature*. New York and London: Greenwood Press, 1986.

Pateman, Carol. 'The Patriarchal Welfare State.' In *Democracy and the Welfare State*, ed. Amy Gutman. Princeton, NJ: Princeton, University Press, 1988.

Pateman, Carol. *The Sexual Contract*. Cambridge: Polity, 1988.

Paulser, Inge Lise. 'Can Women Fly? Vonda McIntyre's *Dreamsnake* and Sally Gearhart's *The Wanderground. Women's Studies International Forum.*' 7 (2, 1984): 103–10.

Pearson, Carol. 'Coming Home: Four Feminist Utopias and Patriarchal Experience.' In *Future Females: A Critical Anthology*, ed. Marlene Barr. Ohio: Bowling Green Press, 1981.

Perutz, Max. 'Forty Years Friendship with Dorothy.' In *Structural Studies in Molecules of Biological Interest*, eds G. Dodson, J. Glusker, and D. Sayre. Oxford: Clarendon, 1981.

Petchesky, Rosalind. 'Reproductive Freedom: Beyond 'A Woman's Right to Choose'.' In *Women, Sex and Sexuality*, eds Catherine Stimpson and E. S. Person. Chicago: University of Chicago Press, 1981.

Petchesky, Rosalind. *Abortion and Women's Choice: The State, Sexuality and Reproductive Freedom*, New York: Longman, 1984.

Pfeffer, Naomi. 'Are British Consent Forms Adequate?' *Bulletin of Medical Ethics* 69 (June 1991): 28–31.

Pfeffer, Naomi and Anne Woolett. *The Experience of Infertility*. London: Virago, 1983.

Phillips, Anne. *Hidden Hands: Women and Economic Policy*. London: Pluto, 1983.

Phillips, Anne and Barbara Taylor. 'Sex and Skill: Notes Towards a Feminist Economics.' *Feminist Review* 6 (1980): 79–88.

Phillips, Patricia. *The Scientific Lady: A Social History of Women's Scientific Interests 1580–1918*. London: St Martin's Press, 1990.

Phizacklea, Annie, ed. *One Way Ticket: Migration and Female Labour*. London: Routledge and Kegan Paul, 1983.

Phizacklea, Annie. 'Gender, Racism and Occupational Segregation.' In *Gender Segregation at Work*, ed. Sylvia Walby. Milton Keynes: Open University Press, 1988.

Piercy, Marge. *Woman on the Edge of Time*. London: Women's Press, 1979.

Pinch, Trevor. 'Reservations about Reflexivity and New Literary Forms, or Why Let The Devil have All the Good Tunes?' In *Knowledge and Reflexivity*, ed. Steven Woolgar. London: Sage, 1988.

Pinchbeck, Ivy. *Women Workers and the Industrial Revolution 1750–1850*. London: Routledge, 1930.

Piven, Frances Fox and Richard Cloward. *Poor People's Movements: Why they Succeed and How they Fail*. New York: Pantheon, 1993 (1972).

Popper, Karl. *Conjectures and Refutations: The Growth of Scientific Knowledge*. Oxford: Oxford University Press, 1963.

Postgate, Raymond. 'Bat's Chance in Hell.' *New Scientist*. 58 (5 April 1973): 12–16.

Potter, Elizabeth. *A Feminist Model of Natural Science*. Haverford: 1986.

Price, Derek De Solla. *Little Science: Big Science*. London: Macmillan, 1963.

Proctor, Robert. *Racial Hygiene: Medicine under the Nazis*. Cambridge, MA.: Harvard University Press, 1988.

Proctor, Robert. *Value Free Science: Purity and Power in Modern Knowledge*. Cambridge, MA.: Harvard University Press, 1991.

Quarashi, M. M. and S. M. Jafar. *Scientific and Technological Development in Profile in Muslim Countries*, Islamabad: Islamabad Natural Science Council, 1988.

Quaraishy, Bashy and Tim O'Connor. 'Denmark: No Racism by Definition.' *Race and Class* 32 (3 1991): 114–18.

Quarrell, O. W. S. and et al. 'Exclusion Testing for Huntington's Disease in Pregnancy with a Closely Linked DNA Marker.' *Lancet* 8545 (1987): 1281–83.

Rapp, Rayna. 'The Story of XYLO.' In *Test Tube Women: What Future for Motherhood?*, ed. Rita Arditti, Renate Duelli Klein, and Shirley Minden. London: Pandora, 1984.

Raverat, Gwen. *Period Piece: A Cambridge Childhood.* London: Faber and Faber, 1960.

Ravetz, Jerome. *Science and its Social Problems.* Oxford: Oxford University Press, 1971.

Raymond, Janice. 'Preface.' In *Man Made Women: How the New Reproductive Technologies Affect Women,* eds Rita Arditti, Gena Corea, and Renate Duelli Klein. London: Hutchinson, 1985.

RCOG (Royal College of Obstetricians and Gynaecologists). *Artificial Insemination.* London: RCOG 1977.

Reid, Robert. *Marie Curie.* New York and London: Saturday Review Press, 1974.

Restivo, Sal. 'Modern Science as a Social Problem.' *Social Problems* 35 (3 1988): 206–25.

Rich, Adrienne. 'When We Dead Waken: Writing as Re-vision.' In *Adrienne Rich's Poetry: Texts of the Poems: The Poet on Her Work: Reviews and Criticisms,* eds Barbara Charlesworth Gelpi and Albert Gelpi. New York: Norton, 1975 (1972).

Rich, Adrienne. *Of Woman Born: Motherhood as Experience.* New York: Norton, 1976.

Rich, Adrienne. *The Dream of a Common Language.* New York: Norton, 1978.

Rich, Adrienne. *On Lies, Secrets and Silence: Selected Prose 1966–78.* New York: Norton, 1979.

Richardson, Ken, and David Spears, eds. *Race, Culture and Intelligence.* Harmondsworth: Penguin, 1974.

Richter, Derek, ed. *Women Scientists: The Road to Liberation.* London: Macmillan, 1982.

Riley, Denise. *Am I that Name? Feminism and the Category of 'Woman' in History.* London: Macmillan, 1988.

Roberts, Helen, ed. *Feminist Methodology.* London: Routledge and Kegan Paul, 1981.

Robertson, Muriel. *Marjory Stephenson, Obituary Notices.* London: Royal Society (1949): 563–77.

Roby, Pamela. 'Institutional Barriers to Women in Higher Education.' In *Academic Women on the Move,* eds Alice Rossi and Alice Calderwood. New York: Russell Sage, 1973.

Rodney, Walter. *How Europe Underdeveloped Africa.* London: Bogle L'Overture Publications, 1972.

Rodriguez-Trias, Helen. 'Sterilisation Abuse.' In *Biological Woman: The Convenient Myth,* eds Ruth Hubbard, Mary Sue Henifin, and Barbara Fried. Cambridge, MA: Schenkman, 1982.

Rohrlich, Ruby and Elaine Hoffman, eds. *Women in Search of Utopia: Mavericks and Mythmakers.* New York: Schocken, 1984.

Roll-Hansen, Nils. 'The Practice Criterion and the Rise of Lysenkoism.' *Science Studies* 2 (1 1989): 3–16.

Rorty, Richard. 'Science as Solidarity.' In *Dismantling Truth: Reality in the Post Modern World,* eds Hilary Lawson and Lisa Appignanesi. London: Weidenfeld and Nicolson, 1989.

Rose, Hilary. 'Hyper-reflexivity: A New Danger to the Countermovements.' In *Countermovements in the Sciences: Alternatives to Big Science,* eds Helga Nowotny and Hilary Rose. Dordrecht: Reidel, 1978.

Rose, Hilary. 'Hand, Brain and Heart: Towards a Feminist Epistemology for the Natural Sciences.' *Signs: Journal of Women in Culture and Society* 9 (1 1983): 73–96.

Rose, Hilary. 'Is a Feminist Science Possible?' MIT Women's Studies Programme, mimeo, 1984.

Rose, Hilary. 'The Miners' Wives of Upton.' *New Society* (29 November 1984): 326–9.

Rose, Hilary. 'Nothing Less than Half the Labs'. In *Agenda for Higher Education*, eds Janet Finch and Michael Rustin, Harmondsworth: Penguin, 1985.

Rose, Hilary. 'Beyond Masculinist Realities.' In *Feminist Approaches to Science*, ed. Ruth Bleier. Oxford: Pergamon, 1986.

Rose, Hilary. 'Women's Work: Women's Knowledge.' In *What is Feminism?* eds Juliet Mitchell and Ann Oakley. Oxford: Blackwell, 1986.

Rose, Hilary. 'Gender Politics in the New Public Health.' In *Health through Public Policy: The Greening of Public Health*, ed. Peter Draper. London: Merlin, 1991.

Rose, Hilary. 'Gendered Reflexions on the Laboratory in Medicine.' In *The Laboratory Revolution in Medicine*, eds Andrew Cunningham and Perry Williams. Cambridge: Cambridge University Press, 1992.

Rose, Hilary. 'Rhetoric, Feminism and Scientific Knowledge: Or, From Either/Or to Both/And'. In *The Rediscovery of Rhetoric*, eds R. Roberts and J. M. Goode. London: Bristol Classical Press, 1993.

Rose, Hilary and Jalna Hanmer. 'Women's Liberation, Reproduction and the Technological Fix.' In *Sexual Divisions and Society: Process and Change*, eds Sheila Allen and Diana Leonard Barker. London: Tavistock, 1976.

Rose, Hilary and Steven Rose. 'The Myth of the Neutrality of Science.' *Impact of Science on Society* 21 (1971): 137–49.

Rose, Hilary and Steven Rose. *Science and Society*. Harmondsworth: Allen Lane and Penguin, 1969.

Rose, Hilary and Steven Rose, eds. *The Political Economy of Science*. London: Macmillan, 1976.

Rose, Hilary and Steven Rose, eds. *The Radicalisation of Science*. London: Macmillan, 1976.

Rose, Hilary and Steven Rose. 'Radical Science and its Enemies.' In *The Socialist Register 1979*, eds R. Miliband and J. Saville. London: Merlin, 1979.

Rose, Steven, ed. *Against Biological Determinism*. London: Alison and Busby, 1982.

Rose, Steven. *The Making of Memory*. London: Bantam, 1992.

Rose, Steven, Richard Lewontin, and Leo Kamin. *Not in Our Genes*. Harmondsworth: Penguin, 1984.

Rosser, Sue. *Female Friendly Science: Applying Women's Studies Methods and Theories to Attract Students.*Oxford: Pergamon, 1990.

Rossi, Alice. 'Women in Science: Why So Few?' *Science* 148 (3674 1965): 1196–1202.

Rossi, Alice. In *Women in the Scientific Professions*. New York: New York Academy of Science, 1965.

Rossiter, Margaret. *Women Scientists in America: Struggles and Strategies to 1940*. Baltimore: Johns Hopkins University Press, 1982.

Roszak, Theodore. *The Making of a Counter Culture*. London: Faber, 1971.

Roszak, Theodore. *Where the Wasteland Ends*. London: Faber, 1973.

Rothman, Barbara Katz. *The Tentative Pregnancy: Prenatal Diagnosis and the Future of Motherhood*. London: Pandora 1988.

Rothman, Barbara Katz. *Recreating Motherhood: Ideology and Technology in a Patriarchal Society*. New York: Norton, 1989.

Rothschild, Joan. *Teaching Technology from a Feminist Perspective*. Oxford: Pergamon, 1988.

Rowbotham, Sheila. *Hidden from History? Three Hundred Years of Women's Oppression and the Fight Against It*. London: Pluto, 1973.

Rowbotham, Sheila. 'The Trouble with Patriarchy.' *New Statesman* (21–28 December 1979): 970–1.

Rowbotham, Sheila, Lynn Segal, and Hilary Wainwright. *Beyond the Fragments: Feminism and the Making of Socialism*. London: Merlin, 1979.

Rowland, Robyn. 'The Meanings of Choice in Reproductive Technology.' In *Test*

Tube Women: What Future for Motherhood?, eds. Rita Arditti, Renate Duelli Klein, and Shirley Minden. London: Pandora, 1984.

Rowland, Robyn. *Living Laboratories: Women and Reproductive Technologies*. London: Lime Tree Press, 1992.

Ruddick, Sara. 'Maternal Thinking.' In *Rethinking the Family: Some Feminist Questions*, eds Barrie Thorne and Marilyn Yalom. New York and London: Longman, 1982.

Ruddick, Sara. *Maternal Thinking: Towards a Politics of Peace*. Boston, MA: Beacon, 1989.

Ruin, Olof. 'Reform Reassessment and Research Policy: Tensions in the Development of University Research.' In *The University Research System: The Public Policies of the Home of Scientists*, eds. Björn Wittrock and Aant Elzinga. Stockholm: Almquist and Wiksell, 1985.

Ruizo, Beatrice. 'The Intellectual Labour Market in Developed and Developing Countries: Women's Representation in Scientific Research'. *International Journal of Science Education* 9 (1987): 385–91.

Russ, Joanna. *The Female Man*. London: Women's Press, 1979 (1975).

Russ, Joanna. 'The Clichés from Outer Space.' *Women's Studies International Forum* 7 (2 1984): 121–4.

Russ, Joanna. 'What Can a Heroine Do? Or, Why Women Can't Write.' In *Courage and Tools: The Florence Howe Award for Feminist Scholarship 1974–1989*, eds Joanne Glasgow and Angela Ingram. New York: Modern Language Association of America, 1990.

Russell, Dora. *Hypatia, Woman and Knowledge*. London: Cape, 1925.

Sadler, Bernice. 'Patterns of Discrimination and Discouragement.' In *Report of New York City Commission on Human Rights*. New York: Avon, 1972.

Salimi, Khalid. 'Norway's National Racism.' *Race and Class* 32 (3 1991): 111–13.

Sargent, Pamela, ed. *Women of Wonder: Science Fiction Stories by Women about Women*. New York: Vintage, 1986.

Sarton, George. *A History of Science*. New York: Norton, 1952.

Saville, John. 'Marxism Today: An Anatomy.' In *The Socialist Register 1990*, eds Ralph Miliband, Leo Panitch, and John Saville. London: Merlin, 1990.

Saxton, Josephine, 'Goodbye to All That' . . .'. In *Where No Man has Gone Before*, ed. Lucy Armitt. New York: Routledge, 1990.

Sayers, Janet. *Biological Politics: Feminist and Anti-feminist Perspectives*. London: Tavistock, 1982.

Sayers, Janet. 'Review of S. Harding, The Science Question.' *New Statesman* (31 July 1987): 30.

Sayre, Ann. *Rosalind Franklin and DNA: A Vivid View of What it is Like to be a Gifted Woman in an Especially Male Profession*. New York: Norton, 1975.

Scheibinger, Londa. 'The History and Philosophy of Women in Science: A Review Essay.' *Signs: Journal of Women in Culture and Society* 12 (2 1987): 305–32.

Scheibinger, Londa. *The Mind Has No Sex: Women in the Origins of Modern Science*. Cambridge, Ma: Harvard University Press, 1989.

Science for the People. *Biology as Destiny?* Cambridge, MA: Science for the People, 1984.

Science for the People. *Sociobiology as a Social Weapon*. Cambridge, MA: Science for the People, 1977.

Scott, Joan Wallach. *Gender and the Politics of History*. New York: Columbia University Press, 1988.

Searing, Susan, ed. *The History of Women in Science, Technology and Medicine: A Bibliographic Guide to the Disciplines and Professions*. Madison: University of Wisconsin Women's Studies Library, 1987.

Sechenov, I. M. *Autobiographical Notes*. Washington, DC: American Institute of Biological Sciences, 1965 (1952).

Shannon, T. A. *Surrogate Motherhood: The Ethics of Using Human Beings*. New York: Cross Road, 1989.

Shapin, Steven. 'Following Scientists Around.' *Social Studies of Science* 18 (3 1988): 533–50.

Shelley, Mary. *Frankenstein: A Modern Prometheus*. Harmondsworth: Penguin, 1987 (1818).

Shiva, Vandana. *Staying Alive: Women, Ecology and Development*. London: Zed Books, 1989.

Siegel, Patricia and Kay Thomas Finley. *Women in the Scientific Search: An American Bio-bibliography 1724–1979*. Metuchen, NJ: Scarecrow Press, 1985.

Siim, Birte. 'Towards a Feminist Rethinking of the Welfare State.' In *The Political Interests of Gender*, eds Kathleen Jones and Anna Jonasdottir. London: Sage, 1988.

Simonen, Leila. 'Caring by the Welfare State and Women Behind It – Contradictions and Theoretical Considerations.' In *Finnish Debates on Women's Studies*, ed. Leila Simonen. Tampere: University, 1990.

Simonen, Leila. *Contradictions of the Welfare State: Women and Caring*. Tampere: University of Tampere Press, 1990.

Singer, Peter. *Animal Liberation*. New York: 1975.

Singer, P. and D. Wells. *The Reproductive Revolution*. Oxford: Oxford University Press, 1984.

Sivanandan, A. N. *A Different Hunger: Writings on Black Resistance*. London: Pluto, 1982.

Sivanandan, A. N. *Communities of Resistance: Writings on Black Struggles for Socialism*. London: Verso, 1990.

Smith, Dorothy. *The Everyday World as Problematic: A Feminist Sociology*. Milton Keynes: Open University Press, 1988.

Smith, Dorothy. *Texts, Facts and Femininity: Exploring the Relations of Ruling*. London: Routledge and Kegan Paul, 1990.

Smith, Dorothy. *The Conceptual Practices of Power: A Feminist Sociology of Knowledge*. Boston, MA: Northeastern University Press, 1990.

Smith, Joan. 'Sociobiology and Feminism: The Very Strange Courtship of Competing Paradigms.' *Philosophical Forum* 13 (1982): 224–43.

Smith, Joan. 'Feminist Analysis of Gender: A Critique.' In *Women's Nature: Rationalizations of Inequality*, eds M. Lowe and R. Hubbard. New York: Athene Pergamon, 1983.

Smith, Paul. 'Domestic Labour and Marx's Theory of Value.' In *Feminism and Materialism*, eds Annette Kuhn and Anne Marie Wolpe. London: Routledge and Kegan Paul, 1978.

Snow, C. P. *The Two Cultures and the Scientific Revolution*. Cambridge: Cambridge University Press, 1959.

Snowden, R. and G. Mitchell. *The Artificial Family: A Consideration of Artificial Insemination by Donor*. London: Allen and Unwin, 1981.

Sohn Rethel, Alfred. *Intellectual and Manual Labour*. London: Macmillan, 1978.

Soper, Kate. *Troubled Pleasures*. Brighton: Harvester Wheatsheaf, 1991.

Spender, Dale. *Men's Studies Modified*. Oxford: Pergamon, 1981.

Spender, Dale. *Invisible Women: The Schooling Scandal*. London: Readers and Writers, 1982.

Spender, Dale. *Man Made Language*. London: Routledge and Kegan Paul, 1983.

Spender, Dale. *For the Record: The Making and Meaning of Feminist Knowledge*. London: Women's Press, 1985.

Spivak, Gayatri Chakravorty. *In Other Worlds: Essays in Cultural Politics.* New York: Methuen, 1987.

Spivak, Gayatri Chakravorty. 'Feminism in Decolonization.'*Differences* 3 (3 1991): 139–75.

Stanley, Liz and Sue Wise. *Breaking Out: Feminist Consciousness and Feminist Research.* London: Routledge and Kegan Paul, 1983.

Stanley, Liz and Sue Wise. *Breaking Out Again: Feminist Epistemology and Ontology.* Manchester: Manchester University Press, 1992.

Stanworth, Michelle. 'Reproductive Technologies and the Deconstruction of Motherhood.' In *Reproductive Technologies: Gender, Motherhood and Medicine,* ed. Michelle Stanworth. Cambridge: Polity, 1987.

Stearns, R. P. 'Colonial Fellows of the Royal Society of London 1661–1788.' *Notes and Records of the Royal Society of London* (8 1951): 178–246.

Stepan, Nancy. *The Idea of Race in Science: Great Britain 1800–1960.* Hamden, CT: Archon Books, 1982.

Stepan, Nancy. *The Hour of Eugenics: Race, Gender and Nation in Latin America.* Ithaca, NY: Cornell University Press, 1991.

Strachey, Ray. *The Cause: Short History of the Women's Movement in Britain.* London: Virago, 1978 (1928).

Struik, Dirk. *A Concise History of Mathematics.* New York: Dover, 1967.

Sunday, Suzanne and Ethel Tobach, eds. *Violence Against Women: A Critique of the Sociobiology of Rape.* Genes and Gender series. New York: Gordian Press, 1985.

Sutton, A. *Prenatal Diagnosis: Confronting the Ethical Issues.* London: Linacre Centre, 1990.

Suzuki, David and Peter Knudtson. *Genethics: The Ethics of Engineering Life.* London: Unwin, 1991.

Swann Report. *Report of the Committee of Inquiry into the Education of Children from Ethnic Minority Groups.* London: HMSO, Cmnd 9453, 1985.

Swinbank, D. 'Japanese Gynaecology: Gender Selection Sparks Row.' *Nature* 321 (1986): 720.

Taylor, Barbara. *Eve and the New Jerusalem: Socialism and Feminism in the Nineteenth Century.* London: Virago, 1984.

Taylor, D. G. et al. *Mental Handicap: Partnership in the Community.* London: Office of Health Economics, 1986.

Thomas, Kim. *Gender and Subject in Higher Education.* Milton Keynes: Open University Press, 1990.

Thompson, E. P. *The Making of the English Working Classes.* New York: Vintage, 1963.

Thompson, E. P. *The Poverty of Theory and Other Essays.* London: Merlin, 1978.

Tobach, Ethel and Betty Rossoff, eds. *Genes and Gender: On Hereditarianism and Women.* New York: Gordian Press, 1978.

Tobach, Ethel and Betty Rossoff, eds. *Genetic Determinism and Children.* New York: Gordian Press, 1980.

Touraine, Alain. *Post Industrial Society: Tomorrow's Social History: Classes, Conflicts and Culture in the Programmed Society.* New York: Random House, 1971.

Touraine, Alain. *The Voice and the Eye: An Analysis of Social Movements.* Cambridge: Cambridge University Press, 1981.

Townsend, Peter and Nick Davidson. *Inequalities in Health.* Harmondsworth: Penguin, 1982.

Traweek, Sharon. *Beamtimes and Lifetimes: The World of High Energy Physics.* Cambridge, MA: Harvard University Press, 1988.

Traweek, Sharon. *Particle Physics Culture.* Cambridge, MA: Harvard University Press, 1989.

Troemel-Ploetz, Senta. 'Mileva Einstein Maríc: The Woman who did Einstein's Mathematics.' *Women's Studies International Forum* 13 (5 1990): 415–32.

Tuana, Nancy, ed. *Feminism and Science.* Bloomington: Indiana University Press, 1989.

Turkle, Sherry. 'Computational Reticence: Why Women Fear the Intimate Machine.' In *Technology and Women's Voices: Keeping in Touch*, ed. Cheris Kramarae. London: Routledge, 1988.

Turkle, Sherry and Seymour Papert. 'Epistemological Pluralism: Style and Voices within the Computer Culture.' *Signs: Journal of Women in Culture and Society* 16 (1 1990): 128–58.

Uglow, Jennifer and Frances Hinton, eds. *The Women's Biography.* London: Macmillan, 1982.

Ungerson, Clare. *Policy is Personal: Sex, Gender and Informal Care.* London: Tavistock, 1987.

Van Sertina, Ivan, ed. *Blacks in Science: Ancient and Modern.* Brunswick and London: Transaction Press, 1985.

Vanek, Joan. 'The Time Spent in Housework.' In *The Economics of Women and Work*, ed. Alice Amsden. Hamondsworth: Penguin, 1980.

Ve, Hildur. 'Women's Mutual Alliances: Altruism as a Basis for Interaction.' In *Patriarchy in a Welfare Society*, ed. Hamet Holter. Bergen: University of Bergen Press 1984.

Ve, Hildur. 'Women's Experience: Women's Rationality.' In Conference on the Construction of Sex/Gender, Lidingo: Mimeo, 1990.

Ve, Hildur. 'Gender Differences in Rationality: The Concept of Praxis, Knowledge and Future Trends.' In International Conference: Ethics. Gender and Technology, Luleå: University of Luleå, 1992.

Verrall, R. 'Sociobiology: The Instinct in our Genes.' *Spearhead* 10 (1979): 10–11.

Versluysen, Margaret. 'Old Wives' Tales: Women Healers in English History.' In *Rewriting Nursing History*, ed. Celia Davies. London Croom Helm, 1980.

Vickers, Brian. *Francis Bacon and Renaissance Prose.* Cambridge: Cambridge University Press, 1968.

Vines, Gail. 'The Hidden Cost of Sex Selection.' *New Scientist* 138 (1 May 1993): 12–13.

Waerness, Kari. 'Caring as Women's Work in the Welfare State.' In *Patriarchy in a Welfare Society*, ed. Harriet Holter. Bergen: University of Bergen Press, 1984.

Waerness, Kari. 'On the Rationality of Care.' In *Women and the State*, ed. Anne Showstack Sassoon. London: Hutchinson, 1987.

Wajcman, Judy. *Feminism Confronts Technology.* Cambridge: Polity, 1991.

Walby, Sylvia, ed. *Gender Segregation at Work.* Milton Keynes: Open University Press, 1988.

Walby, Sylvia. *Theorising Patriarchy.* Oxford: Blackwell, 1990.

Walker, Alice. *The Color Purple.* London: Women's Press, 1983.

Walker, Harris. 'Did Einstein Espouse his Spouse's Ideas?' *Physics Today* (February 1989): 9–10.

Walker, Martin. *The National Front.* London: Fontana, 1977.

Wallsgrove, Ruth. 'The Masculine Face of Science.' In *Alice through the Microscope*, ed. Brighton Women and Science Group. London: Virago, 1980.

Walters, William and Peter Singer. *Test Tube Babies.* Oxford: Oxford University Press, 1984.

Warnock, Mary. *A Question of Life: Warnock Report on Human Fertilization and Embryology.* Oxford: Blackwell, 1985.

Warnock Report *Report of the Committee of Inquiry into Human Fertilisation and Embryology.* Cmnd 9314. HMSO, 1985.

Watson, James. *The Double Helix.* New York: Atheneum, 1968.

Weisstein, Naomi. 'Adventures of a Woman in Science.' In *Working it Out*, eds Sara Ruddick and Pamela Daniels. New York: Pantheon, 1977.

Westergaard, John and Henrietta Resler. *Class in a Capitalist Society: A Study of Contemporary Britain*. London: Heinemann Educational, 1975.

Whyte, Judith. *Girls Into Science and Technology*. London: Routledge, 1986.

Wilkinson, Doris. 'For Whose Benefit? Politics and Sickle Cells.' *The Black Scholar* 5 (1974): 26–31.

Willis, Paul. *Learning to Labour*. London: Methuen, 1977.

Wilson, E. O. *On Human Nature*. Cambridge, MA: Harvard University Press, 1978.

Wilson, Elizabeth. *Women and the Welfare State*. London: Virago, 1975.

Wingerson, Lois. *Mapping Our Genes: The Genome Project and the Future of Medicine*. Dutton, NY: Penguin, 1990.

Winner, Langdon. *Autonomous Technology: Technics-out-of-Control as a Theme in Political Thought*. Cambridge, MA: MIT Press, 1977.

Wittig, Monique. *Les Guérilleres*. London: Oven, 1971.

Wolf, Christa. *The Reader and the Writer*, Berlin: Seven Seas Books, 1977.

Wolpert, Lewis. *The Unnatural Nature of Science*. London: Faber and Faber, 1992.

Woolgar, Steve. 'What is the Analysis of Scientific Rhetoric For?: A Comment on the Possible Convergence between Rhetorical Analysis and Social Studies of Science.' *Social Studies of Science* 14 (1 Winter 1989): 47–9.

Wright, Susan. 'The Military and the New Biology.' *Bulletin of the Atomic Scientists* 41 (May 1985): 10–16.

Wright, Susan, ed. *Preventing a Biological Armsrace*. Cambridge, MA: MIT Press, 1990.

Yalow, Rosalyn. *Le Prix Nobel*. Stockholm: Nobel Foundation, 1977.

Young, Iris. 'Beyond the Unhappy Marriage: A Critique of Dual Systems Theory.' In *Women and Revolution*, ed. Lydia Sargent. Boston, MA: South End Press, 1981.

Young, M. F. D., ed. *Knowledge and Control*. London: Collier Macmillan, 1971.

Young, Michael and Peter Wilmott. *The Symmetrical Family: A Study of Work and Leisure in the London Region*. Harmondsworth: Penguin, 1975.

Young, R. M. 'Science is Social Relations.' *Radical Science Journal* 5 (1977): 65–131.

Yoxen, Ed. *The Gene Business: Who Should Control Biotechnology?* London: Crucible, 1982.

Zarit, Steven, Pamela Todd, and Judy Zarit. 'Subjective Burden of Husbands and Wives as Caregivers: A Longitudinal Study.' *The Gerontologist* 26 (3 1986): 260–6.

Zilsell, Edgar. 'The Sociological Roots of Science.' *American Journal of Sociology* 47 (41 1942): 545–60.

Zimmerman, Bonnie. 'Reading, Seeing, Knowing: The Appropriation of Literature.' In *Engendering Knowledge*, eds Joan Hartman and Ellen Messer Davidott. Knoxville: University of Texas, 1991.

Zipper, Juliette and Selma Sevenhuijsen. 'Surrogacy and Feminist Notions of Motherhood.' In *Reproductive Technologies: Gender, Motherhood and Medicine*, ed. Michelle Stanworth. Cambridge: Polity, 1987.

Zita, Jacqueline. 'A Feminist Question of the Science Question in Feminism.' *Hypatia: Journal of Feminist Philosophy* 3 (1 1988): 157–68.

Zita, Jacqueline. 'The Premenstrual Syndrome: Diseasing the Female Cycle.' *Hypatia: Journal of Feminist Philosophy* 3 (1 1988): 77–100.

Zuckerman, Harriet. *The Scientific Elite: Nobel Laureates in the United States*. New York: Free Press, 1977.

Zuckerman, Harriet, Jonathan Cole, and John Bruer, eds. *The Outer Circle: Women in the Scientific Community*. New York: Norton, 1991.

Index

Index

Women's Social and Political
  Union, 119
Woodward, Clare, 273 n.58
Woolf, Virginia: quoted, 1, 4, 17,
  45, 115, 140, 236; *A Room of One's
  Own*, 52, 211, 273 n.65
Woolgar, Steve, 88–9
Wootton, Barbara, 126
World Health Organization (WHO),
  172, 204, 253 n.40
Wright, Joseph (of Derby), 100
Wrinch, Dorothy, 15, 122
Wrong, Dennis, 110

X-chromosomes, 19

Yalow, Aaron, 159
Yalow, Rosalyn, 136–7, 158–60, 161,
  164, 167, Plate 10
Yardley, Kathleen, *see* Lonsdale
Young, Iris, 46
Young, Michael, 98, 111
Young, R. M., 89
Yugoslavia, women's peace groups,
  54

Zihlman, Adrienne, 92, 262 n.16
Zimbabwe, iron smelting, 9
Zuckerman, Harriet, 130, 138, 149,
  161, 270 n.2